DARE ALL DANGERS

DARE ALL DANGERS

The 741st Tank Battalion in World War II, D-Day to VE-Day

Bill Warnock

Published in the United States by Castlehurst Books.

CASTLEHURST is a registered trademark of Castlehurst Books LLC, Ohio.

Names: Warnock, Bill, author.
Title: Dare All Dangers: The 741st Tank Battalion in World War II, D-Day to VE-Day / Bill Warnock.
Includes maps, appendices, bibliographical references, endnotes, and an index.
Identifiers: ISBN 9798990584402 (hardcover)
Subjects: United States. Army. Tank Battalion, 741st | World War, 1939-1945—Regimental histories—United States | World War, 1939-1945—Campaigns—Western Front

Cataloging-in-Publication (CIP) data can be obtained from the Library of Congress under the control number 2025915739.

Printed in Canada on acid-free paper.

9 8 7 6 5 4 3 2 1

FIRST EDITION

To learn more about the author, visit his website: etohistory.com. Here you will find an online archive with rare primary-source material from libraries, government archives, and private collections around the world. There is also an option to sign up for a monthly newsletter.

On the front cover: Sergeant Mayne Youngblood's M4A1(75) named Asp II crosses a treadway bridge at Dümpelfeld, Germany, on March 9, 1945. Photo by Louis Nemeth, 165th Signal Photo Company (NARA).

Author photo by Kelly Tooman.
All maps created by the author using Adobe Illustrator.
Dust jacket design by Hassan Bikourrane and Ian Pamplona.

For Val, Jack, Verdie, Nelson, and Will,
Finally

Contents

Guide to Acronyms

A&M	Agricultural and Mechanical
AA	Antiaircraft
ABMC	American Battle Monuments Commission
ADC	Army Depository Copy
AGO	Adjutant General's Office
AGRC	American Graves Registration Command
AP	Armor Piercing
ASR	Adjusted Service Rating
AT	Antitank
ATEA	Amphibious Tank Escape Apparatus
AUS	Army of the United States
AWOL	Absent Without Leave
BSM	Bronze Star Medal
CBS	Columbia Broadcasting System
CCB	Combat Command B
CIC	Counterintelligence Corps
CP	Command Post
DSC	Distinguished Service Cross
DD	Duplex Drive
DUKW	6-wheel-drive amphibious truck
ECB	Engineer Combat Battalion
ESB	Engineer Special Brigade
EOD	Explosive Ordnance Disposal
ETO	European Theater of Operations
FA	Field Artillery
FLAK	Flugabwehrkanone (Antiaircraft Gun)
FO	Field Order
FOV	Field of View
GAT	Gap Assault Team
GHQ	General Headquarters
GI	General Issue or Government Issue *

GM	General Motors
GMC	General Motors Corporation
HASAG	Hugo Schneider Aktiengesellschaft-Metalwarenfabrik
HE	High Explosive
HHC	Headquarters and Headquarters Company
HMS	His Majesty's Ship
HVAP	High Velocity Armor Piercing
HVSS	Horizontal Volute Spring Suspension
KIA	Killed in Action
LCA	Landing Craft, Assault
LCI	Landing Craft, Infantry
LCM	Landing Craft, Mechanized
LCT(5)	Landing Craft, Tank, (Mark 5)
LCT(6)	Landing Craft, Tank, (Mark 6)
LCT(A)	Landing Craft, Tank (Mark 5, Armored)
LCT(HE)	Landing Craft, Tank (Mark 5, High Explosive)
LCVP	Landing Craft, Vehicle, Personnel
LST	Landing Ship, Tank
MD	Medical Department
MG	Machine Gun
MGP	Military Government Police
MIA	Missing in Action
MP	Military Police
MS	Mail Steamship
NARA	National Archives and Records Administration
NPRC	National Personnel Records Center
OB	Order of Battle
OCS	Officer Candidate School
OD	Olive Drab
PAK	Panzerabwehrkanone (Antitank Gun)
PFC	Private First Class
POM	Preparation for Overseas Movement
PW	Prisoner of War

QT	Quiet
RA	Regular Army
RCT	Regimental Combat Team
ROTC	Reserve Officers' Training Corps
RMS	Royal Mail Steamship
SCR	Signal Corps Radio
SS	Schutzstaffel
SS	Steamship
SSM	Silver Star Medal
ST	Small Tug
TD	Tank Destroyer
TNT	Trinitrotoluene
USA	United States Army
USAT	United States Attack Transport
USO	United Service Organizations
USS	United States Ship
VA	Veterans Administration
VE	Victory in Europe
VFW	Veterans of Foreign Wars
WD	War Department
WN	Widerstandsnest (Resistance Nest)
WP	White Phosphorus

* The acronym GI is believed to have originated from the letters "GI" stamped on metal garbage cans that were once ubiquitous on US military installations. The letters stood for galvanized iron but were misunderstood as an acronym for general issue or government issue. The author has found this usage in American newspapers as early as September 1941. By late 1942, the meaning of GI had evolved to include US servicemen themselves. Throughout the war, the term "doughboy" was more prevalent when referring to combat soldiers. The same was true of derivative terms like "dough" and "doughfoot." After the war, those words fell into disuse, replaced by GI and various postwar terms. Today, doughboy is largely associated with the First World War. Its use is therefore limited in this book.

Guide to German Military Unit Nomenclature

Abteilung	Battalion or Detachment
Abteilungsstab	Battalion Headquarters
Aufklärungszug	Reconnaissance Platoon
Batallion	Battalion
Eisenbahn-Batterie	Railway Artillery Battery
Fallschirm-Aufklärungs-Abteilung	Parachute Reconnaissance Battalion
Fallschirmjäger-Division	Parachute Division
Fallschirmjäger-Regiment	Parachute Regiment
Feldersatz-Bataillon	Replacement Training Battalion
Flak-Abteilung	Antiaircraft Artillery Battalion
Flak-Division	Antiaircraft Artillery Division
Flakstellung	Flak Emplacement
Grenadier-Regiment	Rifle Regiment or Infantry Regiment
Heeres-Panzerjäger-Abteilung	Army Tank Destroyer Battalion
Kampfgruppe	Battle Group or Task Force
Kompanie	Company
Luftwaffe-Festungs-Bataillon	Air Force Fortress Battalion
Magenbataillon	Stomach Ailments Convalescence Battalion
Panzer Armee	Tank Army
Panzer-Ausbildung-und Ersatz-Regiment	Armored Training and Replacement Regiment
Panzerbrigade	Armored Brigade
Panzer-Division	Armored Division
Panzergrenadier-Division	Motorized Division
Panzergrenadier-Regiment	Armored Infantry Regiment
Panzerjäger-Abteilung	Tank Destroyer Battalion
Panzer-Regiment	Armored Regiment
schwere Panzerabteilung	Heavy Tank Battalion
Sturmgeschütz-Brigade	Assault Gun Brigade

Volksgrenadier-Division	People's Infantry Division (People's was an honorary title)
Waffen-SS	SS Armed Forces (military branch of the Schutzstaffel, the Nazi elite guard)

When expressing ordinal numerals, Germans write the digit(s) followed by a period, for example, 12.SS-Panzer-Division, with 12. meaning 12th.

German quotation marks appear in this book with certain unit names, e.g., „Das Reich" and „Hitlerjugend."

PROLOGUE

This excursion into the past chronicles the 741st Tank Battalion during the Second World War. Most soldiers described in this book spent two years training before entering combat. The chasm between those two worlds was broad and stark. None of the rehearsal adequately prepared the men for the horrors ahead. Nothing could.

Even so, survival and success in battle hinged on the skills ingrained during the training that began soon after the US Army created the 741st. The unit originated as a medium tank battalion but one unassigned to an armored division.

Separate or non-divisional battalions like the 741st emerged at the same time as armored divisions. The army created both when establishing its Armored Force in 1940.

Before that time, the few American tank units in existence operated under infantry control and served as little more than mobile artillery. That changed after Adolf Hitler's Panzer divisions slashed across Poland and Western Europe. The Germans created a new battlefield paradigm. Instead of infantry leading the way, armor set the pace of advance.

The father of the Armored Force, Major General Adna R. Chaffee Jr., adopted many ideas from the Germans when building his armored divisions. He designed them to be self-sufficient, entirely mechanized, and able to mount prolonged strikes deep into enemy territory. He also built separate tank battalions to provide close support for infantry divisions much like prewar American armor. Chaffee had a simple reason: if each infantry division had tanks readily available, there would be no need to borrow from armored divisions and sap their strength.

Separate tank battalions operated with infantry divisions on a temporary or semipermanent basis. More than five dozen separate battalions existed by the time hostilities ended. Only fifteen of these pre-dated America's entry into the war. The 741st was the first one created after Pearl Harbor. The unit originated on March 15, 1942, when Headquarters Armored Force published its General Orders Number 14, the birth certificate of the battalion.

During its first week, the "Seven-Forty-First" existed only on paper, until cadre arrived from the 751st Tank Battalion at Fort Jackson, South Carolina. This infusion of experienced blood reached Fort Meade on March 23 and totaled thirteen officers and 134 enlisted men.

Lieutenant Colonel Jacob R. Moon ranked highest among the cadre. The Alabama native graduated from West Point in 1924 and now commanded the 741st.[1] His battalion grew when 523 enlisted men arrived from the Armored Force Replacement Training Center at Fort Knox, Kentucky. These fledgling troops had only thirteen weeks of basic training.

At Fort Knox, the men had learned how to look and behave like soldiers. Now they discovered what the army expected from them as members of a tank battalion.

The 741st had five companies: Headquarters Company, Service Company, and three medium tank companies—A, B, and C, also known as Able, Baker, and Charley. Each trainee joined a company and received a specific role. Within tank crews, each man mastered his own job before learning the duties of his fellow crewmen.

To everyone's chagrin, they had too few tanks and no medium tanks, just M2A4 light tanks. The battalion also lacked fuel, not even enough to fire up every motor daily.

"Although gasoline was conspicuous by its absence," one enlisted man remembered, "a little inconvenience like that could never stop the army from carrying out its training. The five-man crews 'played' tank and ran about the field practicing tactics until their tongues hung out. We really looked silly."[2]

The men and officers soon traveled to specialist schools and advanced courses at Fort Knox. The subject matter ranged from gunnery, to radios, to tank maintenance. Medics departed for the Medical Field Service School at Carlisle Barracks, Pennsylvania. Kitchen personnel attended the Cooks and Bakers School at Fort Meade.

After everyone returned to the 741st, the battalion packed up for the West Coast and combined-arms training that involved armor, artillery, and infantry.

On July 31, 1942, the 741st boarded railcars bound for the Desert Training Center at Camp Young, California. Two trains transported all personnel, equipment, and wheeled vehicles, but no tanks. New armor waited at Camp Young.

The journey to California lasted until August 5, when the trains reached a dusty, dismal town called Indio. Trucks hauled the men twenty-seven miles through the Little San Bernardino Mountains and into an arid valley flanked by barren hills. Here, in the Mojave Desert, pyramidal tents stood in rows under a blazing sun. This canvas city became home base for the 741st.

The preparation for combined-arms training began when the men inherited a dilapidated collection of medium tanks—M3s and early-production M4s—formerly belonging to the 757th Tank Battalion. Many sat inoperable, some without tracks. The ensuing maintenance headaches lasted several days before the crews had most of the tanks running, at least temporarily.

From August 21 to October 15, the battalion took part in tank-infantry maneuvers with the 7th Motorized Division. Mock battles in the Mojave, near the California towns of Rice and Needles, helped the men master the instruments of killing.

Amid shimmering heat waves, the tankmen rolled across the dead and defeated landscape, a place with no vegetation for shade or camouflage. Sandstorms, frigid nights, and sweltering daytime temperatures challenged their fighting skills. The men learned to work for long periods without sleep. The battalion rationed water, which meant little to drink and no bathing. Sand and sweat created a crusty layer on everyone. Meals were the same day after day: crackers, sardines, chili con carne, and Fig Newtons.

In late October, with maneuvers finished, several tank crews from Company B volunteered for a unique project. Paramount Pictures had begun production of a spy thriller called *Five Graves to Cairo*, and its director needed armor for scenes depicting British Eighth Army tanks. The movie set lay in the desert near Yuma, Arizona. One platoon with five tanks motored there and participated in nine days of filming. The men spent their time waiting for the few minutes each day when the correct sunlight prevailed for photography. In the interim, the platoon members enjoyed drinking with the leading man, actor Franchot Tone.

Only four tanks appeared on camera. Army censorship restricted the fifth one from public view, the machine being an experimental M3A4. The tank had a Chrysler "multibank" power plant: five commercial truck engines joined together. The combination worked well when all the motors ran simultaneously, but that often failed to happen. Most tankers regarded the M3A4 as a joke, not as a military secret.

By the end of 1942, the 741st had wrapped up its training at Camp Young. Orders came for the battalion to pack up and depart for Camp Polk, Louisiana, where the Third US Army planned to conduct maneuvers starting in February.

The move to Louisiana began from the Indio railhead on January 10, 1943. This time, the battalion loaded all of its tanks onto flatcars. The personnel making the trip included one warrant officer, two civilian technicians, forty-eight officers, and 881 enlisted men.

Shortly after reaching Camp Polk, the battalion gave up five officers and 134 enlisted men as cadre for the 701st Tank Battalion forming at Camp Campbell, Kentucky. Those who remained in Louisiana spent eight weeks on maneuvers designed to measure combat readiness. The tanks provided infantry support. Their cannons and machine guns pinned down opposition troops and suppressed their fire, enabling friendly soldiers to advance and overcome resistance. Subjected to an attack, the tanks beat back two infantry battalions, inflicting 50 percent losses as determined by the umpires, who presided over the mock battle.

When the maneuvers ended on March 28, the 741st returned to Camp Polk and garrison life, which included maintenance, marksmanship training, and platoon exercises. The men also waterproofed their vehicles in compliance with an order from corps headquarters. This spawned a rumor that the 741st would depart for an overseas destination. The rumor proved false, but a station change did occur: the battalion left for Camp Pickett, Virginia, on June 13, 1943.

As the train moved east, the men rode in Pullman cars or sometimes atop the tanks. People along the route waved from their homes. While passing through Tennessee, one battalion member saw his mother waving from the back porch of his house, but she stood too far away to recognize him. Homesick, he broke down in tears.

After four days, the train reached its destination. The need for waterproofing then became apparent when the battalion began amphibious training.

The first part involved no vehicles. Navy instructors lectured about shipboard life and landing techniques. Disembarkment practice followed. In full combat gear, the men lumbered down a cargo net slung from a tall wooden wall. The structure replicated a troopship hull. To imitate heaving swells, sailors rocked the wall back and forth.

Training then shifted to Camp Bradford, a naval base along the coast at Norfolk, Virginia. More disembarkment practice took place but this time at sea. The men also rehearsed loading vehicles aboard a large vessel known as an LST (Landing Ship Tank) resembling Noah's Ark. It all ended with a mock invasion at mosquito-infested Solomons Island, Maryland.

In August 1943, a large party from the battalion traveled by rail to Cape Cod, Massachusetts, for antiaircraft training at Camp Wellfleet. Twelve officers and 322 enlisted men lived in pup tents pitched on the beach.

For nine days, the tankers learned antiaircraft gunnery. They started by firing caliber .30 machine guns at dummy mortar shells launched into the air. Training then moved to the shoreline, as each gunner fired a caliber .50 machine gun at a canvas sleeve pulled across the sky by a Lockheed Hudson. Several gunners struck the tow cable, sending the canvas fluttering down into the ocean. Others blazed away at the cable until the angle of fire became so extreme that bullets nearly hit the aircraft, and the airmen threatened to stop flying.

After the antiaircraft training, the men headed to Fort Dix, New Jersey, a pre-staging area for units moving across the Atlantic. The entire 741st assembled there on September 2.

Questions abounded. When and where would the battalion board a troopship? Was the battalion destined to participate in the coming invasion of continental Europe?

Many men enjoyed their first furlough since joining the army—and their last until after the war. Those who remained behind spent time writing letters, reviewing previous training, and working on POM requirements (Preparation for Overseas Movement). Among the manifold burdens, the men packed all organizational equipment in watertight crates.

On September 18, the battalion posed for the official Fort Dix photographer, who made a panoramic photo of each company. Such pictures had become an army tradition during the nineteenth century. The men purchased prints and mailed them to their families for safekeeping.

Fort Dix remained home base for the 741st until the second week of October, when the battalion moved to Camp Shanks, at Orangeburg, New York, the final overseas staging area. All tanks and trucks stayed at Fort Dix. New ones waited abroad.

Orders restricted everyone to camp while the battalion completed its POM requirements. The men underwent physicals, inoculations, and dental exams. "Show-down" inspections identified missing clothing and personal equipment. Additional orders required all personnel to mail home radios, cameras, and electric razors. For security's sake, enlisted men unpinned the 741st crests from their lapels and garrison caps, and officers removed them from their epaulets.

On October 19, the battalion journeyed by train and ferryboat to the New York Port of Embarkation. Weighed down with bedrolls, backpacks, and overstuffed duffel bags, the men struggled up a narrow gangway to the MS *Capetown Castle*. This converted ocean liner with its British crew would carry the 741st across the North Atlantic. The destination remained hush-hush, but everyone guessed it with spectacular accuracy—England.[3]

CHAPTER 1 **PRELUDE TO INVASION**

The *Capetown Castle* remained at the pier for two days, taking on more troops and tackle before weighing anchor. She steamed past the Statue of Liberty and onto coastal waters toward a rendezvous with Convoy UT-4, which formed near New York City and grew larger after reaching the Boston area. The final count stood at twenty-three ships. Their cargo included mail, troops, aircraft, frozen meat, and petroleum products.

Under the aegis of twelve destroyers and the battleship USS *Texas*, the convoy weaved along like a snake to evade U-boats. Aviators patrolled above looking for any hint of the enemy, and enlisted men from Company B kept watch while manning antiaircraft guns. Lifeboat drills occurred daily and offered the only physical activity other than walking the decks. Craps and poker provided an escape from boredom.

For most enlisted personnel, bland British rations comprised their meals—prunes, oatmeal, and boiled sausages. Mutton frequently appeared on the supper menu. Officers had dibs on American favorites like steak and eggs, but heavy seas left many soldiers too nauseated to eat.

The only calamitous event occurred the first night out, when the petroleum tanker *Bulkoil* accidentally struck the destroyer USS *Murphy*. The tanker sheared off the warship's bow and bridge section, which sank, taking thirty-five sailors to the bottom. The rest of the *Murphy* remained afloat and was later towed away. The *Bulkoil* returned to port under its own power.[4]

After ten days at sea, UT-4 reached the British coast. The convoy dispersed, and its ships steamed for different ports. The *Capetown Castle* dropped anchor near Liverpool.

On November 2, the ship pushed up the River Mersey and docked with help from tugboats. Late that night, all but two 741st companies disembarked. Liverpool lay blacked out and enveloped in dense fog. Many buildings stood roofless and crumbling, bombed out during the Blitz. Through the misty gloom, the tankmen groped along cobblestone streets to the nearest railway station. Red Cross ladies served coffee and doughnuts as the men boarded a troop train for Ogbourne Saint George in southern England.

When the train chuffed into the station after sunrise, the GI travelers found themselves in a storybook hamlet with thatched cottages and dainty flower gardens. This small English community in Wiltshire became their new home. Most battalion members lived in Nissen huts—prefabricated structures that cluttered every corner of the once-orderly countryside. The huts bunked sixteen men each, and each had two coal-burning stoves.

Back at Liverpool, Company A and Service Company had remained aboard the *Capetown Castle*. These men disembarked at noon on November 3 and boarded a southbound train. Late that night, the two companies reached Ogbourne Saint George.

Like all newly arrived units, the battalion received a code name ("Vitamin" in this case) and underwent an orientation program to acquaint its men with English money, customs, and vernacular. The soldiers also learned about the wartime hardships endured by the civilian population: blackouts, food shortages, and enemy air raids. The horrific years under German aerial bombardment meant that many housewives in aprons and kids in knee pants had survived more high explosives than anyone in the 741st.

Once orientation ended, the men acquired new tanks and trucks from the ordnance depot at Tidworth. This sprawling facility was the principal depot for armor in the European Theater. The 741st received brand-new tanks for the first time in its short existence. These included the medium

M4, which the British had nicknamed the "General Sherman," a name officially adopted by the US Army later in the war.[5] Factory workers and other well-wishers had emblazoned several tanks with messages. One Able Company crew found a simple statement chalked on the side of their machine: "Good luck boys. I hope you win soon—I am a little girl."[6]

It took sweat and elbow grease to prepare the tanks. Hand tools, periscopes, fire extinguishers, and dozens of other items required unpacking, cleaning, and installation. The men stripped off Cosmoline, a petroleum product applied to inhibit corrosion. They peeled away tape that covered apertures, bearing caps, and other nooks where moisture might accumulate.

The new armor obtained at Tidworth also included M5A1 light tanks, nimble little machines almost twice as fast as any medium tank. Named the "General Stuart" by the British, the M5 series was part of an army-wide change. The War Department had restructured all medium tank battalions, giving each a light tank company for reconnaissance, flank security, and roadblocks. Thus emerged Company D.

In the 741st, the formation of Dog Company took place on December 1, 1943. Its personnel primarily came from other companies in the battalion. The remainder arrived from the 3rd Tank Group, including men formerly part of the American garrison protecting Iceland.

First Lieutenant John F. Sicks, a former Company B officer, commanded the new company. The lieutenant had graduated from the journalism school at the University of Missouri and worked at the *Chicago Daily News* before being drafted in 1941. He later completed an officer training course at Fort Knox before joining the 741st. As he organized Company D, the rest of the 741st departed Ogbourne Saint George on December 18. The men—minus their tanks—traveled by rail and trucks to southwest England for a training maneuver called Exercise Duck.

Yet unknown to anyone in the 741st, Allied high command had chosen to invade Hitler's Fortress Europe by way of French beaches in Normandy. During early planning, the need for large-scale amphibious exercises became apparent. Exercise Duck was the first. It encompassed all stages of an actual invasion: marshalling, embarkation, seaborne transit, and beach assault.

The British War Office allocated a training area at Slapton Sands, where the beach consisted of shingle and tiny pebbles resembling the Normandy coast. Exercise activities commenced when troops began arriving at campsites in southwestern England. The 741st split its companies between two locations: Ladock in Cornwall and Brixham in Devon.[7]

At Ladock, the men found pyramidal tents waiting in a hayfield. Boots and truck tires quickly turned the place into a muddy morass. At Brixham, the men stayed in tents near Lupton House, a Georgian mansion with wooded grounds. The tankers did little but eat, sleep, and loaf time away for two rain-drenched weeks.

Exercise Duck culminated on January 3, 1944, when approximately 26,400 troops moved from their camps to embarkation points at Falmouth (near Ladock) and Dartmouth (near Brixham). Too few naval transports meant that less than half the soldiers assaulted Slapton Sands. The remainder never left the water's edge, including the 741st men at Ladock and Brixham. Amid a cold drizzle, they simulated embarkation procedures.

Afterward, the battalion re-boarded trucks and returned to Ogbourne Saint George, where they found Christmas parcels and letters waiting for them. The men passed around boxes overflowing with cakes, cookies, and other homemade delicacies.

Besides mail, the first month of 1944 brought Jacob Moon's departure. Due to illness, the 741st commander left on January 11 for the 2nd General Hospital. He subsequently joined the First US Army Group on March 1 by way of a replacement depot. The former battalion commander became executive officer for the group's Special Troops headquartered at Bryanston Square in London.[8] According to scuttlebutt, his new posting occurred because he was deemed physically unfit for combat. At age forty-four, he was older than most battalion commanders, and he suffered from chronic degenerative arthritis. His job passed to a younger officer, Major Robert Newell Skaggs, the thirty-six-year-old battalion executive officer.

Raised in California by his widowed mother, Skaggs was short in stature with a bulldog physique. He joined the enlisted ranks in 1926 and became a reserve officer three years later through the ROTC at the University of Arizona. While attending the school, he worked for a bank in Tucson, managed a polo team, and also met his first wife, Patty Newton.

They married in 1932, but the union ended after two years. Immediately following the divorce, he wed Anna Mae Lowery of Dallas, Texas. At the time, Skaggs commanded Company 808 of the Civilian Conservation Corps at Platt National Park, Oklahoma. In 1940, the army called him to active duty as a first lieutenant, and he landed a position with the newly created 2nd Armored Division. When the division formed a cadre for the 751st Tank Battalion, Skaggs transferred to the new battalion and became commanding officer of its Headquarters Company. Reassigned to the 741st as part of its cadre, he joined as a captain, first serving as battalion motor officer. Before the unit left the United States, he advanced to executive officer and received a promotion to major.

Soon after Skaggs took charge of the 741st in England, Companies B and C began training with the British Army in Valentine tanks. Formally known as the Infantry Tank Mark III, the Valentine paled in comparison to any medium tank in the American arsenal. It had thinner armor, lighter weaponry, and moved slower, despite weighing less and having only three crewmen. The men from Baker and Charley received no explanation for the shift to Valentine tanks, but everyone assumed the worst. They expected to trade their Shermans for Valentines.

After learning to operate and maintain the British tanks, the two companies transferred to the port city of Great Yarmouth. The tankers, numbering 160 men, arrived on February 1 and quartered in a row of houses on a side street.

The move to Great Yarmouth and the Valentine training related to a high-level decision. Allied commanders had selected the 741st and two other US tank battalions—the 70th and 743rd—to receive amphibious armor. The British named these super secret weapons, DD tanks.

The letters DD stood for "duplex drive," meaning the tank could drive on land and traverse water under its own power. A watertight canvas screen surrounded its upper hull. This permitted the vehicle to attain buoyancy by sinking deep enough to displace its own weight in water. Pneumatic pillars raised the screen from a collapsed position, and metal struts held it taut. When deployed, the screen resembled an oversized washtub. To provide propulsion, the tank had two bronze propellers mounted at its rear and geared to the idler wheels for power. The propellers rotated in opposite directions and turned to either side like a rudder. After emerging from the water, the crew would disengage both propellers and collapse the screen.

The British produced the first DDs in England using Valentine tanks. Their performance impressed Allied commanders, who opted to use amphibious armor for the coming invasion but with one fundamental change: American medium tanks would replace the Valentines. To furnish enough DDs, manufacturing took place on both sides of the Atlantic.

No American-built DDs reached England during the first two months of 1944, which forced the three US tank battalions to begin training with Valentine DDs.

The first instruction topic at Great Yarmouth was underwater escape. All crewmen learned how to use the Davis Submarine Escape Apparatus, a breathing device developed by the Royal Navy to give sailors a thirty-minute oxygen supply when abandoning a sunken submarine.

Training occurred in a twelve-foot-deep pool. After donning the apparatus and entering the water, each man descended a steel ladder, walked across the pool floor, and climbed out using a ladder on the opposite side. This helped the tankers develop confidence in the breathing device and overcome any aversion to being under tons of water. Not everyone followed the program. Several men descended partway and lost their nerve. They pushed away from the ladder, hurriedly swam across the pool, and burst to the surface.

The trainees also learned to rise from the bottom without a ladder. On the pool floor, each man inflated a life-preserver belt worn under his arms, which popped him to the surface. The instructor warned against rising too fast in deep water. Each man learned how to reduce his ascent rate and control decompression.

Besides using the bulky Davis apparatus, the men trained with a smaller, lighter device that provided up to seven minutes of oxygen. Manufactured in England by Siebe Gorman & Company, the Amphibious Tank Escape Apparatus (ATEA) became standard in all American DDs. After learning to breathe with it on dry land, the men practiced in water.

To begin, they climbed into an empty pool with a Valentine hull resting on the bottom. One tank crew at a time, the men sat inside the steel shell as an instructor flooded the pool. With water gushing in, the tankers began using their breathing devices. When the pool reached full, the men squirmed out of the hull and swam to the surface. Most trainees spent less than three minutes submerged. After that test, the crews moved to a deeper pool with an M4A1 hull.

Throughout escape training, the instructors emphasized two points: crewmen should always attempt to flee a foundering DD before it went under, and, if that failed, nobody should exit a submerged tank until it hit the seabed and the interior filled with water. Crewmen who did otherwise would expend all of their energy and oxygen fighting the water.

Once the tankers finished preparing for the worst, they began working with actual DDs. Each morning, trucks carried the men from their quarters in Great Yarmouth to nearby Fritton Lake. A canvas fence surrounded its perimeter, screening the lake from view, and armed guards patrolled around the clock. All the doings there stayed on the QT. The men received a stern lecture on secrecy, and each man signed a document pledging his silence.

While training at Fritton, each crew had a Valentine DD and a British sergeant, who served as an instructor. The soldiers practiced deploying and collapsing the flotation gear. They first entered the lake from the shore and then learned how to launch from a ramp like the one on an LCT (Landing Craft, Tank).

One rainy morning, before Companies B and C departed Great Yarmouth, a British general visited to expound on the DD training. Both companies stood in ranks for his speech, which climaxed with a heady announcement: "You have the honor of being chosen to invade the shores of France."[9] At that precise moment, a man from Company B passed out, falling flat on his face into a puddle. Overindulgence the previous evening caused the dramatic moment, rather than the general's words.

After Great Yarmouth, the two companies motored to Barton Stacey in Hampshire. This rural community stood alongside the Salisbury Plain, the great chalk downs that served as the main peacetime maneuver area for the British Army. The Baker and Charley men lingered less than a day before heading on to Fort Monckton.

The aging brick-and-stone citadel guarded the approach to Portsmouth Harbour, while the nearby shoreline served as a training area. Quartered in ramshackle barracks, the men woke each day before sunrise to eat breakfast and begin saltwater training with Valentine DDs. They cruised a section of coastal water and practiced assault landings. The DDs traveled as far as Queen Victoria's bathing beach on the Isle of Wight.

After two weeks of "swim" practice, everyone returned to Barton Stacey. The place was now home station for the 741st, and the whole battalion billeted there, except for Company A.

Able Company had departed Barton Stacey for Puddletown in Dorset. There its men prepared for Exercise Fox, a training maneuver involving a seaborne assault at Slapton Sands.

Since the company's tanks had to wade ashore, each crew outfitted their vehicle with a deep-water fording kit that included two sheet metal fording stacks. One stack attached to the engine's air intake and the other connected to its exhaust. Together, they worked like a snorkel, permitting the vehicle to operate in water up to its turret.

The crews waterproofed their tanks with Bostik, a tar-like substance used to seal crevices where water might seep in. The men caulked large cracks with felt strips or rags before applying the black goop. Rivets, bolt heads, and drain plugs also received a coating. An asbestos compound waterproofed periscopes and their turntables. The men pasted water-repellent canvas over ventilators, gun shields, and gun muzzles.

Company A drove its waterproof tanks south to the port city of Weymouth. They boarded LCTs at Portland Harbour and sailed toward Slapton Sands. After two nights at sea, the armored vehicles landed with the first wave at 8:30 A.M. on March 11. They hit the beach along with the 16th Regimental Combat Team and the 116th Regimental Combat Team.

Late on March 12, Company A departed Slapton Sands and headed to Barton Stacey. The men arrived the next morning but without tanks. Their armor remained on the beach, awaiting transport to Barton Stacey, via tractor-trailer.

The entire 741st stayed at Barton Stacey until March 24. Early that morning, four companies from the battalion began a two-day trek to Wales and a firing range located at Camp Merrion near Pembroke. (Company A and Headquarters Company stayed behind at Barton Stacey. Both units had spent most of February at the range. Company D and Service Company were also there in February but now returned for more training.)[10]

At Camp Merrion, the men used a variety of weapons, ranging from handguns to tank cannons. They also practiced antiaircraft gunnery at an indoor facility where a projector flashed enemy-aircraft silhouettes on a dome-shaped ceiling. The trainees fired at the silhouettes with dummy weapons, while a machine tabulated the hits and ammunition expended.

Once marksmanship practice ended, everyone returned to Barton Stacey. Ten days later, Companies B and C headed to a DD school at Torcross in Devon. Until this time, they had only operated Valentine DDs. They now received American-built M4A1 DDs, sixteen per company.

The new Torcross residents changed the town's name to Camp MacDevon. The "Mac" portion alluded to Colonel Severne S. MacLaughlin, commander of the 3rd Armored Group, to which the 741st belonged. Major William D. Duncan from the 743rd Tank Battalion served as school commandant. All the men attending his school occupied civilian dwellings vacated by their owners in December 1943, when Torcross and the surrounding coastal region became a place for secret maneuvers. Most 741st troops lived in cottages at nearby Slapton Sands.

Training at MacDevon involved "sailing" in the calm waters of Slapton Ley, a freshwater wildlife preserve. Here the tank crews familiarized themselves with their new DDs, especially the steering and control mechanisms, which differed from the Valentine DDs. The training also encompassed practice LCT launchings in nearby Start Bay. Occurring several thousand yards off Slapton Sands, most launchings took place during the day, but some transpired at night under blackout conditions. The entire program lasted less than two weeks. Afterward, Companies B and C remained at MacDevon to participate in a dress rehearsal for the invasion.

The 741st took part in Exercise Fabius I.[11] Warships shelled mock enemy fortifications at Slapton Sands, and Army Air Force planes dropped bombs. Foot soldiers splashed ashore from landing craft, while DD tanks "swam" in and LCTs unloaded regular tanks outfitted for deep-water fording. After securing a beachhead, the infantry captured MacDevon before continuing on to other objectives. The tanks provided heavy, well-aimed fire. Without them, the infantry had nothing more potent than mortars, machine guns, and bazookas.

The vital need for tanks led to the inclusion of DDs in the Normandy assault force. Army planners believed these unique vehicles offered advantages. They presented small, unthreatening targets bearing no resemblance to tanks during their journey to shore. Moreover, they could land in a well-dispersed formation. The enemy might also suffer psychological shock upon seeing DDs crawl from the surf and reveal themselves as tanks. Navy planners saw another advantage. Because the vehicles launched at sea, their LCTs could remain well away from shore, where they were less vulnerable to German artillery fire.

Despite the perceived benefits, doubts lingered about the amphibious contraptions, necessitating the addition of regular tanks to the assault force.

★★★

Fabius I concluded all the training and rehearsals. In accordance with invasion plans, the 741st commander split his battalion into three echelons: assault, support, and residue.

Attached to the 1st Infantry Division, the assault echelon included the DDs and regular Sherman tanks. The DDs would land five minutes before H-hour, followed by the regular tanks at H-hour. One minute later, the first wave from the 16th Regimental Combat Team would touch down. The disembarkment area spanned two Omaha Beach sections: Easy Red and Fox Green.

Also with the 1st Division, the support echelon would begin landing thirty minutes after H-hour and continue arriving over the next fourteen hours. Its troops included maintenance men, supply personnel, and a skeletal headquarters staff led by the battalion commander. The support echelon possessed five reserve tanks, five tank-recovery vehicles, four quarter-ton trucks (peeps or jeeps), one halftrack, and three 2-½ ton trucks, each carrying 750 gallons of gasoline.[12]

Major Jack H. Browder, the battalion executive officer, led the residue echelon. Its men moved from Barton Stacey to the coastal city of Bournemouth. They would begin crossing the English Channel nine days after the invasion.

Only one part of the residue had a role in the invasion, albeit an indirect role. The battalion Medical Detachment joined an elaborate casualty evacuation network. The detachment's commander, Captain Michael R. Godett, a physician from Pasadena, California, established an aid station near the Bournemouth waterfront. There he and his men would treat casualties received from transport ships. The team included Captain Harold E. Lippman, assistant battalion surgeon, and Captain Henry I. Stanger, battalion dental officer.

On May 16, Baker and Charley began the final countdown to D-Day. They loaded their DDs on eight unarmored Mark 6 LCTs at Slapton Sands, four vehicles per vessel. The following day, they sailed to Portland Harbour with a contingent of tankmen on each boat. All other personnel left Camp MacDevon on May 25 and journeyed to Camp D-2 outside the Dorset town of Lytchett Minster.

The men enjoyed comfortable conditions inside D-2, one of fourteen concentration camps in Marshalling Area "D." Everyone had ample food, most of it top quality. Special Service troops presented feature films and distributed magazines and paperbacks. Civilian radios abounded. The Radio Berlin broadcasts of Mildred E. Gillars ranked among the most popular programs. She called herself "Midge," but her American listeners knew her as "Axis Sally" or the "Bitch of Berlin." Gillars played pop music and pitched not-so-subtle propaganda, including warnings about the gruesome fate awaiting any Allied soldiers who dared cross the Channel.

Activities at D-2 included final preparations for the invasion. Everyone received French phrase books and invasion currency—two hundred francs per man, instant gambling money for some. Quartermaster troops doled out Mae West life-preserver vests, as well as socks, leggings, and coveralls impregnated with a chemical to protect against blister agents like mustard gas. Each soldier also received a black rubberized bag containing a gas mask.

The tank crews attended a briefing on their role in the invasion. They learned about terrain features, enemy defenses on the far shore, and heard the code name "Omaha" for the first time. Visual aids included maps, dioramas, and aerial photographs. The briefing officer revealed that H-hour was 6:10 A.M., but he withheld the D-Day date and the town names along the invasion beach. The men would receive those details only after setting sail.

Orders from Allied headquarters kept all troops quarantined in camp. They had become privy to some of the most important secrets of the war. The camp population remained inside the perimeter fence. Armed guards enforced the lockdown.

Like the DD crews at D-2, the 741st men who operated the regular tanks received their own briefing. They too remained under quarantine at Camp D-14, near Weymouth, from April 23 until June 2, when everyone moved to Portland Harbour.

The crewmen loaded sixteen Able Company tanks, each towing armored trailers, onto eight armored Mark 5 LCTs. The trailers held tank ammunition and Bangalore torpedoes for the infantry. The vessels had a special modification: a temporary wooden platform that permitted two tanks to fire shells over the bow at land-based targets while heading toward the beach.

Besides regular tanks, every landing craft took aboard a single tankdozer with an eight-thousand-pound blade for earth moving. These special vehicles belonged to an ad hoc platoon assigned to Headquarters Company and led by the Company A maintenance officer. His little unit had a special mission to assist army and navy demolition men in clearing safe lanes through the enemy's beach obstacles. The platoon included men from all five 741st companies, although two machines and their crews originated from the 610th Engineer Light Equipment Company.[13]

On June 2, the DD crewmen left Camp D-2 for Portland Harbour. Red Cross women welcomed the men with coffee and doughnuts. Besides a quick snack, each man received a leaflet with a message from the Allied Supreme Commander, General Dwight D. Eisenhower, proclaiming a "great crusade" ahead.

Leaflets in hand, the men boarded small boats that shuttled everyone to the LCTs carrying their DD tanks. There they waited to receive the green light.

The assault echelon left port shortly after 2 A.M. on June 4, 1944. D-Day was June 5.

One after another, LCTs headed out to join the invasion fleet. Violent swells met the small craft once they passed the protective breakwater surrounding Portland Harbour. It became impossible for many to stay on course. Seasickness overcame countless soldiers, and, as conditions worsened, the possibility of a successful landing in France seemed to fade.

At 6:30 A.M., all vessels received a prearranged signal: "POST MIKE ONE." General Eisenhower had postponed the invasion and recalled the armada to port.

His staff quickly rescheduled D-Day for June 6 with H-hour at 6:30 A.M.

Weather forecasts indicated a chance for improved conditions starting on June 5 and lasting for thirty-six hours. Broken clouds and moderate seas seemed likely. Afterward, meteorologists predicted a return to stormy weather.

The situation wrapped Eisenhower in worry. Delay meant a huge security risk. Not only did thousands of soldiers know details about the invasion, but Luftwaffe reconnaissance aircraft might spot the massive buildup. The troops themselves also required consideration. How much longer could they stay in suspense with seasickness and nervous tension wearing them down?

Despite these factors, half of Eisenhower's senior commanders argued against going ahead. The weather bothered them too much. With his commanders split, the decision fell to Eisenhower alone. "I'm quite positive we must give the order," he announced. "I don't like it, but there it is."[14]

By midnight, the invasion force received Eisenhower's order and began leaving port. Minesweepers led the way, followed by the slowest vessels in the fleet, the LCTs. These included the craft carrying the 741st assault echelon.

At 3 A.M. on June 5, the tankmen and their armor cleared Portland Harbour. The weather and sea had improved but remained ugly. And more than 120 miles lay ahead.

The support echelon also rode in LCTs, as well as in larger and faster ships. The attack transport USS *Samuel Chase* carried the battalion commander, Lieutenant Colonel Skaggs, and his staff. He had risen to lieutenant colonel on May 10.

The *Chase* sortied from Portland Harbour at 5:27 P.M. on June 5 and steamed toward Normandy. The anticipated weather break had arrived, with the sun cutting through the clouds now and again, but the water remained choppy, an unremitting procession of whitecaps.

In his cabin, Colonel Skaggs chatted with his roommates, including *Life* magazine photographer Robert Capa. They also had a visitor, Colonel George A. Taylor, gritty commander of the 16th Regimental Combat Team (or 16th RCT for short). Capa and Taylor had known each other since the campaign in North Africa. Skaggs listened as the two veterans reminisced. They had no illusions about the rugged task ahead.

The struggle had already begun for Skaggs's men riding aboard LCTs. The turbulent sea punished their flat-bottomed boats, heaving them up and slamming them down. Waves broke over the gunwales, and the decks ran ankle deep in salt water. Most navy crewmen had grown accustomed to such conditions but not so for the tankmen. They hurled stomach fluid and undigested food. That sight and stench begot more sickness, as did diesel fumes and the stink from overflowing toilets. But one factor lessened the torment—everyone endured it together.

Since ancient times, shared adversity has created cohesion among soldiers. It transformed the 741st. The battalion began in 1942 as a discordant mishmash: Yankees, Johnny Rebs, New Yorkers, Westerners, Midwesterners. These parochial identities, with their inherent antipathies, eroded under rigorous training and the strictures of army life. A collective identity evolved and formed the bedrock of each man's ability to survive in combat.

The first test of that ability stood hours away. Ahead, beyond the horizon, lay that beach called Omaha—the razor's edge of Hitler's Fortress Europe.

CHAPTER 2 ABANDON SHIP!

June 6, 1944, 2:00 A.M.
The sea erupted like a powder keg.

Technician Fifth Grade Valentine Fister stuck his head out from the turret of the Company A command tank chained to the landing craft's deck.

What he saw jolted him wide awake.

"The damn ship is sinking!" Fister screamed, turning to his buddies inside the tank, "Get the hell outta here!" He, Technician Fourth Grade Francis G. Thomas, and Private First Class Herman W. Mayo fumbled for their life-preserver belts and flew out of the vehicle, their hearts pounding.[15]

The dark, swelling English Channel crashed over the deck. The landing craft rode the waves at an odd angle, as if starting to capsize.

Frigid sea spray pelted their faces as the men struggled across the tilting deck now awash in swirling water. "Look out!" someone yelled over the deafening gusts of wind. A tankdozer ripped free from its chains, and its rear end crunched into the vessel's side.[16]

Their boat, LCT 2049, lay twenty miles off the French coast, reeling from a jagged hole torn in its starboard side by an enemy mine. The explosion also ripped away the starboard gunwale. Sailors frantically attempted to pump out a torrent of black water gushing into the engine room. Yet, within minutes, one of the diesel engines sputtered and quit.

On deck, Val Fister and his comrades found their gunner, Corporal Adolf C. Rodriguez, and the tank commander, Captain Cecil D. Thomas. The captain commanded Company A, and his moniker, "Smiley," had inspired the name "Always Smiling" painted in white letters on the command tank. Though lacking his ready grin today, Thomas remained levelheaded as he gathered the crews of all three tanks aboard the vessel.[17]

His fifteen men joined the other passengers, twenty-seven soldiers from Company A, 299th Engineer Combat Battalion, some on temporary assignment from the 2nd Infantry Division. There were also thirteen sailors from Navy Combat Demolition Unit 46.

Like Cecil Thomas, the skipper of 2049, Lieutenant j.g. Raymond E. Eberhardt, stayed composed and focused on the job at hand. With his craft listing twenty-five degrees, he knew the situation was untenable, and only one course of action remained. Eberhardt barked out the command, "Prepare to abandon the ship!"—the worst order he had ever given.[18]

The doomed vessel towed a smaller landing craft laden with explosives for the engineers and demolition men. Eberhardt signaled the coxswain of the smaller craft and directed him to steer alongside 2049. The plan: everyone on the sinking boat would then jump aboard—a perilous undertaking, as both vessels rolled and pitched with each swell. The men risked being smashed between the two boats or landing in the water and drowning in the darkness. Yet nobody hesitated. "We jumped in the little boat, and away we went," Fister recalled years later. "All I had on were a pair of coveralls, my shoes, and my gasmask. No gun, no nothing. I had a handkerchief that I put over my head to keep it warm."[19]

He and the others looked back at 2049 and saw it bobbing like a cork. It slowly vanished in the darkness.

Around 4:00 A.M., the gray silhouette of a landing ship loomed ahead—LST 317. Eberhardt received permission to board. The crew from 2049 and the tankers climbed a rope ladder to the ship's main deck. The engineers and navy demolition men remained on the little craft and headed off toward their assigned landing area to destroy beach obstacles.

After ascending the ladder, Fister, dripping wet, shook himself like a dog. He encountered a couple of sailors, who asked if he had fallen in the water.

"No," Fister said.

"Are you sure?" one of them said.

"No, I didn't fall in."

"Do you swear to it?"

"I *didn't* fall in."

"Well," one of them replied, "I'm sorry, you said the wrong thing. We have a double shot of booze here for anybody who fell in."[20]

The sailors walked away. Fister stood there, thinking how much he wanted that drink.

After awhile, he retreated below deck with Private Clarence Olson, the driver of Sergeant George Cora's tank, which had also been aboard 2049. The two men found cots and flopped down.

As they lay there, warm and comfortable, Fister said to Olson, "Why the hell don't we just stay right here, and when this ship goes back to England, we'll go back with it?"[21] Olson agreed to the idea, but soon both men felt guilty. Had they come all this way to contribute nothing to the invasion?

The would-be stowaways gave up their cozy accommodations and emerged topside. They gazed at the panorama spread out before them. The sea and sky had merged into a single shade of gray, an abyss above and below. The water brimmed with ships, their bows pointed toward France and a nightmare yet to unfold.

CHAPTER 3 DISASTER AT SEA

Sixteen landing craft from LCT Group 35 crested swell after swell, riding up and down. The LCTs carried DD amphibious Shermans from the 741st and 743rd Tank Battalions.

In the predawn darkness, the vessels reached their rendezvous area. The LCTs split into a pair of columns, one carrying the 741st and the other with the 743rd.

The sleek hull of a control boat sliced through the columns, and confusion arose. The senior navy officer in the 741st column had instructions to trail the little boat, but it cut through too fast and turned too sharp. His column had difficulty following.

Two LCTs loaded with 741st DDs inadvertently tagged onto the 743rd column. Nobody realized the mistake.

On June 6 at 4:00 A.M., the LCT skippers sounded general quarters aboard their craft. The tank crews sprang into action and removed the chains locking their vehicles in place. Each tank driver turned a valve, opening two compressed-air cylinders, which inflated thirty-six rubber pillars. These pneumatic pipes raised his tank's canvas flotation screen. The other crew members checked the screen, ensuring it had deployed correctly and had no damage.

Engines at full speed, the LCTs scudded along toward the launching point for the tanks, an imaginary line three miles from the beach.

Force Four sea conditions prevailed, and they created three- to five-foot swells. The undulating water presented no problem for landing craft but spelled big trouble for DD tanks.

Previous training had shown that DDs could withstand Force Three conditions. Anything greater posed an extreme risk. With its screen raised, a DD tank had three feet of freeboard at its rear and four feet at the front. Waves any higher would crash over the screen, swamping the tank.

Before the invasion, the army and navy recognized that rough seas might preclude the launching of DDs. In that event, the landing craft should deliver the DDs to the beach.

For each tank battalion, the launch decision rested with one person, the senior army officer assigned to its DD contingent. For the 741st, Captain James G. Thornton Jr. had that responsibility.

Jimmie Thornton commanded Company B and had earned his men's respect during two years of training. Even though he lacked a commanding presence, they felt confident in his leadership. At five foot four and weighing 140 pounds, the little captain possessed a boyish face and a high-pitched voice to match. His coveralls, too large for him, dragged on the ground around his ankles. Jimmie grew up in Wilmington, North Carolina, and graduated from The Citadel in 1940, after his father enrolled him at the prestigious military school to set him straight. The young Thornton played running back and tight end on the school's football team and proved himself a fearless, almost reckless, ball carrier. "He would not have hesitated to run through a brick wall," recalled one family member.[22] During his freshman year, Jimmie suffered a nasty hip injury. Surgery to mend the break resulted in his left leg being slightly shorter, a handicap insufficient to disqualify him from military service. He obtained a commission two months after Pearl Harbor and joined the 741st.

On D-Day, as the launching point drew near, Thornton rode aboard one of the LCTs that had joined the 743rd column by mistake.

Far away from Thornton, his naval counterpart, Lieutenant j.g. John E. Barry, skippered LCT 549 and led the 741st column. Barry's landing craft swung left and ran parallel to the beach in preparation for launching the DDs fifty-five hundred yards from shore.

But the final decision to launch rested with Thornton.

He had radio contact with the Company C commander, Captain Charles R. Young, a tall, athletic man with a dominating personality. Young rode aboard LCT 598, two vessels behind Barry's lead craft. The army officers debated whether to launch.

By now, the skipper of Thornton's LCT realized he had mistakenly joined the 743rd column. He altered course to port, as did the LCT behind him, which also carried 741st DDs.

Around that time, word came down the 743rd column that its senior army officer had decided against launching. He expected Thornton to follow suit.

All the LCTs in the 741st column turned toward the beach, ready to launch.

The driver in Young's tank, Private First Class Ralph A. Woodward, from Columbus, Wisconsin, listened as Young and Thornton spoke over the radio. Young argued against launching, recommending instead that the LCTs ferry the DDs to the beach.

"He got 'no' for an answer," Woodward remembered.[23] Thornton decided to launch.

Young relayed the decision to the skipper of his LCT, and several sailors dropped the ramp. At 5:31 A.M. the first DD tank plunged into the water, followed by the remaining three. The sailors raised the ramp, and the vessel turned away, heading out to sea. The navy men last observed the DDs afloat and underway toward shore.

On Thornton's order and following Young's lead, DD tanks on the other LCTs also launched. The situation quickly deteriorated. Swells crashed against the DDs, and water gushed over their canvas skirts.

Ralph Woodward described the effect upon his tank's flotation apparatus. "Pretty soon the pipes started kinking. The struts started snapping back in the center. The canvas started tearing."[24]

Captain Young shouted for his crew to abandon their tank. The men inflated their yellow Mae West life preservers.

Woodward scrambled out through the turret, but his Mae West snagged on a caliber .50 machine gun mounted outside. Frigid water rushed in around him as the gun pulled him under the waves. The sea turned darker and darker while the tank descended into a watery tomb. Woodward held his breath and frantically wrestled to free himself, dislocating his left arm in the process. The tank hit the seabed with a muffled thud, and sand swirled up in his face. One awful thought crossed his mind: his parents would never know what happened to him.

With a desperate will to survive, the twenty-one-year-old tank driver finally broke free from the gun, but he thought he had lost his Mae West, as its neck loop had slipped off his head. Unaware that the life-saving device remained secure around his waist, he floated upward, feet first. Even though he had an escape apparatus strapped to his chest with seven minutes of oxygen, Woodward continued holding his breath as the sea around him grew lighter and lighter.

At last, he burst through to the surface, coughing and spitting blood as he gasped for air, his chest heaving. Blood vessels in his lungs and throat had ruptured. The tank driver spun around in the water and spotted two other men from his crew, Captain Young, and, farther away, Private Francis P. McKellar.

A landing craft outfitted with rocket launchers swiftly approached, and sailors hauled Woodward out of the drink along with Captain Young.

McKellar had inflated a life raft stowed aboard the tank. The dinghy floated upside down in the water. Alone, McKellar had no way to flip it over, so he crawled on top of it—but not for long. Several hundred yards away, the USS *Arkansas* opened fire with its twelve-inch guns in a thunderous roar. With every blast, a shock wave raced across the water, and the raft quivered like gelatin. McKellar slid off each time, then kicked and tugged to boost himself back onto it again.

Elsewhere on the water, other DD crews struggled to survive. The current swept some away from their assigned landing area, so their drivers turned to adjust for the drift. The swells hit those machines broadside and engulfed them. The soldiers had never learned to keep their tanks perpendicular to the swells, and they feared landing in the wrong area. Still other DDs foundered soon after launching, poor seamanship playing no part.[25]

Lieutenant Barry watched the first three tanks plunge off his LCT and sink within a minute, all from Charley Company. While driving forward to launch, the fourth tank ripped a hole in its canvas screen after catching a 20 mm gun barrel. The tank's commander was Staff Sergeant John R. Sertell, a vigorous, can-do soldier, who had reenlisted three days after Pearl Harbor at age thirty-three, having already served three years as an infantryman. Navy crewmen begged him to stay aboard the vessel. They had instructions to beach a damaged tank, but Sertell refused, convinced that his bilge pump would compensate for the tear.

On another LCT, Private First Class Philip L. Fitts, from Danville, Virginia, served as bow gunner aboard the Company C tank commanded by Sergeant John G. Suydam. After launching, their DD motored along for twenty minutes before white-capped waves crushed its canvas skirt with stunning force. All five crew members escaped and pulled themselves into a raft that Suydam inflated.

Life rafts offered the best chance for survival, as they provided protection from the cold sea. Any man floating in just a life preserver faced death from hypothermia.

Sailors on a control boat sighted the raft carrying Suydam's crew. The helmsman attempted to maneuver alongside it, but the high swells thwarted him. He persisted until reaching the hapless soldiers.

Once aboard the boat, Phil Fitts gazed in dismay at the surrounding sea. Here and there, tankers huddled in bright-yellow rafts, while others floated alone in life preservers, some with their heads hanging under the waves. He and another man fished out one floater, catching him by his belt with a hooked pole. Fitts recognized a familiar face. It was John Sertell, and he had stopped breathing. Fitts tied him to the deck, and himself, too. He feared them both washing overboard as waves broke against the boat.

"I worked on him for about an hour, giving him artificial respiration, but I never could bring him around," Fitts recalled. "I got bubbles coming out of his nose, but I never could really say he was breathing. I didn't take a pulse reading. I was so anxious to get artificial respiration going that I didn't want to waste any time."[26]

A medic later put a stethoscope to Sertell's chest and pronounced him dead.

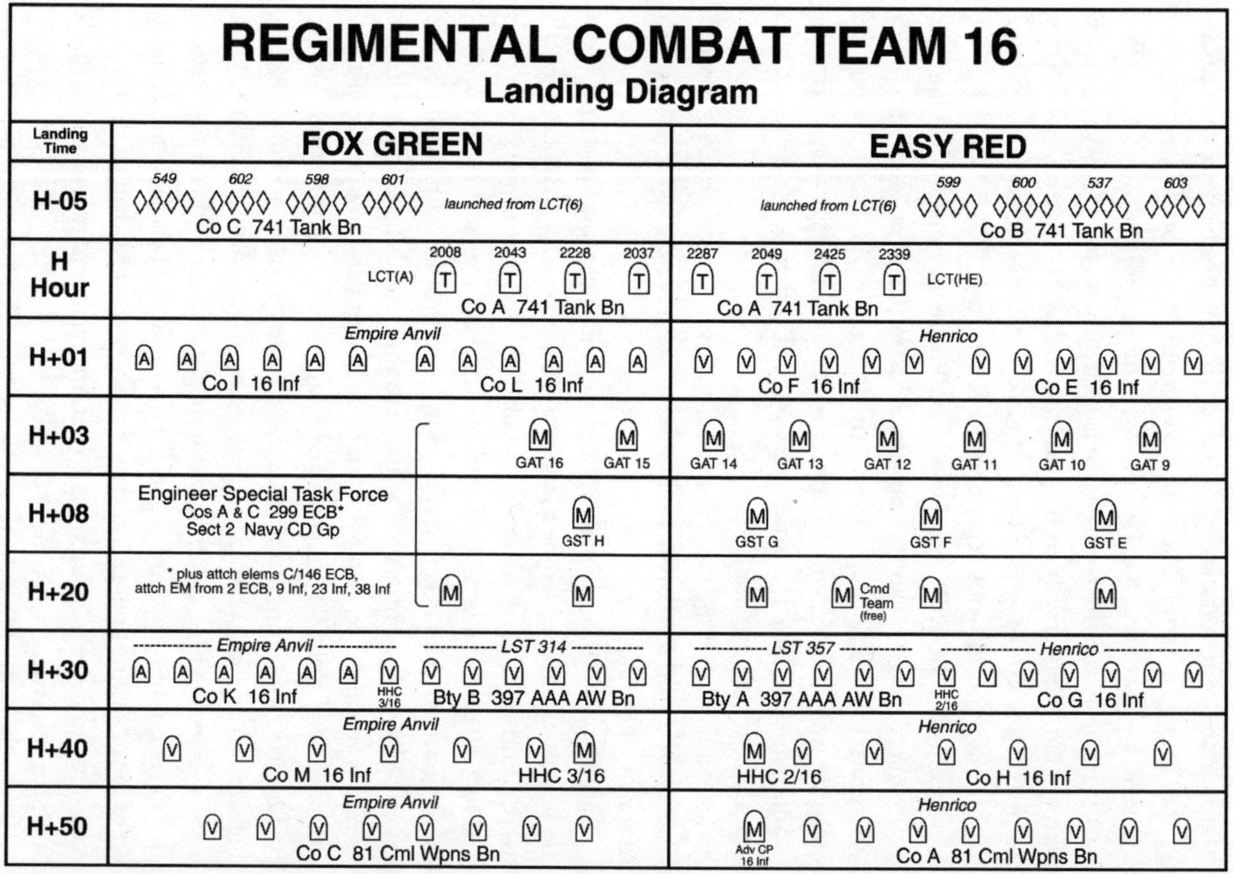

REGIMENTAL COMBAT TEAM 16

Landing Diagram

Landing Time	FOX GREEN	EASY RED
H-05	549 ◊◊◊◊ 602 ◊◊◊◊ 598 ◊◊◊◊ 601 ◊◊◊◊ *launched from LCT(6)* Co C 741 Tank Bn	*launched from LCT(6)* 599 ◊◊◊◊ 600 ◊◊◊◊ 537 ◊◊◊◊ 603 ◊◊◊◊ Co B 741 Tank Bn
H Hour	LCT(A) 2008 T 2043 T 2228 T 2037 T Co A 741 Tank Bn	2287 T 2049 T 2425 T 2339 T LCT(HE) Co A 741 Tank Bn
H+01	*Empire Anvil* A A A A A A — Co I 16 Inf A A A A A A — Co L 16 Inf	*Henrico* V V V V V V — Co F 16 Inf V V V V V V — Co E 16 Inf
H+03	M GAT 16 · M GAT 15	M GAT 14 · M GAT 13 · M GAT 12 · M GAT 11 · M GAT 10 · M GAT 9
H+08	Engineer Special Task Force Cos A & C 299 ECB* Sect 2 Navy CD Gp M GST H	M GST G · M GST F · M GST E
H+20	* plus attch elems C/146 ECB, attch EM from 2 ECB, 9 Inf, 23 Inf, 38 Inf M · M	M · M Cmd Team (free) · M · M
H+30	*Empire Anvil*: A A A A A A — Co K 16 Inf; V — HHC 3/16 *LST 314*: V V V V V V — Bty B 397 AAA AW Bn	*LST 357*: V V V V V V — Bty A 397 AAA AW Bn *Henrico*: V — HHC 2/16; V V V V V V — Co G 16 Inf
H+40	*Empire Anvil* V V V V — Co M 16 Inf V V M — HHC 3/16	*Henrico* M V — HHC 2/16 V V V V V — Co H 16 Inf
H+50	*Empire Anvil* V V V V V V V V — Co C 81 Cml Wpns Bn	*Henrico* M — Adv CP 16 Inf V V V V V V V V — Co A 81 Cml Wpns Bn

H+60	5 Engineer Special Brigade (first echelon: 37 ECB, Co C 6 NBB, misc units) — 544 (T), 543 (T), 545 (T) — Co A 20 ECB	541 (T), 539 (T), 538 (T), 540 (T), 542 (T) — Co A 1 ECB
H+65	493 (I) — LST 376: D D D D D D — 459 Amph Trk Co — V	Det V Corps — LST 374: D D D D D D D — 459 Amph Trk Co — 88 (I)
H+70	Begining H+60, landing craft use channels cleared by gap assault teams unless otherwise directed by beachmaster	V V V V V V — Co C 16 Inf; *Chase*; V V V V V V — Co A 16 Inf
H+80		V V V V V V — HHC 1/16, misc engrs; *Chase*; V V V V V V — Co B 16 Inf
H+90	213 (T), 206 (T), 293 (T), 209 (T), 276 (T) — 62 Armd FA Bn	V V V — 5 ESB, HHC 1/16; *Chase*; V V V V V — Co D 16 Inf
H+95		V — Med Sect 16 Inf; *Chase*; M — HHC 16 Inf
H+105	-LST 314-: D D D — HHC 3/16 — V	-LST 376-: D D D — HHC 1/16; -LST 374-: D D D — HHC 2/16; ----LST 357----, ----LST 314----, ----LST 374----: D D D D D D D D D D D — 7 FA Bn
H+110		LST 376: D D D D D D D D — Cn Co 16 Inf, Det 1 Sig Co; *Chase*; M V M — Adv CP 1 Inf Div
H+120	Co B 20 ECB — 83 (I); 305 (T), 18 (T), 195 (T), 20 (T) — 197 AAA AW Bn	547 (T), 25 (T), 200 (T), 201 (T) — 197 AAA AW Bn; 199 (T), 546 (T), 271 (T) — AT Co 16 Inf; 89 (I), 85 (I)
H+130		815 (T), 814 (T), 623 (T), 550 (T), 624 (T), 625 (T), 2487 (T) — 7 FA Bn, 5 ESB

I LCI — T LCT — M LCM — A LCA — V LCVP — ◇ DD tank — D DUKW

Diagram reflects planned landings, not actual. Additional RCT 16 elements to land H+135 through H+880

★★★

To the west, among the landing craft carrying Baker Company, Ensign Henry P. Sullivan commanded LCT 600 carrying four DD Shermans from the First Platoon.

At 5:05 A.M. the platoon leader, Second Lieutenant Patrick J. O'Shaughnessy, reported to Sullivan. O'Shaughnessy had spoken by radio with Captain Thornton and received the go-ahead.

Minutes later, O'Shaughnessy's tank experienced throttle trouble. He wondered whether to proceed and contacted Thornton. The captain deferred to Sullivan, who had authority in the event of a damaged or malfunctioning tank.

Sullivan later recounted his discussion with O'Shaughnessy. "We both felt it was a little too rough to launch. However, after working on the tank, he decided to try launching."[27]

The lieutenant's tank had a sticky throttle, so his driver coasted down the ramp to the water before starting the motor. Just then, an enemy shell exploded off the port beam, and a large wave slapped the landing craft. The three DDs behind O'Shaughnessy lurched backward, sliding into one another. The mishap tore their canvas skirts.

O'Shaughnessy's tank hit the water at 5:35 A.M. As it labored toward shore, Corporal Wardell C. N. Hopper stood on its engine deck, attempting to maneuver the machine using its auxiliary steering system, a wooden tiller. On either side of him, O'Shaughnessy and Private First Class William S. Merkert struggled to hold the canvas upright, pushing against it with all their strength. Sheets of water fell on them until the heavy swells overpowered the duo.

One side of the skirt caved in, and the DD sank. O'Shaughnessy and Merkert clung to a life raft, while Hopper and two other crewmen went down with their vehicle.

Hopper felt the tank settle on the Channel floor. "My cigarettes worked out of my pocket and floated up past my face. I gave up hope when I saw that and started to breathe water."

But Hopper's life vest saved him.

"I opened my eyes again and saw the water getting lighter."[28]

He surfaced, as did the other men who had gone to the bottom. One of them, Private First Class Alfred Snike, swam over to O'Shaughnessy's life raft.

Back on LCT 600, Corporal Jack B. Boardman, from Downers Grove, Illinois, agonized over the torn canvas. He repeatedly warned his tank commander about the problem. The commander, Sergeant Richard L. Maddock, dismissed the warnings. "He kept telling me to shut up," Boardman remembered.[29]

Staff Sergeant Paul W. Ragan commanded the tank parked behind Maddock's machine. Ragan had watched O'Shaughnessy's tank founder. He also eyeballed the damage to the three remaining tanks. As the senior-ranking soldier on the vessel, he decided against launching.

Sullivan leaned from the conning tower and spoke with Ragan. The sergeant instructed Sullivan to ferry the three damaged DDs to the beach. The navy officer ordered the ramp raised.

En route to shore, the crew of LCT 600 rescued the men in O'Shaughnessy's raft. The sailors also tried to recover Hopper but initially lost him in the rough water. The landing craft eventually reached him, his arms and hands now numb from cold. Unable to keep a tight grip on a rope thrown to him, Hopper wrapped the line around his arms, while his rescuers pulled his limp body aboard their craft. Covered with a wool blanket and totally exhausted, he passed out but survived the ordeal.

One other crew member, the driver, Technician Fourth Grade Gordon D. Crouse, later turned up on the USS *Samuel Chase*. Men on another landing craft had rescued him.

The last two LCTs to launch DD tanks arrived late. These vessels—LCTs 537 and 603—had mistakenly joined the landing craft carrying DDs from the 743rd Tank Battalion. After correcting course, the two wayward craft reached the waters off Easy Red sector.

Ensign Robert I. McKee skippered LCT 537. The four DDs on his boat included the one commanded by Captain Thornton. The captain motioned to McKee that he wanted to launch, and McKee's men lowered the ramp. Thornton rolled into the water followed by Sergeant Millard I. Case and his tank.

The swells immediately overwhelmed Case's DD. He scrambled to inflate a life raft, but the vehicle sank during the process, sucking him into the dark abyss. Case desperately latched onto the raft, and it pulled him upward to the surface. His gunner, Corporal Howard E. Riley, popped up nearby. They were the only survivors of their crew.[30]

Staff Sergeant Turner G. Sheppard, from Seneca, South Carolina, departed from LCT 537 just after Case. Behind Sheppard, Sergeant Daniel H. Potcher Jr. launched his machine, and it immediately sank.

Sheppard maneuvered his DD using its wooden tiller. Trying to keep pace with Thornton's tank, Sheppard yelled for his driver to open up the throttle.

"We proceeded toward shore," Sheppard later reported, "but Captain Thornton's tank went down after we had gone about one thousand yards. We were now alone."

Sheppard described what happened next. "A small rocket-launching boat pulled alongside, and its commander asked which beach we were headed for. I replied Easy Red, and he stayed with us until we were about one thousand yards offshore. At this point, he launched his rockets and said, 'There's your beach!' We started to draw fire but made it onto the beach."

The tank emerged from the water near the boundary between Easy Red and Fox Green. Sheppard ordered his driver to "raise propellers" and then to "break struts and deflate." The driver first lowered the front of the canvas skirt, then dropped the rear portion.

Sheppard recalled, "I was then able to get into the turret for the first time. This was the best landing we had ever made."[31]

To his left on Fox Green, he spotted the DD commanded by Sergeant George R. Geddes Jr. from Company B's Third Platoon. His tank had also reached the beach after launching from LCT 599. He and Sheppard shared the distinction of commanding the only DDs from the 741st to reach shore under their own power.

At 6:32 A.M. Sheppard watched LCT 600 drop its ramp on Easy Red. The three tanks with torn skirts rumbled off the vessel and onto the beach.

Of the thirty-two DDs launched by the 741st, only five reached shore.

CHAPTER 4 **ON THE DEVIL'S DOORSTEP**

The men aboard LCT 2425 saw the French coast ahead in the pallid morning light. They picked out landmarks like the church steeple at Colleville-sur-Mer and the mouth of the Ruquet valley, until an unexpected sight grabbed everyone's attention.

"Swimmers in the water!" someone shouted.[32]

The scene made no sense at first. Nobody had witnessed any naval vessels sink or aircraft fall from the sky. The onlookers unraveled the mystery when they noticed what looked like football helmets on the men in the water. Only tank crewmen wore such headgear.

Lieutenant j.g. Donald H. Sandie skippered 2425, which carried two tanks from Company A and one tankdozer from the 610th Engineer Light Equipment Company. Sandie had orders to bypass anyone in the water and reach Easy Red, bar nothing. His coxswain remained on course, pounding past the stranded men. Some of them waved for help in vain.

German shells tossed up geysers in the water as 2425 crossed the final mile to shore. The tanks and tankdozer responded, lobbing high-explosive shells toward the enemy.

The coxswain bore down on the beach. He reduced speed and felt his way through the beach-obstacle maze erected by the Germans. Bullets splashed in the surf and rattled against the landing craft's armored hull.

Sandie reached Easy Red on schedule at H-hour. Smoke drifted across the pebbly beach as the boat's ramp dropped.

Staff Sergeant Thomas R. Fair commanded the first Sherman that lumbered down the ramp and into the breaking waves. His machine sank up to its turret before gaining traction on the bottom and crawling forward. Sergeant James B. Larsen followed in his tank. Behind him, Technician Fourth Grade Donald G. Adkins from the 610th Engineers commanded the tankdozer.

Tommie Fair's crew opened up with their caliber .30 machine guns, peppering the hillside behind the beach. His driver crept forward but stayed partially submerged for protection.

The sergeant and his gunner scanned the terrain through their periscopes, trying to locate enemy bunkers. Fair, a native of Petrolia, Pennsylvania, recounted, "It wasn't long until we spotted one, which no doubt was a machine-gun emplacement. I put my gunner on it, and he fired."[33]

The first shells landed short, missing the bunker's narrow embrasure. The gunner adjusted his aim. "The next shots entered straight through the opening," Fair remembered.[34] The machine gun fell silent.

With a large white "10" painted on its exhaust stack, Don Adkins's tankdozer supported Gap Assault Team 10 in clearing a fifty-yard gap through the beach obstacles. Adkins and his crew used their earth-moving blade to uproot and shove these barriers aside. Army engineers and navy demolition men used explosives to destroy others.

The team scrambled to complete its work before the morning tide rose and submerged the obstacles. All the while, foot soldiers disembarked from small gray landing craft. Many infantrymen sought cover behind the obstacles, disrupting removal operations.

The troops wading ashore made simple targets for German machine gunners, who swept the beach with vicious, hammering fire. Lifeless bodies floated face down, their blood churning in the surf.

"The Germans had full command," Adkins recalled.[35] Bullets and larger projectiles ricocheted off his tankdozer. The assistant driver, Technician Fifth Grade William C. Mitchell, remembered the noise. "It was like being in a house with a tin roof in a hail storm."[36]

The enemy defenders transformed the beach into chaos and carnage, but the Germans failed to incite mass panic. The engineers and demolition men attached explosives to obstacles and tied them together with Primacord for simultaneous detonation.

Meanwhile, Adkins's tankdozer had difficulty maneuvering with infantrymen bunched together behind it for protection. Despite enemy fire and casualties, Gap Assault Team 10 opened a path twice the required width.

Seven other teams worked to remove obstacles on Easy Red and Fox Green. Only two teams succeeded in opening gaps during the first hour on the beach. The men from Team 9 cleared a fifty-yard-wide lane. Team 12 did the same but incurred casualties when a German shell prematurely detonated charges they had planted.

The rising tide eventually drove the gap assault teams ashore. They would return to work after it began receding.

★★★

First Lieutenant Frank A. Klotz from Company A (now assigned to Headquarters Company) commanded tankdozer 12, which assisted in clearing obstacles. His machine was not actually a dozer, being the only one in the dozer platoon without an earth-moving blade. Klotz led the platoon—eight machines numbered 9 to 16 and assigned to the gap assault teams.

Two dozers never reached Omaha that day—tankdozers 11 and 15. Number 11 sank along with LCT 2049. An enemy shell struck 15 as it disembarked from LCT 2043. The explosion broke one of its tracks, rendering the vehicle inoperable. The landing craft itself sustained hits and retracted from the beach with the dozer still onboard. The damaged vehicle, commanded by Sergeant Elmer Middleton, eventually reached the beach on June 8.

Sergeant Dennis C. Holcombe commanded tankdozer 14, which sailed to France aboard LCT 210.[37] Its coxswain beached the craft at 6:33 A.M., making it the first vessel to land on Fox Green. As four sailors dashed forward and lowered the ramp, German bullets pelted the boat, and a shell set fire to a storage locker.

Holcombe's tankdozer and two Company A tanks rolled off the vessel and began pumping shells at enemy positions. LCT 210 withdrew from the beach, losing its ramp and anchor in the process. Sailors onboard later quelled the blaze.

While Holcombe's crew pushed ahead, tankdozer 13 joined the battle, with Private Edward L. Ayers in charge. This Headquarters Company soldier was an unlikely tank commander as a buck private.

Busted to that rank before joining the 741st, Ayers had racked up multiple AWOL convictions with the battalion, but nobody in the unit knew of his troubled past. As a civilian, he spent time in Texas State Prison on a felony conviction for check forgery. Ayers also had arrests and jail time for public loitering as well as drunk-and-disorderly conduct. His superiors in the 741st denied him promotion because of unexcused absences after joining the battalion, but he also demonstrated leadership ability and therefore commanded a tank.[38]

Ayers scarcely had a chance to prove his value in combat when the Germans torched his machine with an armor-piercing shell. The crew escaped and waded over to Holcombe's dozer. Ayers suffered a concussion and a fractured vertebra in his lower back. All but one crewman climbed onto the engine deck of Holcombe's dozer and rode to dry land.

Technician Fourth Grade Bolick "Bill" Smulik, the remaining man from Ayers's crew, dodged German bullets and hustled to an abandoned bulldozer. He climbed into the driver's seat and grabbed the controls. Alone and fully exposed to the enemy, he began clearing obstacles. The son of Polish immigrants, Smulik earned a Distinguished Service Cross for his efforts.

Holcombe's dozer never linked up with its gap assault team. Enemy fire cut up the team, killing army engineers and all the navy demolition men. The engineer officer ordered his men to quit their mission and join the infantry.

Holcombe, a rawboned South Carolinian with a cadaverous face, lost his nerve in the melee. He ordered his crew to "abandon tank."[39] The sergeant climbed out of his machine and sprinted inland, where he dug a foxhole. One of his subordinates, Private John Froehlich, took over the dozer. Hours later, after the tide began receding, he and two crew members commenced removing obstacles. Holcombe shouted from his foxhole and threatened them with imprisonment for ten years.

Froehlich encountered Major Robert F. Hunter, who commanded an engineer battalion.[40] He suggested shooting Holcombe on the spot. Froehlich declined. The crazed sergeant received a demotion to private the following month. Hunter recommended Froehlich and his two comrades for decorations, and they each received a Bronze Star Medal.[41]

Like Holcombe's dozer, number 16 also failed to link up with its gap assault team. The dozer's landing craft had stopped six miles offshore when its ramp unexpectedly dropped. After emergency repairs, the vessel resumed its course, but the ramp dropped again, twice. The craft finally reached Fox Green at 8:15 A.M.

After disembarking, the dozer stood still for several minutes, blasting away at German positions. Its commander, George A. Habib, a husky, dark-haired corporal from Wisconsin, ordered his driver to move out. They searched for Gap Assault Team 16 but ground to a standstill on the beach, its entire length jammed with men and equipment.

After the tide ebbed, Habib and his men set about clearing obstacles. Work abruptly stopped when their dozer threw a track. The corporal and several crew members dismounted to fix the problem. The driver moved the tank backward and forward, hoping to snap the track back into place. He instead threw the other track. Under incessant enemy fire, the crew eventually repaired their disabled machine. Flying debris bruised Habib's fingers in the process. The crew resumed work on the obstacles and helped unload a damaged LCT by pulling equipment from its deck.

Habib, born to Syrian parents, garnered a Purple Heart and a Distinguished Service Cross, while each man in his crew received a Silver Star Medal. They became the most highly decorated crew in the dozer platoon.[42]

Only one man in the platoon died on D-Day: Technician Fifth Grade John W. Farr Jr., who served aboard tankdozer 9 from the 610th Engineers. Staff Sergeant Hubert W. Spooner commanded number 9, which landed at H-hour. Farr and his crewmates eliminated beach obstacles and blasted away at enemy emplacements. As the tide crept higher and higher, the men found themselves restricted to a narrow strip of stony shingle. The smooth, oblong rocks formed an embankment along the tidal plain. The tankdozer crew motored along the shingle until a shell thumped their vehicle, and its engine conked out. The soldiers climbed out, dozer 9 now a stationary target.

While standing behind it, a shell exploded and knocked down three crewmen. Fractured bones protruded from Spooner's legs and another man's foot. The third man clutched a bloody knee penetrated by flying debris. The explosion missed Farr, who stood around side the vehicle to its left.

Nearby, an armored bulldozer chugged along the shingle. The machine belonged to an engineer unit and carried a large metal box filled with explosives. Red-hot shell fragments ripped through the bulldozer's camouflage net, which burned and set off the explosives, perhaps by igniting a fuse. The bulldozer disintegrated in a bright flash, with debris rocketing skyward. Metal hunks plunked onto the shingle, and flaming bits floated down. The blast wave bowled over John Farr, and the young soldier from Maryville, Tennessee, died where he fell.

William Mitchell in tankdozer 10 witnessed the tragedy. He later discovered a broken fuel line in Spooner's machine and found a crewman alive inside. The man, Private Franklin E. Stallings, had no physical injuries, but the terror in his eyes told all. Mitchell coaxed him out.[43]

Mitchell's own tankdozer met its fate afterward by rolling over a German antitank mine.

Staff Sergeant Michael Kark commanded the first Able Company tank knocked out, an M4A1 named "Aide de Camp."

Shortly after debarking, his machine sustained a hit to its gunshield. Concussion from the enemy shell ejected Kark from the turret. He landed on a hatch cover and snapped his spine. The injury left him paralyzed from the waist down.

The stunning blow also cracked several of the tank's waterproof seals. Salt water flowed into the engine compartment, cutting the motor.

Its driver, Technician Fourth Grade Herman W. Sendelbach, climbed out, dizzy and grasping a submachine gun. The gunner, Corporal Frank

M. Elliott, also emerged. He tended to Kark, holding him upright in the water to save him from drowning. The paralyzed tank commander remained conscious while his crew sheltered behind the tank in four feet of water.

"I got hit while we stood there deciding what to do," Sendelbach later recalled. "My gun flew out of my hands, and Frank said, 'Why the hell did you throw your gun away?'"[44]

Several hundred yards away, an antitank shell crashed against Sergeant Robert D. Coaker's tank, named "Adeline II." It had disembarked from LCT 2287 along with Frank Klotz. The projectile broke off a bogie wheel, but the tank remained functional. It hobbled along and assumed a stationary position on the shingle embankment. The vehicle's periscopes turned from side to side as the crew searched for targets.

Kark's crew decided they had to leave the frigid water. Elliott chose to make a run toward Adeline II.

"Here, you take Mike," he said to Sendelbach. "I'm going over by Coaker's tank."

"Christ, don't go over there," Sendelbach said. "They'll be shooting at that tank. That'd be suicide."

"I'm going," Elliott said.[45]

He left, along with two other crewmen, Private Robert J. Figg and Private First Class Leslie L. Robinson.

Raised on a farm in Tiffin, Ohio, Sendelbach had the brawn to haul Kark out of the water. Six foot two and 190 pounds, Kark had played tackle for the Canton McKinley High School football team in Ohio. Sendelbach dragged him onto the shingle, where medics laid him on a litter and in time evacuated him to England. Another medic dressed Sendelbach's bullet wound. He kept the slug, a gift from the Germans. D-Day was his twenty-fourth birthday. Unfortunately, he lost the bullet before the war ended.

Sendelbach, too, departed for England, but not before learning that a shell burst had killed Figg and Elliott. Robinson survived the ordeal and approached Tommie Fair's tank. "I heard someone call, and I looked over the side," Fair later reported. He recognized Robinson, but the roar of battle made it impossible to communicate. Fair dismounted and spoke with Robinson. "He wanted a cigarette and told me to give him morphine."[46]

Fair escorted Robinson to where crewmen from another knocked-out tank had taken refuge. He squeezed morphine from a syrette into Robinson and covered him with a blanket.

Mike Kark's platoon leader, First Lieutenant Edward S. Sledge II, commanded a tank named "Alabama II."[47] His crew engaged Germans operating two Belgian 75 mm field cannons of First World War vintage in casemates at Widerstandsnest 62 (Resistance Nest 62 or WN62). Sledge's gunner fired at the cannons, as did the gunner aboard the "Ace of Spades," First Lieutenant Roger J. McDonough's tank. The American gunners hit both emplacements.

Enemy riflemen and machine gunners beat the area with murderous fire. Sledge had no luck pinpointing the exact source, but it originated somewhere near the Belgian cannons. Minefields blocked the tanks from moving in that direction. Infantrymen began an attempt to move up the Ruquet valley to outflank the enemy.

The two lieutenants and their crews covered the infantrymen, as did tanks commanded by Sergeants Thomas Carlson and Thomas H. Rieman. Sledge's machine detonated an antitank mine and stopped dead. An armor-piercing shell punctured McDonough's vehicle, and it burst into a sheet of flames. Mercifully, the crew escaped. McDonough limped away with a thigh wound caused by a shell fragment. Rieman's tank continued supporting the infantry.

Carlson motored along the shingle to Fox Green and the beach exit nearest Colleville-sur-Mer, where he encountered the DD tank commanded by Staff Sergeant Sheppard.

★★★

Turner Sheppard called out targets to his gunner, Corporal Corlin N. Legler, who battered machine-gun positions at WN61. Through dust and smoke, Sheppard spotted an embrasure, the firing port of a casemate. He later described what happened:

"Legler put two concrete-piercing shells into it. One of them resulted in a large explosion, and I believe it was the ammunition supply for an 88."[48]

Soon afterward, an infantry messenger approached Sheppard's DD tank and asked him to help capture the Colleville beach exit. Low on cannon shells, Sheppard dismounted his vehicle and hurried to confer with Second Lieutenant Gaetano R. Barcellona.

The Company A lieutenant commanded a tank named "Always in my Heart," its hull scarred by five hits from a Soviet-made 76.2 mm howitzer at WN64. Barcellona's gunner pinpointed and knocked out the weapon before his crew relocated to Sheppard's area.

Barcellona gladly gave the DD men ammunition from a trailer towed behind his tank. Sheppard's crew reloaded and pushed forward on a dirt road leading toward Colleville, followed by Carlson's tank. Sheppard soon met an infantry major, the messenger's commander. The officer doubted that a German antitank gun lay ahead, but he made no guarantee.

Sheppard's tank continued along the road to a ruined house at a sharp bend.

"Just as we pulled around the house," Sheppard recalled, "I spotted an antitank gun sitting in the center of the road about six hundred yards away. Before we could get our gun on him, he hit us. His first shot killed my assistant gunner, PFC Arthur Edwards, injured me, and set the tank on fire. We bailed out, and my assistant driver, Woodrow Snyder, helped me back to the beach."[49]

Carlson and his crew thought the shot that hit Sheppard's tank came from a bunker they spotted. Carlson's gunner hammered away at the position. Did he destroy it? He never knew. Carlson's tank later threw a track, and the crew evacuated it under mortar fire.

Back on the beach, an army medic bandaged Sheppard's leg and injected him with morphine. He waited over a day before leaving Omaha on a British LCT.

Sheppard's colleague George Geddes commanded the other DD tank that swam to shore and landed on Fox Green. His gunner bagged two Belgian 75 mm cannons atop a cliff at WN60. The enemy, in turn, tagged his vehicle with several shots. One hit injured the bow gunner, Private First Class Burt A. Moyers. The bilge-pump control jarred loose and smashed against his right knee. He jumped up, intending to open his hatch and bail out. Instead, the driver, Technician Fourth Grade Arthur E. Kinkade, reached over and grabbed him. "No, you don't. No, you don't. Hang on."

No German shells penetrated the tank.

"We kept moving around, backing up, and shifting places," Geddes recounted. "We backed into the water, the tide came in, and we drowned out and had to abandon tank."[50]

So, Geddes and his crewmen became foot soldiers.

Paul Ragan's DD tank rolled up to the Colleville beach exit, but a wide antitank ditch blocked the way. His driver looked left and right for a way around the trench. Meanwhile, Ragan and his gunner hunted for enemy positions.

"I spotted a mortar operating behind a big oil drum," Ragan remembered. "I put my gunner on it, and with the first round we blew it sky high."

Ragan searched for other tanks from his unit and saw only two vehicles. "I tried to get in touch with them by radio," he recalled, "but my radio would not work because of water that had gotten into the control boxes."[51]

Ragan's driver attempted to move down the beach, but the tank had difficulty gaining traction on the shingle. The machine threw a track on the loose stones, and the crew had no luck remounting it. Amazingly, the enemy never scored a hit on the immobilized tank. After the crew exhausted their ammunition, they vacated it.

When Dick Maddock's DD tank landed, his gunner, Jack Boardman, found the beach enveloped in a smoky pall from grass fires touched off by the naval bombardment.

Maddock pounded on Boardman's back, yelling, "Shoot!"[52] Boardman wondered what to shoot at because he saw nothing through his periscope but smoke. He then noticed something dark and rectangular to his left at WN62, on the heights overlooking the beach. He rotated the turret forty-five degrees in that direction and set his telescopic gunsight on a casemate embrasure.

Boardman ordered his assistant, Private First Class Enrico Valente, to load a concrete-busting shell. Valente had four such rounds, and Boardman fired them all. He then switched to armor-piercing shells, aiming each one at the embrasure. Boardman noticed dust flying up behind the casemate. Some of his shots had gone straight through the bunker, which had an open rear entrance. No enemy gunfire emanated from the position, and Boardman ceased firing.

An incredible, almost comical, scene suddenly erupted inside Maddock's tank when Valente and another crewman began bickering about whether Bing Crosby or Frank Sinatra had the better voice. The two soldiers tried to pull their fellow crewmen into the silly squabble.

The debate ended when an enemy shell glanced off the turret and reverberated through the tank. The impact left a groove in the turret, like a scoop mark in a tub of ice cream. Boardman resumed shooting at the casemate. Decades later, while visiting Omaha Beach, he speculated that the shot might have come from an identical casemate slightly lower on the heights, a position he never noticed during the battle.

Maddock received instructions to help open a beach exit. As his tank motored along the shingle, his driver swerved to avoid an American soldier on the ground. The tank scattered stones in all directions and threw both tracks. Maddock and Boardman emerged to fix them.

Almost immediately, wounded soldiers gathered behind the tank, all seeking cover. The two tankers tried to shoo away the injured men. One of them spit curses at Boardman, wishing death upon him. After clearing away the wounded, Maddock and Boardman broke both tracks while trying to remount them. They gave up working, and the crew abandoned the tank.

Sergeant Kenneth V. Williams also commanded a DD tank on Omaha. He narrowly escaped death while operating his caliber .50 machine gun. Shell fragments tore the weapon from his hands, and he fell back into the turret with bleeding fingers. His gunner bandaged them, and Williams continued as tank commander. While driving inland from the beach, an enemy shell severed a track, and the tank lurched to a stop. The crew scurried out and dug in nearby.

Not a single 741st DD remained in operation.

★★★

As H-hour approached on the *Samuel Chase*, the battalion commander, Bob Skaggs, prepared to disembark on an LCM bound for Easy Red. Into his gear, he tucked a phonograph record that had arrived from his wife just before he left port. Perhaps he would find a record player in France. He had failed to find one on the ship, even after asking the ship's chaplain for help. The clergyman dismissed his request.

The battalion commander climbed down a rope net hanging from the *Chase*. The landing craft below rolled with the waves and clanged against the ship's hull. Skaggs dropped into the boat along with his operations officer, Captain William M. King, a tall, square-chested man who graduated from the New York Military Academy and was a former criminal investigator with the Pennsylvania Motor Police.

Three other 741st men also descended the net: Technical Sergeants Malcolm C. O'Brien and William D. McClintock, the operations and intelligence sergeants, respectively. The third man, Technician Fifth Grade Donald O. Momany, carried an SCR 509 radio.

Almost one hundred soldiers packed the boat, mostly regimental headquarters troops from the 16th Infantry. The vessel carried no vehicles. *Life* magazine photographer Bob Capa had a spot on the craft but had opted to depart on an earlier boat carrying First Battalion soldiers. (After the war, Capa falsely stated in his memoir that he landed with Second Battalion troops in the first wave. Those soldiers were aboard the USS *Henrico*, not with Capa on the *Chase*.)

The LCM helmsman pulled away from the mother ship and motored several hundred yards to a rendezvous area. The rough water quickly turned stomachs. Seasick men upchucked eggs and sausage, many soldiers plastering one another with vomit.

At the prescribed time, the helmsman turned the craft toward shore, and the journey resumed. While the men inside crouched down, cold salt water sprayed faces, ran down necks, and saturated clothing. As the little boat neared shore, bullets pinged against its steel hull, which held death at bay, for the moment.

After spending more than two hours on the boat, the soldiers felt it scrape along the sandy bottom. The ramp dropped, and everyone rushed into the briny surf.

Skaggs glanced at his watch—8:20 A.M. He looked up, and his eyes smarted at the sight of dead and torn bodies all around. Nevertheless, the battalion commander charged ahead, trying to ignore the carnage in an effort to keep focused on his job. Bullets and shell fragments sliced down man after man around him, some spinning crazy pirouettes as they fell. Other soldiers dropped in bloody pieces, ripped apart. Horrific was too weak a word for it.

Amazingly, Skaggs and the soldiers from 741st headquarters all reached the shingle embankment unharmed and pressed themselves flat against the rocks. Hundreds of men, living and dead, lay packed together on the embankment. Ahead, beyond a grassy dune, an escarpment overlooked the beach. The steep bluff resembled a fortress wall.

Bill McClintock spotted a soldier lying motionless on his back with the tide lapping at his knees. He crawled over to pull the soldier higher onto the shingle.

"He was still alive," McClintock later wrote, "but with a great gaping hole in his stomach and an expression of such complete wonderment in his eyes. He knew he was going to die—but no fear, just that astonished look."

Skaggs attempted to communicate with his tank crews, but water had seeped into the radio that Don Momany carried, and the device had shorted out. The headquarters group rose from the shingle and, fully exposed to enemy fire, raced from tank to tank, directing their fire and coordinating their movement. Momany, McClintock, and Malcolm O'Brien all fell wounded. McClintock took a bullet through the fatty part of his left thigh, the pain amounting to a sharp, pinching sensation. "If that's all there is to getting hit," he thought, "I've got this war made."

Gimping along, McClintock noticed an officer from the 741st sitting on the shingle, tossing stones into the surf. He appeared oblivious to the furious combat all around.

"As I sat down beside him," McClintock recalled, "I could see he was crying as if his heart would break. He was condemning, between alternate earthly expressions, the failure of American intelligence and the general confounded situation. I joined him in both words and actions."[53]

Omaha looked like a hellish scene from the Final Judgment. Vehicles burned, flames lashing from them. Black, tarry smoke plumes uncoiled into the sky. The vehicle accumulation clogged Omaha, and the navy responded by suspending all landings. Transport craft pulled away from the beach and disappeared beyond the horizon.

Soldiers hugging the shingle misconstrued what they saw as a retreat. It appeared the navy had aborted the landings. Nobody who witnessed it would ever forget the feeling of defeat and abandonment.

CHAPTER 5 **FOOTHOLD IN FRANCE**

One naval vessel that retreated from the beach carried a T2 tank-recovery vehicle from the 741st and four tank mechanics. The T2 and several other vehicles sat aboard an LCT scheduled to land at 7:30 A.M. on Fox Green.

The boat approached the beach, but enemy fire drove it away. Its skipper searched for a more favorable landing spot but found none before receiving word that all vehicle landings had been temporarily halted. The coxswain circled in the water for over an hour before obtaining permission to land, the halt now lifted.

As the vessel neared shore, it bounced up and down on the rolling breakers. When the ramp fell, the T2 driver, Technician Fourth Grade Lytle C. Morton, saw the beach through the glass in his periscope slit. The boat appeared to have stopped too far from shore. Morton feared his T2 would drive off the ramp and plunge nose first to the bottom.

At 9:30 A.M., the vehicle commander, Sergeant Leonard H. Trimpe, tapped Morton on the shoulder, the signal to go.

Their machine immediately sank, and Morton shouted, "I can't see! I can't see!"[54] The T2 landed upright with only its uppermost portion above the waves. With zero visibility, Morton gripped the steering levers, keeping the machine headed straight. After several minutes, he finally saw gray sky through his periscope, and then the battle unfolding ahead.

Morton found a beach soaked in blood and horror. Soldiers hauled wounded men, while medics fumbled with bandages and tourniquets. One GI crawled on his belly, dragging a leg almost severed above the knee. Morton winced as enemy fire struck dead men again and again. Limbs, torsos, and innards lay across the shoreline.

When he steered the T2 out of the water at a place free of dead and wounded, the vehicle immediately became a target. Morton heard someone outside shout, "Let's get the hell out of here! That damn thing is drawing fire."[55]

Sergeant Trimpe alerted Morton to an immobilized Able Company tank sitting nearby, and he drove toward it. The Sherman had turned sideways on the shingle and thrown both tracks. The crew remained inside, all hatches shut. Trimpe noticed that an armor-piercing shell had penetrated the hull, flown through the engine compartment without causing damage, and lodged in the right rear corner. The shell's nose protruded several inches from the hull.

Trimpe and fellow crewman Raymond E. Markby worked on the tracks, while Morton remained in the T2 along with Carl A. Rudhman, the winch operator. The Germans continued to hurl fire in their direction.

While struggling with the tracks, the men saw a figure running toward them, arms flailing as he yelled, "Trimpe! Trimpe!"[56] The four tank mechanics recognized their battalion commander, Colonel Skaggs, who suddenly fell flat when he heard an incoming mortar round. The projectile exploded just ahead of him. Undaunted, the colonel jumped up and dashed over to ask if the men could repair the tank. Trimpe assured him it was possible. Skaggs darted away before Trimpe could show him the armor-piercing shell lodged in it.

As promised, Trimpe and his team repaired the tank and put it back in action. The arduous job required the entire morning. Miraculously, none of the mechanics fell dead or wounded. All four men later received medals for heroism.

While Trimpe's crew worked on the tank, Captain Bill King, the battalion operations officer, acted as a liaison between the 741st and the 16th

Infantry. He heard the regimental commander, Colonel George Taylor, exhorting his troops, "If you stay on the beach, you're dead or about to die."[57]

The men on the shingle had violated two cardinal rules—never bunch up, and never stop moving. Their mistakes made them ripe targets for German mortars and artillery.

Even before Taylor uttered those words, small groups of intrepid infantrymen had penetrated minefields, breached barbed wire, and ascended the bluff. They slipped into the German trench network and began digging out the defenders with grenades and bayonets. Those men required no coaching from their colonel.

Taylor dispatched King to round up the few operable tanks and launch an assault against the Colleville beach exit, the widest exit on Omaha. The captain set out to notify each crew and direct them to a rally point.

He sent Sergeant Bob Coaker's tank in that direction. When he reached Staff Sergeant Walter J. Skiba's machine, King found Skiba dead, killed by a shell splinter. His crew removed his body from the tank, and Corporal Steve J. Hoffer took command.

King also contacted Sergeant John W. Call, who was receiving first aid after a shell fragment cut a lurid gash across his cheek. King took over Call's tank and directed its driver toward the rally point.

On the way, the driver swerved to avoid a group of vehicles and wounded men. When he did, he struck an antitank mine, which severed a track and damaged the tank's suspension system. King ordered everyone to dismount, and he led them away. Shell fragments struck the stalwart officer a half-dozen times that morning, but none injured him.

Everyone in the crew followed King, except John Call. The injured sergeant stayed to help Hoffer's crew repair their thrown track, a task they completed the following day.

Meanwhile, an enemy shell crippled Coaker's tank, knocking off a bogie wheel. The machine limped forward but could not climb over the shingle embankment. The crew assumed a stationary position and kept firing. After several hours, enemy artillery fire became too intense and fell too close. The crewmen abandoned their tank and sought cover behind a grounded LCT.

The effort to crack open the Colleville exit failed.

American attack plans called on tanks and infantry to unlock the beach exits by frontal assault, but those points proved stubborn impediments. The Germans had fewer defenders on the bluff between the exits and no concrete bunkers there.

The tank crews watched as more and more foot soldiers ascended the bluff. The tankers waited for the infantry to outflank the exits and captured them from behind.

That happened about 11:30 A.M. at the Ruquet valley, the first exit to fall.

★★★

Sergeant Robert W. Compton from Headquarters Company commanded a reserve tank, which landed about 1:00 P.M. He and his crew joined engineers and tankdozers in clearing beach obstacles. The tide had just begun receding.

At around 3:00 P.M., an LCT unloaded a T2 from Headquarters Company and a halftrack commanded by the battalion communications officer, Captain John D. Sterrett.

Thirty minutes later, four more reserve tanks landed: one from Company B, one from C, and two from Headquarters Company. Sergeant Frederick A. Neidrich commanded the Charley Company tank. He found no other tanks from his company on the beach, a sobering discovery.

★★★

Val Fister finally reached Omaha. He and the other Able Company men aboard LST 317 hitched a ride to shore on a barge made from pontoon sections. Resembling a floating football field, the navy called it a rhino ferry.

The ferry touched down on the west side of Omaha, well away from Easy Red, and disgorged its cargo. Fister strode up the beach before glancing back to see nobody following from his company, including Captain Thomas. Fister dashed back, looking for them.

German artillery interrupted his search. He dove into a runnel, drenching himself.

"When I came out," Fister recalled, "the artillery continued to pound away at us. God, that was really hell. I would run then fall down when I heard more coming."

Fister spied a piece of timber and dove for it, but he suddenly jumped up. "Something said no, get the hell out of here. I no sooner hit the ground again when another barrage came in. Boom! There was a big hole right where I had been."

He ran stride for stride with another soldier, whose legs suddenly buckled when enemy fire hit him. Fister kept running until he reached a dirt road beyond the beach. He flopped down, and a lieutenant hollered at him.

"Are you coming along?"

"No, I'm a tanker," Fister replied.

"Thanks for what you did," the lieutenant said.

Only later did Fister realize what the lieutenant meant. "I was the point man running across the beach. Where I went, they went. When I fell down, they fell down."[58]

He looked out to sea and watched an artillery shell strike the rhino ferry. The explosion killed the ferry pilot, obliterating him. The unfortunate sailor had befriended Fister earlier that morning. He had given him a blue watch cap to wear.

Fister sat beside the road, exhausted and shaking with fear, when a military policeman approached and started talking.

"I've been watching you all along. What are you, a Seabee?"

"No, hell, I'm a tanker," Fister replied.

"You got any cigarettes?" the MP asked.

Fister pulled a soaking-wet pack from his shirt pocket. The policeman smiled and offered Fister a smoke instead.

"He gave me one and handed me a match," Fister recalled. "Hell, I couldn't light the thing. I was shaking so much. He lit the cigarette for me, and we talked for awhile."[59]

Fister stopped trembling by the time he finished the cigarette. The policeman, who had introduced himself as Gene King from Indianapolis, offered to escort Fister up the bluff to where the MPs had a command post. King needed to finish his duties on the beach, but, later that afternoon, he returned carrying an M1 rifle and a helmet for the unarmed tanker. The two men scrambled up the bluff together. Once on top, Fister discovered that sand had clogged the bore of his rifle. Originally trained as a medic, he had no idea how to clear the blockage. He swapped the infantry weapon for a caliber .45 pistol that King carried.

Both men sat down for a rest as German shells whooshed overhead now and again. Fister suddenly heard a tank motor. He looked up and saw Lieutenant Barcellona's machine ascend the bluff on a dusty road, but the lieutenant was absent. His gunner, Private First Class Lawrence G. Sweeney, had temporary command. Fister hustled over to him.

"Where you goin,' Sweeney?" Fister asked.

"We're goin' up there," Sweeney replied.

"You got room for me?"

"Yeah, we got room for one more."[60]

Fister climbed inside and sat in the gunner's seat. He inquired about Lieutenant Barcellona, and Sweeney said he would return shortly. Fister listened as Sweeney named the Able Company men killed that morning: Figg, Skiba, Elliott, Whited, and Waldkoetter. Another man had a grievous abdominal wound, Corporal Lavander D. Mizer.

Sweeney gave Fister dry clothes, and Fister jumped out of the tank to put them on. After he had climbed back inside, Barcellona arrived. "Okay, Fister," he said, "get the hell out." The corporal dismounted, and the tank roared away, the lieutenant waving goodbye.

Fister took a seat on a fallen tree trunk. Unknown to him, Colonel Taylor had issued an order at 5:00 P.M. for the 741st to move through the Ruquet valley exit and assemble near Saint-Laurent-sur-Mer, not far from where Fister sat.

The corporal spotted Captain Thomas walking with three or four other men.

"He didn't have a happy look on his face," Fister remembered. "I thought he was going to chew me out, but he didn't say a word. I didn't say a word either, leaving well enough alone. A few more people came along, then a few more."

The group continued to grow, mostly tank crewmen without vehicles. Night descended before Thomas finally spoke. "Well, Fister, this is where we're gonna settle down for the night. Do you know where we are?" Fister shook his head. "This is exactly where we were supposed to meet when we got off the beach."[61]

The battalion assembled there, and its survivors started digging foxholes. Fister borrowed an entrenching tool. Antiaircraft guns banged away nearby, jarring the earth and illuminating the darkness. They eventually quit, and Fister claimed a long day's reward.

Sleep.

CHAPTER 6 **DEATH'S TOLL**

Frank Elliott from Able Company had a family in New Castle, Pennsylvania. His wife, Pauline, and their young daughter, DeRonda, spent June 6, 1944, at home. Pauline knew nothing about her husband's role in the invasion when she penned a letter to him.

"All day the radio has broadcast invasion news—constantly—all regularly scheduled programs have been canceled, and I have been virtually 'glued' to the radio. News broadcasts and prayers led by eminent clergymen have occupied most of the time.

"Naturally none of us here at home can think of anything else. People took the news calmly and soberly—how else could it be taken."

Pauline went on. "Little DeRonda was the only one not affected by the D-Day news. She went about her happy little business of living, as usual, entirely unaware of the great event. I hope and pray that she will never remember any of this but only the happiness of the hours that will follow her Daddy's homecoming step on the porch. Good luck to you, darling, wherever you are. We are waiting for you and loving you with all our hearts."[62]

She dropped the letter in her mailbox.

No reply ever arrived. None could.

Her husband lay among the fallen on Omaha Beach. She learned about his death when a belated War Department telegram arrived on August 6. DeRonda would always remember her mother's heartrending screams on that Sunday afternoon.

★★★

The 741st troops who remained in England learned about the Omaha bloodbath from injured comrades returning by ship. The battalion Medical Detachment had an aid station in a building near the Bournemouth waterfront.

Technical Sergeant Leonard E. Hatfield, originally from Company B, transferred to the detachment in the United States. He later recalled processing "about forty patients" on D-Day.[63] Many belonged to the 741st, but the group included infantrymen, too. Many tank crewmen arrived with broken eardrums, which had burst when they ascended from sunken DD tanks. The men also suffered from exposure after their ordeal in the frigid water.

Hatfield encountered two Company B survivors who served in the same tank crew with his best friend, James D. Longmire. Hatfield inquired about his buddy.

"How's Longmire?"

"Sorry, he didn't ..." one of the Company B men replied.

"What happened?" Hatfield asked.

"Shot through the eye."[64]

The aid station remained in operation for several days until the casualty flow tapered off, and ambulances dispersed the patients to hospitals in England.

Death claimed hundreds of lives on Omaha. Two temporary cemeteries sprang up there, one on the bluff and one on the beach itself.

The 741st lost thirty-seven men, all but four serving in the DDs. The water proved deadlier than the Germans. More tank crewmen died from hypothermia than enemy fire or drowning.

British sailors aboard the HMS *Aristocrat* sighted a floating body. The dead man was Second Lieutenant Florus W. Laird, a native of Auburn, Illinois, who served as a platoon leader with Company B. The sailors pulled him aboard their vessel and removed his personal effects, including an identity card and invasion currency totaling three hundred francs.

The Royal Navy men transferred Laird's body to ST-758, a US Army tugboat. The little vessel's crew buried Laird at sea, making the English Channel his grave.

Sailors on the HMS *Northern Reward* retrieved another floating body, Technician Fourth Grade Donald E. Ford from Company B. He also received a burial at sea, as did Private Grady E. Colvin from the same company. Army crewmen aboard tugboat ST-782 interred Colvin.

Graves Registration troops accounted for all the 741st dead except for three men: Corporal Alberto Lopez, Technician Fifth Grade Hoover Fugate, and Private Raymond Dahlman.

★★★

D-Day left the 741st unrecognizable as a tank battalion. It reported losing forty-eight Shermans, though some returned to action following repair work.

On the morning of June 7, Leonard Trimpe and his crew worked on a tank at the battalion assembly area atop the bluff above Omaha. As Trimpe labored on the machine, Lieutenant General Omar N. Bradley arrived, having recently come ashore. One of his aides approached Trimpe and said, "General Bradley wants to speak with you. Keep working. Don't salute."

The general walked over.

"How many tanks do you have besides this one?" he asked.

"I think five, sir," Trimpe said.

"Will you get this one back in combat?"[65]

Trimpe replied yes. He and Bradley talked awhile longer before the general departed. He spoke to nobody else from the 741st.

Bradley's presence gave Trimpe a feeling of reassurance. The overall situation must have been improving for an officer of his rank to be in the area. Trimpe also saw another hopeful sign. Engineers had constructed an airstrip on the bluff. Small monoplanes had begun using it, even as the engineers worked to expand the runway. The following week, a B-17 "Flying Fortress" landed carrying officers and newsmen.

As he rode a captured German bicycle, the sergeant observed the bomber as it taxied to a stop. He pedaled over to the area where it parked, and a reporter asked him for an interview. Trimpe agreed, and the journalist switched on a portable recording device. The young soldier said he felt fine, and the reporter commented on his three- or four-day growth of beard. Trimpe mentioned his current tank-repair duties but said nothing about the horrors he had witnessed.

The following day, June 15, residents in Trimpe's hometown, Seymour, Indiana, heard the interview during a radio broadcast. They telephoned his family with the news. His parents and siblings felt relieved. Nobody had heard from him in over a month.

★★★

Captain Thomas led the five tanks that Trimpe had mentioned to General Bradley. One belonged to Able Company, one to Baker, one to Charley, and two to Headquarters.

At 7:00 A.M., June 7, the captain received orders to support the 16th Infantry, but the plan changed when an 18th Infantry officer flagged down Thomas and requested tank support. Thomas radioed the 741st command halftrack and obtained permission from Captain King to join the 18th.

Thomas and his tanks engaged the enemy, pounding machine-gun positions, including a farmhouse converted into a strongpoint. They even blasted three German bicyclists, killing two. The infantry advanced.

By 2:00 P.M., the tanks reached Engranville and shelled the village. They soon bypassed the town but later doubled back to bombard it again after the infantry received enemy fire from its direction. The crews returned to the bluff when their ammunition ran low.

One tank returned to action that evening to smash an enemy artillery piece. The tank gunner lobbed shells at the weapon, his shots directed by a forward observer. Together they scored a direct hit on the German gun.

The following morning, Colonel Skaggs (colonel being the term of address for a lieutenant colonel) organized a salvage operation in an effort to recover all repairable Shermans and ammunition trailers. Five T2 recovery units worked on the beach, including Trimpe and his crew. Skaggs also took steps to obtain replacement tanks for those lost at sea or damaged beyond repair.

The battalion had eight tanks operational on June 8. They joined soldiers from the 16th Infantry but saw little combat. Lieutenant Barcellona led three tanks, and Captain Jimmie Thornton led five. Thornton had rejoined the battalion the previous day after a small landing craft rescued him and his crew from the water on D-Day. The men transferred to a larger vessel, where they received dry clothing. Thornton wangled his way onto a landing craft headed for the beach, and he joined an infantry platoon until locating the 741st.

Crewmen from other sunk DD tanks also drifted back to the battalion. In addition, replacement troops started joining during the first week in France. New armor also began arriving that week, and, by mid-June, the battalion fielded thirty-two Shermans.

As the 741st reorganized, it remained assigned to the 1st Infantry Division but had little contact with the enemy. The tankers spent most of their time in reserve or in defensive positions.

The men wondered how long the respite would last.

CHAPTER 7 **IVANHOE**

June 15. Warning orders reached battalion headquarters shortly before 4:00 P.M. They relieved the 741st from its assignment to the 1st Infantry Division and alerted the battalion for movement to an undisclosed location. Something was afoot.

More information trickled in two hours later when Colonel MacLaughlin called from 3rd Armored Group headquarters. He issued a verbal order attaching the 741st to "Ivanhoe," code name for the 2nd Infantry Division. Soldiers of this famed unit wore an eye-catching shoulder patch with a red-and-blue Indianhead embroidered on a white star. Pleased to join the division, Colonel Skaggs reported to its command post, or CP, hidden among tall, shady beech trees near the Château de la Boulaye in the countryside outside Cerisy-la-Forêt.[66]

He met the division commander, a bespectacled major general, who graduated from the US Military Academy in 1912. With his quiet reserve and scholarly appearance, the general resembled a college professor or university president. At age fifty-six, Walter M. Robertson, known to many as "Robbie," was older than his superiors: Gerow, Bradley, and Eisenhower. Age would end his career five years after the war, but in 1944, it scarcely mattered. Stability and intelligence *did* count, and he had both.

Robertson brought his first troops ashore the day after the invasion began, and his men made even progress for the next three days. Against patchy opposition, they pushed sixteen miles inland to the Elle River, making the deepest penetration into France of any Allied unit to date. However, the drive stopped there when German resistance stiffened.

Paratroopers from the 3.Fallschirmjäger-Division held the line. These elite enemy soldiers shoveled deep into the French countryside with no notion of retreat or defeat. They had orders to attack toward the coast, but, for the moment, they dug in, while their division closed ranks after moving from Brittany to Normandy.[67]

This was the time for the Americans to regain the initiative. Wait too long, and the Fallschirmjägers would sit snug in earth-and-timber bunkers—phone lines strung, land mines sown, and their array of weapons tied together in a cohesive and deadly defense.

Robertson understood the need for quick action as he spoke to Skaggs and other officers gathered at his CP. They learned about an attack plan handed down the previous evening from corps headquarters.

The scheme called on the 2nd Division to grab four objectives. The first was an oblong hill, ripe with apple orchards. An observation tower at the top enabled enemy observers to peer deep into the American rear, as far as Utah Beach in clear weather. The hill also dominated the main highway in the area, Route Nationale 172. French maps gave the place no name, so the army called it Hill 192, the number denoting its height in meters. Hedgerows and sunken lanes crisscrossed the second objective, high ground near Saint-Germain-d'Elle. The other two objectives lay a couple of thousand yards farther behind enemy lines.

The officers at the CP studied circles and tactical symbols drawn on a map overlay. These illustrated the attack plan. The burden of conquering Hill 192 would fall to the 38th Infantry Regiment, while the second objective belonged to the 9th Infantry Regiment. The plan also called for the 23rd Infantry Regiment to capture Bérigny, thus driving a wedge between the two objectives.

What part would the 741st play? General Robertson wanted one tank company ready to attack in the morning. In preparation, the 741st set course for an assembly area near the eleventh-century abbey at Cerisy-la-Forêt. Night fell as tank traffic roared through the town. The crewmen parked their machines and sacked out in the darkness. H-hour was 8:00 A.M.

★ ★ ★

American artillery began blasting away fifteen minutes before the infantry attacked. Some ninety minutes after H-hour, Robertson ordered Company A into action. The tankers drove their fifteen vehicles to the division CP to receive instructions. Second Lieutenant Barcellona's platoon joined the 23rd Infantry. The two other platoons stayed at the CP with Captain Thomas.

Barcellona's tanks left Cerisy-la-Forêt and rumbled to Colonel Hurley E. Fuller's headquarters. The colonel had first served in combat as an infantry captain during the First World War, earning a Purple Heart in France. He now commanded the 23rd Infantry, and his troops had barely inched forward since attacking thirty minutes after H-hour. Fuller ordered Barcellona's tankers to attack in column formation down a country lane and to cross the Elle River at a small bridge above Bérigny. The colonel permitted no reconnaissance but gave Barcellona an infantry platoon for protection. The lieutenant felt uneasy about ambling blindly down a narrow road in single file. His tanks would make easy targets, but orders were orders.

After reaching the Elle, the lead vehicle tried to cross the bridge, and its thirty-five tons broke the rustic structure. The tank crunched through, coming to rest on its side. Luckily, nobody inside suffered injury, and the crewmen crawled out.

Commanding the second vehicle, Barcellona ordered his driver to veer off the road. They would attempt to extricate the tipped-over tank. Technician Fifth Grade John H. Barner, the lieutenant's driver, steered down into a neighboring field. Crewmen scrambled between the two tanks, attaching tow cables, and Barner wrenched the incapacitated machine upright.

The Elle ran narrow and shallow, so the attackers could easily ford it, but German paratroopers had observed the bridge fiasco. As the tankers unhooked the cables, mortar shells burst around the tanks. Two rounds struck Barcellona's vehicle while he, Barner, and their gunner, Lawrence Sweeney, stood outside with the officer leading the infantry platoon. The explosions showered steel fragments on all four men and several others, including Second Lieutenant George M. Kovachy Jr., who had joined the 741st two days earlier.

Heedless of his own shell-torn leg, Barner helped load the casualties onto the rear of his tank before driving the machine toward an aid station. The tank, with its groaning cargo, lumbered to the destination.

The welcome was less than warm as medics ran out, waving their arms, and shouting, "Get that tank outta here! This is an aid station!"[68] Their tone suddenly changed when they saw the wounded passengers. Barner spent sixty-three days hospitalized and later received a Silver Star Medal.

Back at the collapsed Elle bridge, the American attack foundered against enemy mortar rounds and small-arms fire. All but one tank withdrew. The corpses of infantrymen lay in the field, olive-drab lumps scarcely discernible as human beings.

★ ★ ★

Captain Thomas and his two remaining platoons joined the 23rd Infantry shortly after Barcellona's tanks departed from the division CP. An officer from 23rd headquarters guided the armor to a point near Bérigny, where Thomas met the commander of Company E, 23rd, Captain William A. Smith. Easy Company had orders to grab Bérigny from Fallschirmjägers, who occupied bunkers and fortified houses there, and the two captains sketched out a plan to seize the village.

At noon their joint force crashed into Bérigny, and a violent reception followed. The enemy played machine guns on the attacking infantrymen with murderous effect. Men dropped and died as the tanks kept advancing, firing all the while.

Captain Thomas commanded four vehicles, and First Lieutenant Earl H. Asker led four. Thomas quickly realized they had lost their infantry support, and he sent an order over the radio. His crews broke formation to battle enemy strongpoints that had pinned down the infantry.

Thomas and his tankers maneuvered on individual enemy positions, each assault a perilous mission without riflemen to reckon with Fallschirmjägers wielding handheld antitank weapons. The terrain also presented difficulties. Several tanks drove for the Elle to clobber emplacements on the other side, but marshy ground brought one machine to a halt.

After Lieutenant Asker's tank sank in the muck, he abandoned it and made for a vehicle commanded by Technician Fourth Grade Robert W. Compton. The lieutenant climbed aboard and took charge, but minutes later a shell burst left him with a concussion and an injured hand. Compton dressed Asker's wound, relinquished him to medics, then returned to the fight, until bullets smashed his main gunsight and several periscope lenses. Thomas ordered Compton's tank to the division CP for repairs, along with Sergeant James V. McCoy's tank after its electrical system failed.

At 5:00 P.M. Captain Smith from Easy Company relayed an order from regiment calling on Thomas to assist Company G, which was attacking west of Bérigny. Thomas sent two tanks in that direction, led by Staff Sergeant Robert A. Nicol.

Meanwhile, Thomas and his crew drove to assist Sergeant Lloyd C. Ball, whose tank stood in a field protecting foot soldiers pinned down by machine-gun fire. To help extricate the lot, Thomas's tank stormed into the field and raked the enemy with bullets and high explosives, allowing the infantrymen to escape. When Ball's tank turned to leave, a Panzerfaust projectile found its target. Flames snapped from the engine compartment, while all five crewmen bolted out and safely retreated.

Having failed to seize Bérigny or to achieve any gain against the enemy, the tanks and infantry companies withdrew to their start line.

That same afternoon, Colonel Skaggs and Captain King visited the 23rd Infantry CP, accompanied by Lieutenant Sledge. The three men encountered Colonel Fuller and Brigadier General Thomas L. Martin, the assistant division commander. According to an official report, Fuller was in a "highly emotional state and incoherent."[69] He stammered something about his orders that morning to Barcellona, and that he had not seen him since. Nobody understood what the colonel meant. He had lost control, so Robertson sacked him.

The general had long been at odds with the cantankerous Fuller, who never withheld his opinion or temper and often challenged his superiors. This had provoked red-faced quarrels and melodramatics, but no more.[70]

The 38th Infantry also attacked on June 16. Its push began at H-hour with a frontal assault up Hill 192. Two battalions attacked abreast and fought partway up the hill before faltering against a lethal sleet of lead and steel.

The regimental commander told his operations officer, Major Tom C. Morris, to telephone General Robertson for reinforcements. Morris called for help and heard the general's dispiriting words: "Morris, we don't have it to give!"[71] Another plea drummed up the 2nd Engineer Combat Battalion. Its commander disdained using his engineers as infantrymen but nevertheless joined the fight. Fourteen died, and thirty-one fell wounded, including the battalion commander, but the engineers helped the 38th Infantry maintain a foothold on the hill.

Thus far, Robertson had withheld the 741st from Hill 192. The battalion was too weak, owing to its D-Day losses, to support the entire 2nd Division. For the time being, the infantry and engineers were the only ones beating their brains out against the hill.

Away from the battle zone, additional 741st personnel landed in France during the final hours of June 16—Headquarters Company men who had spent D-Day in England. They arrived aboard LST 345 and included Major Browder and Captain Jack W. Duke, the battalion intelligence officer.

Sergeant Nelson W. Hall, a Californian from Headquarters Company, accompanied them and journeyed by motor convoy to the 741st CP, located in a Gothic building on the abbey grounds at Cerisy-la-Forêt. The gray, timeworn edifice included a chapel, gatehouse, justice room, and a dungeon. Hall found the operations halftrack parked nearby at a farm building, duffel bags stacked alongside the vehicle.

He reported to Master Sergeant Barnette Ates, the battalion sergeant major, a straitlaced professional soldier from Louisiana. Hall learned that he was now the operations sergeant, replacing Technical Sergeant O'Brien, who had been a D-Day casualty.

Like many men, the new operations sergeant wasted no time in exploring the abbey next door. Its stone walls and spire soared into the sky. On close inspection, the soldiers found a structure that had long ago descended into disrepair. German troops had recently used it as a stable, and the interior smelled of horse urine. Dirt, straw, and manure fouled its tile flooring.

While Hall and battalion headquarters set up shop, Service Company and Company D embarked from Key Docks at Southampton on June 17 and set sail for Mulberry A, an artificial harbor at Omaha Beach. After nearly twelve hours at sea, LST 527 reached the harbor and opened its whale-sized mouth. Company D rolled out with its seventeen light tanks, as did the battalion Medical Detachment and headquarters personnel from the medium tank companies. Service Company lagged behind, having embarked nearly six hours after the others, aboard the liberty ship SS *Stephen B. Elkins*.

DD tank survivors from Companies B and C arrived from England the following day and reported to the battalion CP before reuniting with their units.

The 741st was becoming whole again but lacked the strength called for in War Department tables of organization and equipment. The battalion needed more replacement crewmen and twice the number of Sherman tanks.

★★★

Less than two weeks after D-Day, the staff at 741st headquarters drafted a Distinguished Unit Citation proposal honoring the battalion for its performance on Omaha Beach.

Sergeant Robert B. McMullen from Service Company helped prepare the documentation. The Saint Louis native landed with the assault echelon on D-Day and had since undertaken administrative chores with Headquarters Company pending his company's arrival.

Around noontime on June 17, McMullen set out for 1st Division headquarters to complete the proposal. He also planned to fetch mail for the battalion at the 1st Division post office. Letters and parcels addressed to 741st personnel still arrived there, a change to the 2nd Division post office slow to take effect.[72]

To reach 1st Division headquarters, McMullen had a chauffeur, Private Thomas D. Chastain from Company A, who had been driving a captured German truck since losing his tank on Omaha Beach. With paperwork in hand, the two men headed east from Cerisy-la-Forêt toward the 1st Division sector.

Their journey took an errant turn at a seven-way crossroads in the Cerisy Forest, when an MP sent them heading southwest and into enemy territory. The two men awakened to the stupendous blunder when bullets ripped through their vehicle. One slug struck McMullen above the right hip. German infantrymen converged on the road, and Chastain braked to a stop. He snatched the paperwork and concealed it under his clothes before the enemy soldiers flung open the doors.

After yanking out the Americans, several Germans discussed McMullen's condition. He needed medical help. Chastain was in good shape and could march.

The enemy transported McMullen to a prisoner-of-war hospital in Rennes, France. He arrived at the former Catholic girls school two days after his capture and recuperated in a bed until July 12, when the Germans moved him to Frontstalag 221, also in Rennes.[73]

After McMullen reached the camp, the Germans evacuated Rennes before advancing American troops arrived. Guards forced the internees onto a train, which departed the city on August 3, 1944. The locomotive and boxcars embarked on a slow, halting journey along the Loire River. The procession temporarily stopped at Langeais due to a bombed-out bridge. Then tragedy struck from above, when American aircraft pounced on the stalled train, their bullets killing twenty-three persons outright and injuring another seventy.

McMullen was among the wounded rushed by truck to the Bretonneau Hospital in Tours, where a doctor amputated his left leg. American troops from the 83rd Infantry Division liberated the city on September 1, 1944, and McMullen's captivity ended.

Chastain ended up at la Chappelle-sur-Vire, a collecting point for American prisoners. There he found other captives, many wearing US paratrooper garb. He also met a sergeant from the 2nd Ranger Infantry Battalion and a musician from the 83rd Infantry Division Band who became a prisoner while working as a litter bearer. They had little to eat, and the men dubbed the place "Starvation Hill." On one occasion, Chastain made a flavorless meal from the paperwork he carried on his person. The Germans had failed to search him, and he still had the unit citation documents. Enemy guards eventually prodded the disheveled prisoners into trucks and hauled them to Alençon and later across France to Chartres and then to a railhead at Châlons-sur-Marne. The ensuing train trip ended in Germany at Stalag XIIA.

Chastain and McMullen were the first 741st members to fall into German hands. At battalion headquarters, their disappearance remained unexplained. The personnel officer reported them missing in action, but McMullen's status changed after his release at Tours. Chastain's status remained unchanged until December, when word arrived through the International Red Cross that he was a prisoner in Germany.

On June 18, the 2nd Division again requested support from the 741st. Company B joined the 9th Infantry Regiment, and its commander, Colonel Chester J. Hirschfelder, ordered Captain Jimmie Thornton to send seven tanks to the Second Battalion for a special mission to recover two 57 mm antitank guns abandoned in a muddy field under enemy observation. Previous recovery attempts had ended in casualties.

Second Lieutenant George W. Wilson led the seven tanks. He claimed Cheswick, Pennsylvania, as his home and had joined the company only the day before as a replacement platoon leader. Wilson now reported to the Second Battalion commander, Lieutenant Colonel Walter M. Higgins. The colonel poked his finger at a map and picked out two covering positions, ordering that a pair of tanks should occupy each spot with another machine in reserve. The two remaining tanks would enter the field to retrieve the guns. Wilson readily agreed. The colonel seemed to know his stuff. Captain Buford Arms offered to help the tankers reconnoiter the area, as the abandoned guns belonged to his unit, Antitank Company, 9th Infantry.

The vehicles commanded by Sergeant Richard Maddock and Technician Fourth Grade Lee O. Jenkins would make the trip for the guns. Arms and the two sergeants reconnoitered the area on foot. The tankers studied the scene, crafted a plan, and returned to brief their men. One soldier from Antitank Company would ride with each crew and, at their destination, crawl out to hook the guns to the tanks.

Artillery from the 2nd Division warmed up the area before Maddock and Jenkins took to the field at 2:00 P.M. Their tanks rumbled up to the guns while the enemy harassed them by dropping artillery shells.

Corporal Jack Boardman was Maddock's gunner, and he watched as the antitank man in his vehicle exited through the escape hatch on the floor. The lone man slithered through the muck until emerging behind the machine. Boardman described what happened next:

"He hooked the thing on, and then he popped back into the tank. He was supposed to crawl back across the field, but he popped back in, and we started off. Well, the hitch came undone. We backed up, and he went out and reattached it. We pulled out again, and it dropped again. Apparently, there was something wrong with the clevis."[74]

Maddock's tank returned with no gun. Jenkins's crew and its antitank man brought back their quarry but all for naught. The Germans had detonated a grenade inside the barrel, ruining it.

Captain Arms wanted to send Maddock's tank back for the remaining gun. Captain Thornton objected. He saw no point in risking a single life to retrieve a weapon that was surely in the same useless condition, but the decision had already been made. Maddock's tank ventured back out and returned twenty minutes later with the gun.

The Germans had indeed destroyed its barrel.

Second Lieutenant Earnest L. Gray joined Able Company on June 17. He preferred his middle name, Lynn, and had worked as a salesman for Sears, Roebuck and Company in Lafayette, Indiana, before being drafted in 1941 and later receiving an OCS commission. The tall, young officer drew a tank platoon as his first combat assignment and found the unit attached to the 23rd Infantry in a hamlet called Touzé.

Colonel Jay B. Lovless now commanded the 23rd, and he had instructions from division headquarters to storm Hill 192 with First Battalion. The division also assigned a supporting role to Company A, the first use of armor on the hill.

On June 19, Lieutenant Gray received orders to lead five tanks and one tankdozer to an assembly area several hedgerows from Saint-Georges-d'Elle. The group departed at 7:50 A.M., but mechanical difficulties reduced their number by one. The remaining vehicles reached the assembly area and awaited direction from First Battalion commander, Lieutenant Colonel William S. Humphries.

Throughout the morning, rain and wind lashed the Norman countryside as Gray and his tankers waited. About noon, a runner from the 759th Light Tank Battalion appeared. His unit was also supporting the attack, and he brought orders for Gray's platoon to move out. The runner also mentioned that enemy shells had struck the First Battalion CP, causing fatalities.

Although the tactical situation lacked clarity, the tanks moved ahead. They passed through Saint-Georges-d'Elle and drove south into a gully. The crews machine gunned suspicious-looking hedgerows and shelled other targets. Bill Smulik's tankdozer led the way, followed by Lieutenant Gray and tanks commanded by Sergeants James Larsen, James McCoy, and Donald L. Hecox.

Smulik became a soldier during the Depression, finding greater opportunity in uniform than in the Pennsylvania coal mining town where he grew up. His fellow tankers called him "Helmet," as the scrappy little warrior always wore his steel helmet in France, even while asleep.

The attack plan called for Smulik and his tankdozer to clear roads and plow gaps in the hedgerows for those behind, but the rain turned the low ground to mush, and the dozer sank up to its hull. Gray's tank also sank. The vehicles lacked adequate flotation due to their narrow tracks—a design weakness, one of many.

All five crews blasted their cannons uphill toward a house, eliciting small-arms fire in return. Colonel Humphries watched from afar and realized the tank crews were aiming at his infantrymen, but he lacked communication with the tanks, a maddening predicament. The frustrated officer tore down into the low area, fully exposed to enemy fire. Darting from tank to tank, he beat on each with a stick until the men inside heard him and stopped shooting. Humphries's valor resulted in a Silver Star Medal, but he did not live to receive it. Machine-gun bullets mortally wounded him in the head and spine on June 21. He died four days later.

Sergeant Larsen tried to hook onto Gray's vehicle and drag it free, but the soft ground refused to release the tank. He radioed Sergeants McCoy and Hecox and told them to hook onto the machine while he covered them, but that failed when the enemy intervened with mortar shells and 88 mm rounds.

Gray ordered the mired crews to abandon their vehicles, while he and McCoy crawled along a hedgerow, guiding Hecox's tank away from the area. As rain and mortar fire poured down, shell fragments nicked McCoy but penetrated Gray's skull, cutting into his brain. The twenty-seven-year-old lieutenant died from the wound, having survived in combat only two days.

Later that afternoon, McCoy and Larsen reported the story to Colonel Lovless, who withdrew the leaderless platoon.

Smulik hitched a ride in a Headquarters Company jeep and reported to the battalion CP. He described the day's events and received permission to go back for his tankdozer. First Lieutenant Romolo P. Legnini, the battalion reconnaissance officer, directed the recovery operation, which included rescuing Gray's tank. The pair left for the Able Company bivouac at 11:00 P.M. and borrowed that company's T2 tank-recovery vehicle along with its crew. Smulik led everyone to the gully near Saint-Georges-d'Elle, where he had been forced to leave his dozer, but it was too dark to see.

The team withdrew and waited until 4:15 A.M., when daylight at last tinged the horizon. Water had pooled inside the vehicles and killed their batteries, but their gasoline-powered Homelite generators still functioned. They supplied enough current to fire up both tank motors. With tank tracks churning and the T2 pulling, the team extracted both machines from the mud.

The Germans never interfered.

CHAPTER 8 **COMBINED ARMS**

Hedgerow country waylaid the Americans. They came to drive Hitler from the fields and forests of Northwest Europe but found themselves crawling in jungle—shades of the Pacific.

Just like the Japanese, the Germans used the dense vegetation to complement their killing power. The ancient hedges and sunken lanes in Normandy became a tangled nightmare for US troops, their superior mobility halted, their numerical advantage undermined.

Created as far back as Roman times, the hedgerows fenced in cattle, marked property lines, and reduced soil erosion caused by English Channel winds. Each hedge consisted of an embankment three to five feet wide and up to nine feet tall (sometimes higher) topped with a leafy mass of saplings, brambles, and mature trees. When viewed from above, hedgerow country, or bocage, looked like an asymmetrical mosaic of pastures and apple orchards.

Within this green patchwork, the Fallschirmjägers burrowed deep into the embankments and chiseled out firing slits at ground level, allowing them to spray the pastures and orchards with death. Mortar crews launched shells from sunken lanes nestled between hedgerows. Unseen and always moving, the mortarmen made difficult targets for counter-battery fire.

The Americans needed new ideas. Tanks no longer constituted the wonder weapons that shook the world in 1940, their supremacy challenged by new weapons: bazookas, Panzerfäuste, and high-velocity antitank guns. Warfare had changed.

The leadership at 2nd Division headquarters needed time to dope out a better tactic. The heavens obliged by forcing a rainy respite.

The clouds and downpour of June 19 blew into a force-seven gale the following day. Catastrophic damage befell the artificial harbor at Omaha Beach. Its piers and caissons snapped apart. The storm smashed five hundred ships and landing craft and beached another eight hundred vessels.

Service Company was the only residue element caught at sea, but its transport ship stayed afloat. The gale caused less damage inland, where it amounted to a soggy interlude rather than a catastrophe. Jack Boardman rode it out in an improvised sleeping bag.

"It was made from the same canvas used for the DD tanks. We had a bunch of it for repairing and such. I glued the edges with Bostik, then turned the thing inside out. I stuck three or four blankets in there and spent the whole rainy time reading a book in that sleeping bag with the flaps up so the rain wouldn't come in."[75]

Major General Leonard T. Gerow, the corps commander, visited the battalion CP in Cerisy-la-Forêt on June 20, to decorate four battalion members for their actions on D-Day. Lynn E. Hurte, Arthur W. Mains, Charles Lombardo, and Anthony E. Annunziata stood at attention while the general pinned on their Silver Star Medals. He shook their hands, spoke a few words of praise, and departed to repeat the performance at another CP.

That night, enemy shells rocked the 741st CP. The explosions caused no injuries but resulted in immediate plans to relocate the headquarters (Vitamin Rear) to orchards west of town, a venue less likely to draw fire than the abbey and its grounds.

The move took place after daybreak the next morning.

★★★

The gale abated on June 21, as the headquarters men labored to set up their new CP. Staff officers at division headquarters provided a method to end the hedgerow agony. The prescription for success involved a tight-woven partnership between armor, infantry, and combat engineers, a combined-arms approach.

Team exercises began on June 28, when 2nd Division infantrymen joined tankers from the 741st for secret training at a quiet rear area. Soldiers from the 2nd Engineer Combat Battalion also participated. Each team included two tanks, an infantry squad, and four combat engineers. They rehearsed a carefully scripted process:

The squad leader deployed his dozen infantrymen along a hedgerow that served as the departure line. With the engineers stationed nearby, the lead tank pulled up to the hedgerow. The second tank parked directly behind to protect the team from the rear.

When firing commenced, the lead tank and infantrymen hammered the next hedgerow past the departure line. The tank focused on the hedgerow corners where the Germans usually placed heavy machine guns. While the enemy crouched for cover, the squad leader sent his men forward, sprinting along the hedgerows running perpendicular on either flank.

Once the infantrymen reached their new position, and routed any lingering opponents with bayonets and hand grenades, the squad leader selected a new firing position for the lead tank and signaled the vehicle ahead.

The engineers sprang into action if the tank failed to punch through the hedgerow at the departure line. They hauled out two satchel charges from a wooden rack on the tank's engine deck. Connected together, the charges resembled a saddle, which the engineers flung over the hedgerow. The demolition device blew both sides at once, the dual explosion tearing open a passageway for the tank.

The engineers, armed with rifles, kept near the tank to protect it from tank hunters armed with a Panzerfaust, Panzerschreck, or magnetic mine. The support tank moved up to the hole, and the team repeated the process.

To facilitate communication, the tankers mounted an ordinary EE-8 field telephone on the rear of each tank, permitting discussion between the crew and those outside.

Troops from the 2nd Division rotated to allow all infantry units an opportunity to practice this hedge-busting technique. Exercises occurred daily, until each tank, infantry squad, and engineer party demonstrated the requisite timing and collaboration.

Second Lieutenant Charles D. Curley Jr., a platoon leader with Company E, 38th Infantry, participated in the training. He felt uneasy about the engineers stowing three hundred pounds of TNT on each tank. That seemed like a swift way to make dead Americans. "I can remember asking what would happen if something hit the charges and was told they wouldn't explode unless a fused detonator set them off. The instructor demonstrated by firing an M1 into a block of TNT on the ground to show that it wouldn't explode."

The demonstration alleviated Curley's concern, but he spotted another problem when engineers placed TNT satchels against a hedgerow and cautioned everyone away. The enormous blast stunned him. "I thought the world had come to an end. I knew we were in trouble when I saw how much smoke and debris flew up in the air," Curley recalled.[76] The soaring pillar of smoke would attract enemy artillery, something nobody had considered.

While the Americans rehearsed, the Germans kept watch on the frontline. They disappeared into their holes among the greenery, remaining invisible to American artillery observers and photo-reconnaissance aircraft.

Fallschirmjägers careless enough to show themselves received a sharp rebuke from American artillery. Nevertheless, casualties among the Germans remained inconsequential.

Intelligence officers like Jack Duke at 741st headquarters faced difficulties figuring out the enemy's order of battle and troop deployments. The problem resembled a thousand-piece jigsaw puzzle. Prisoner interrogations sometimes provided missing pieces, but prisoners were scant and most had few words. The captured Fallschirmjägers held tight to their vision of German superiority and expected the Allies to suffer a towering defeat in Normandy.

Their division commander, Generalleutnant Richard Schimpf, viewed recent events as having bolstered the self-confidence of his paratroopers. They sharpened their war skills against the feckless attacks made by the 2nd Division at Bérigny and Hill 192.

Their tenacity tested the 2nd Division from general to private. The Fallschirmjägers proved rugged opponents. "If the same type of soldier had faced us between the beach and Forêt de Cerisy," commented General Robertson, "we would have had considerable difficulty maintaining our beachhead."[77]

★★★

On June 22, Robertson summoned Colonel Walter A. Elliot, a pudgy officer with basset-hound jowls, who commanded the 38th Infantry Regiment. The regimental operations officer, Major Tom Morris, remembered the summons to division headquarters:

"I grabbed the map board, then Elliot and I took off thinking we were going to get orders to attack. 'Robbie' met us at the top of his dugout. He said, 'Morris, you wait here.' They went below, and within minutes, Colonel Elliot came out, got into the jeep, and said that he had been relieved. He never got over being relieved and didn't know why."

The division personnel officer shipped Elliot to England for reclassification, though he kept his rank. Morris added, "He was a gentleman of the highest order, and I loved and respected him like a father. But he just didn't have the background and ability to command a regiment. I noticed that he grew mentally tired during the long hours in the first days of Normandy."[78]

His replacement took over on June 23. Colonel Ralph W. Zwicker graduated from West Point in 1927 and had recently served as a War Department observer assigned to V Corps. Three days earlier, he had received a Silver Star Medal from General Gerow. Now he took over the 38th Infantry, his first regimental command.

After just three weeks in combat, the 2nd Division had burned through two regimental commanders, Elliot and Fuller, plus the assistant division commander. The latter had countermanded an order from Robertson. The general fired his assistant and reduced him to colonel.

Years later, Morris asked Robertson why he waited until entering combat to relieve senior officers who he knew were problematic. Robertson replied, "If I had attempted to have them reclassified before combat, I would probably have ended up being relieved myself. But once in combat, as the commander, all I had to do was say you're being relieved, go to the rear."[79]

While Robertson rearranged at the top, Colonel Skaggs rearranged at the bottom. Skaggs shuffled men from Headquarters Company to the medium tank companies, mostly to Companies B and C to mend their D-Day losses. Replacement troops continued to arrive as well.

Charley Company required the most time to rebuild. Captain Young accomplished this one platoon at a time, while also restoring morale, which had sunk along with the DD tanks. His aggressive personality and imperishable optimism helped spark confidence among his men. He was a leader with no false front who knew his worth and had nothing to prove. Young lived true to his own credo, and everyone in the company recognized it and followed him.

Known to his friends and relatives as Dick or Richard, the captain spent his early years in Boston before his family moved to Hackensack, New Jersey. His father supervised printing operations for the local newspaper, the *Bergen Evening Record.* Young worked for his dad at the paper after finishing high school and played semi-pro football for the Hackensack Steam Rollers. He also attended Columbia University, where he met his future wife, Isabelle Hardman.

Young's civilian pursuits ended in May 1941 with an induction notice. While training at Pine Camp, New York, with the 4th Armored Division, he volunteered for Officer Candidate School at Fort Knox. He possessed an aptitude for command and completed the course in April 1942, becoming a second lieutenant. The following month, he married Isabelle, and the newlyweds took up residence in Baltimore. The bridegroom commuted to Fort Meade, where he led a tank platoon with the 741st. His wife was pregnant when he departed for Europe, and she delivered a son in March 1944, several weeks after Young took over Company C.

In the waning days of June, he wrote to his wife and recounted his part in the invasion, though omitting any mention of DD tanks. Isabelle also learned that his unit was near Saint-Lô and preparing for the next big push against the enemy.

Dry runs for that offensive continued on July 1. The unit journal of the 741st summed up the day's work: "Unit continued to engage in training activities in rear areas in connection with tank-infantry-engineer teams. All medium companies participating."[80]

For the training to change anything on Hill 192, the attack plan required attention to detail. Engineers from the 2nd Division prepared topographic maps (1:5,000 scale) showing every dirt trail, apple tree, hedgerow, sunken lane, farmhouse, and cow pasture. The mapmakers assigned a number to each pasture for easy reference, and all company officers and tank commanders received a copy. The officers reconnoitered the hill, sketched their observations on the maps, and penciled in assault lanes for the tank-infantry-engineer teams.

The overall situation provided a lesson. When an enemy entrenched himself, the attacking force needed topographic maps detailing every field and all manner of minutiae. With an enemy on the run, a Michelin road atlas sufficed.

Independence Day 1944. The staff at corps headquarters published Field Order #10, which outlined the long-anticipated attack. The order called for a push to secure positions along the Bayeux-Saint-Lô highway in the 2nd Division sector. The single objective was explicit: "Elements of the 2nd Inf Div will attack on Corps order to secure and hold Hill 192."[81]

With the big push in the offing, word came from "Vestibule," code name for the 3rd Armored Group, that six assault guns awaited pickup. Captain Hurdle E. McDaniel Jr., from Headquarters Company, took a lieutenant and six enlisted men to receive the new vehicles from the 83rd Ordnance Battalion.

Equipped with 105 mm howitzers, these modified Shermans provided more powerful infantry support, especially against fortified positions. Each medium tank company received one, and the Assault Gun Platoon of Headquarters Company received three. The newly assigned crews trained west of le Bois d'Elle under Second Lieutenant John A. Bruck, from Headquarters Company. They had several days to master their machines before the attack.

As preparations ramped up, General Robertson blew in for an inspection. He toured the battalion headquarters area and found that many men lacked overhead protection for their slit trenches, making them vulnerable to air bursts from enemy artillery. Orders from the general sent the men scrounging for material to roof their holes. Top cover was in place the next day when he returned to dole out medals.

Four battalion members received Bronze Stars for heroic achievement on D-Day: Robert W. Compton and Clarence W. Blankenship from Headquarters Company, Robert A. Nicol from Company A, and George R. Geddes from Company B.

Two days later, the general returned with four more D-Day Bronze Stars. He pinned them to Charles D. Anderson and Carl Sergent from Headquarters Company, Robert D. Coaker from Company A, and Arthur W. Barker from Company C.

While the general dispensed awards, corps headquarters sent down target dates and times for the impending attack. These initially called for offensive operations to start at 6:00 A.M. on July 9, but a change pushed everything back twenty-four hours. Another revision postponed the attack until 6:00 A.M. on July 11, with artillery preparation beginning twenty minutes prior.

CHAPTER 9 HILL 192

When Field Order #10 arrived from corps headquarters, subordinate units rushed to issue their own orders defining their particular roles, including the 2nd Division and the 741st.

The three-page document produced by the 741st explained each tank company's assignment during the attack. The companies had regained sufficient strength to support four infantry battalions. Mortar Platoon from Headquarters Company would also participate.

Captain Duke, the intelligence officer, typed a report and drew a map overlay summarizing the enemy situation and identifying the various German units opposite the 2nd Division. He named Fallschirmjäger-Regiment 9 as the unit holding Hill 192, with two battalions in frontline positions on its slopes. Duke also commented on the landscape:

"Highest elevation, 192 meters. Many small fields, hedgerows, and orchards, tending to hinder tank operation. Enemy has well-defended emplacements for MG, bazookas, 88 mm and mortars. Terrain definitely favors the enemy."

Duke went on to stress the grit and youth among the enemy's lower-enlisted ranks: "Personnel consist of men ranging in age from sixteen to twenty-one years. Morale of these troops is very high. They are fanatics. They will attack a tank with pistols and rifles."[82]

Not every Fallschirmjäger was so gutsy. Some intended to toss up their hands at the first chance, but these men were in the minority. Most enemy defenders brimmed with confidence.

★★★

July 10, 8:00 P.M.: H-hour was less than twelve hours away when the medium tank companies began leaving the training area. They crept forward to the start line in small groups under cover of night. Each machine drove in low gear and at idling speed to diminish motor noise and the squeaking sound of tank tracks.

When the vehicles arrived on line, the tank-infantry-engineer teams married up, forming six assault teams for each infantry battalion. Aside from the teams, two infantry squads in each battalion would attack without armor or engineers. Routine foot patrols, meanwhile, skulked about the hedgerows, conducting normal operations for the benefit of enemy eyes and ears.

The battalion operations officer joined Charley Company to monitor events at close range. Colonel Skaggs moved to his advanced CP (Vitamin Forward) set up near several stone-and-clay farm buildings called la Giffardière.[83]

★★★

5:00 A.M., July 11: Cold, drizzling rain fell from the morning sky as shell fire erupted along the front. The tumult boomed with deafening intensity as salvo upon salvo drummed enemy positions. Shells checkered the hill with bright flashes and shook the earth. Here and there, phosphorus rounds burst into fat white incendiary blossoms.

Ten minutes before H-hour, the artillery focused its destructive power on the first line of enemy-held hedgerows. It seemed nothing could survive the nerve-breaking drumfire.

★★★

6:00 A.M.: In the 23rd Infantry sector, the artillery preparation transitioned to a rolling barrage as the tank-infantry-engineer teams moved out. The bombardment preceded them, advancing one hundred yards every four minutes. Enemy counterfire soon fell among the attackers.

Fragments from a German shell killed two Company C men. The dead belonged to Second Lieutenant Jerry B. Brown's crew. Splinters wounded a third crewman and Brown himself, cutting his hand. The lieutenant also had a concussion but refused evacuation. Born to Jewish immigrants from Russia, Brown had survived the DD tank debacle and chafed to wage battle against the enemy. He rushed to find replacement crewmen.

The Fallschirmjägers quickly dashed all hope that artillery alone might quash resistance. Their subterranean fortifications safeguarded them during the bombardment.

All attacking units reported murderous opposition, especially one rifle platoon from Company A, 23rd Infantry. Its soldiers launched a frontal assault into the gully near Saint-Georges-d'Elle, now nicknamed "Purple Heart Draw." The Germans watched these troops descend to the bottom and walloped them with well-registered mortar and artillery fire. Rifles and machine guns spit bullets at them from houses and hedgerows overlooking the draw. The hapless infantrymen dug in with their entrenching tools, and, according to an official account, "refused to retire," but no chance for retreat existed. The trapped Americans could only shout for help.[84]

Since tanks were unable to motor straight across the precipitous draw, four Charley Company machines from Warrant Officer (jg) Quentin J. Swan's platoon lined up along the draw and fired over it to cover the embattled platoon.[85]

Staff Sergeant Malcolm A. Reynolds tried to skirt the draw with four of Brown's tanks, but they quickly halted, lacking infantry support and facing a steep grade. They held up in an orchard where the Germans pelted them with mortar rounds. The vehicles inched ahead, shelling the hedges and houses overlooking the draw.

Mister Swan ("mister" being the term of address for a warrant officer) left four of his Shermans behind, while he led another two around the draw accompanied by the reserve platoon from Company A, 23rd. They successfully rolled up the German flank. The tanks pumped high explosives into the enemy-held houses until their foundations disintegrated.

With that, the pressure let up in Purple Heart Draw, but only thirteen of the trapped men came through unscathed.

Reynolds's tanks finally left the orchard and swung behind the draw as the infantry pursued the uprooted enemy and drove toward the regimental objective.

Company C, 23rd attacked to the right of Purple Heart Draw. Finding the enemy's defenses less stout, the company hastened forward with six tanks from Second Lieutenant Edwin H. Jacobs's platoon and two from Brown's, including Brown's own tank.[86]

Rain and low visibility prevailed throughout the morning and thwarted the planned air strikes, but General Robertson ordered an exception. He requested one dive-bombing mission, and, after that, P-47s dipped from the heavens, plunging toward red smoke that American artillerymen had laid to mark the target. The planes leveled off short of the smoke and released their bombs. Phil Fitts, the replacement driver in Brown's tank, witnessed the results:

"The air force came over and bombed the blazes out of the area. They started bombing even before they got to the frontline. One of the bombs just about turned my tank over. It hit a log cabin or barn right next to my tank."[87]

That structure held enemy prisoners, who had surrendered to the 23rd Infantry. Those who survived the blast wobbled out, their senses swaying. Robertson canceled further bombing, the risk of fratricide too great.

Sergeant Henry J. Zeniewicz, a former gasoline-station attendant from Detroit, Michigan, commanded a machine in Jacobs's platoon. The sergeant and his crew smashed through two hedge-bound fields along with engineers and infantrymen from Company C, 23rd. The tankers silenced a machine-gun nest, but the uphill climb became too difficult for their vehicle. They left their designated assault lane and detoured through another field, finding the grade less steep. Before long, four more fields passed beneath their tracks.

Fallschirmjägers emerged to stop them with antitank rockets and magnetic mines, but American infantrymen trounced the Germans with rifle grenades. Several more fields fell, and the tanks demolished three farmhouses in the process.

Another threat arose when a lone Sturmgeschütz appeared. It belonged to a Luftwaffe assault gun brigade supporting the 3.Fallschirmjäger-Division.[88] Zeniewicz and his crew thumped the vehicle with cannon shells and wrecked it.

Short on gasoline and ammunition, the sergeant and his men withdrew, and a support tank took over the lead.

Close support by the infantry and engineers prevented tank losses that morning, but there was a narrow escape when an enemy rocket struck Lieutenant Brown's vehicle from behind. The projectile hit near the crank hole, according to Phil Fitts, who had been driving. An explosion reverberated through the tank, but it kept moving, though several cylinders on its motor conked out, and a fire started among satchel charges stowed on the engine deck. Nobody inside knew about the fire as Fitts drove to the rear.

"We made it back to the bivouac area, and I wondered why everybody was running when they saw us coming," Fitts later stated.[89] Lieutenant Brown climbed out of the turret, saw the flaming satchels, fainted or lost his balance, and fell off the tank. The next day, an order from battalion headquarters sent him to Company A to replace a wounded officer. Brown never felt well, and medics eventually sent him to the 24th Evacuation Hospital, where he received treatment for a concussion. He spent a week hospitalized and never returned to the 741st.[90]

Fitts came out to assess the situation and reacted without hesitation. "I stomped out the fire. I thought—like in hand grenade training—you can't outrun a grenade that's had its pin pulled. You have to throw it. I applied the same philosophy here, and it worked."[91]

Fitts counted himself lucky, though he learned later that TNT merely burns in a fire, but had the rocket detonated among the satchels, nobody would have survived.

★★★

The Germans facing the 23rd Infantry lost the east wing of the hill, while their comrades on its crest battled the First Battalion, 38th Infantry.

Combat there began twenty minutes after H-hour, when two companies from the 38th attacked alongside six Able Company tanks. They struggled up the steepest part of the hill as rain fell on the grimy, unshaven troops.

Defiant Fallschirmjägers ignored falling shells and slipped ahead several hedgerows. They adjusted shell fire on the assault wave as it started forward.

Two tanks anchored the right flank. Sergeant Donald W. Hecox commanded the lead tank for Assault Team One, while Staff Sergeant Thomas R. Fair led Team Two. They drove uphill from la Carosserie, an abandoned farmstead, and crossed the departure line. Fair's tank plunged into an apple orchard, the machine's tracks carving twin furrows in the sod.

Private First Class Charles R. Buchanan served as Fair's bow gunner. Through his periscope, Buchanan saw the blood-red trails of tracer bullets streaking through the air as the infantry charged forward to a hedgerow on the far side of the orchard.

Then it happened. A hammering blow thundered through the hull.

An antitank projectile punched into a fuel tank on the right side and sent fire surging through the crew compartment. The brain-jarring impact threw Buchanan into unconsciousness, but his blackout was only momentary. When he came to, blood poured from his nose and mouth. Searing heat washed over him in waves as flames licked at his face and hands. He squinted through double vision, trying to see his fellow crewmen in the blistering inferno.

To his left, the driver, Willis Nixon, sat slumped in his seat, unconscious. Behind him, thrashing in the flames, Tommie Fair screamed, "Let me outta here! Let me outta here!"[92] He beat wildly on his hatch, but it never opened.

Buchanan jumped up to escape the hellish crematorium, throwing open his hatch. He climbed out, slid down the front of the tank, and staggered away, his clothes smoldering. He was making an easy target for enemy sharpshooters until a soldier from the 38th Infantry sprinted forward and tackled him. Buchanan attempted to stand as his benefactor dragged him away, but the infantryman admonished, "You don't walk. You're bleeding too much."[93]

All the while, Lieutenant Sledge looked on from a support tank and reported the grim details over the radio. He broke into tears as he described the fiery scene, his words reminiscent of the radio journalist who famously reported the Hindenburg disaster.

As Sledge struggled to speak, an enemy mortar shell hit satchel charges stowed atop the tank's engine deck. The detonation shattered the rear hull and threw the turret into the air. Earth, rocks, and bits of tank rained down for what seemed like minutes. Some nine hundred yards away, a track segment from Fair's tank thudded against a tree trunk, wrapping around it and dumbfounding GIs standing nearby. The enormous blast ended Fair's agony and the lives of three others: Corporal Frank Pasqualini, Technician Fourth Grade Willis E. Nixon, and Private Raymond L. Perkins. The corporal was the only married man in the crew. He left behind a wife and a seven-year-old son.

Well away from the blazing pile of steel, litter bearers loaded Buchanan on a stretcher and carried him to a farmhouse called les Antiers, where the First Battalion, 38th had its aid station. Medical personnel infused him with blood plasma before evacuating him.

Bandaged, tranquilized, and his limbs tied down, Buchanan later took flight in a C-47 bound for England. He received care at the 1st General Hospital near London before earning bed space on a Liberty ship bound for the United States. His journey ended in San Antonio, Texas, at Brooke General Hospital, a treatment center for burned tankmen from Europe and the Pacific. After 179 days of hospitalization and numerous skin grafts, Buchanan's service ended with a disability discharge.

Before leaving the battlefield that dreadful day, he witnessed the demise of Sergeant Hecox's tank. It struck an antitank mine (or mines) on a muddy road just to the right of Fair's tank and shuddered to a halt. Fire immediately erupted.

The sergeant and his gunner, Corporal Clyde Rohrbaugh, quickly escaped, as did the assistant gunner, Private Peter DeNota. The driver, Technician Fifth Grade Russell M. Bradsher, and his assistant, Private First Class Andrew G. Anderson, remained inside. Hecox tore back and pulled out Anderson, while Rohrbaugh struggled with an unconscious

Bradsher. The flames burned Rohrbaugh's neck, and he finally gave up, the intense heat driving him away.

Minutes later, the tank's ammunition exploded, blowing the turret sky high and embedding the motor in the road. Satchel charges on the tank may have contributed to the blast as had been the case with Fair's machine.

As morning light filtered through the drizzling rain, both machines burned, their fires casting an eerie glow visible for miles.

The four other Company A tanks that led the assault also encountered enemy fire. Second Lieutenant Calvin B. Meyers's cannon suffered a damaging hit, and an enemy rocket crippled Staff Sergeant Nicol's tank. Both vehicles withdrew. The tanks commanded by Sergeants Henry P. Kult and George Cora also sustained battle damage and pulled back for repairs.

The company lost Lieutenant Sledge when a shell fragment pierced the front of his helmet. The jagged sliver tore a hole in the steel shell, traveled around the liner, and gashed the back of his neck. His driver quickly reversed their machine in urgent search of medical help. The lieutenant's bloody wound required eight sutures and hospital time.

Support tanks commanded by Sergeants Ball and Coaker remained in the fight without having to withdraw. Sergeant Larsen's Sherman fell back with mechanical difficulties. Two tankdozers, one commanded by Bill Smulik, found the terrain impossible. The rainy weather turned Hill 192 into a muddy slick for tracked vehicles. Many had smooth-rubber pads on their track links, which sent the tanks sliding back twenty feet at a time.

At 8:05 A.M., an urgent message from Captain Thomas arrived at the 741st forward CP. He requested permission to reorganize his company and shift it to more favorable terrain. Skaggs dispatched his intelligence officer to make a firsthand assessment. An hour later, the officer reported a litany of Able Company woes. Skaggs immediately approved Thomas's request, and Colonel Zwicker gave final assent.

Thomas spent forty-five laborious minutes locating two alternate routes, while the infantry slogged uphill alone and secured a diamond-shaped forest code-named "Ford Woods." Despite falling short of their objective, the foot soldiers produced a significant result by seizing the highest point on Hill 192, located in a copse of tall trees called "Dodge Woods." There they found an observation tower, a mishmash of barn ladders and old planks nailed to tree trunks. The spindly structure afforded a view clear to the coast.

With it in American hands, the enemy no longer controlled the high ground.

Baker Company of the 741st fared better during the attack. Two tank platoons led the assault, along with an engineer company and two rifle companies from the Second Battalion, 38th Infantry, specifically Company E on the right and Company F on the left.[94]

The attackers slinked ahead through rain and mist as a rolling barrage fell on the enemy. About thirty Fallschirmjägers jumped up on the left with their arms raised. They readily surrendered, believing American propaganda leaflets that promised food and safety. With these prisoners bagged, Company F and its tanks sliced south and uphill to crack open a weak portion of the German line. The hilltop soon lay within rifle-grenade range.

On the far right, Company E faced a prickly obstacle, an enemy strongpoint dubbed "Kraut Corner." Artillery from the 2nd Division had blasted it for several days before the attack and again that morning. Its defenders held firm in bunkers, prepared for whatever malice came their way.

Lieutenant Charles Curley's platoon from Company E faced a less-defended area alongside Kraut Corner. His men straddled a muddy road known to the French as D390. According to plan, Curley deployed one tank and one rifle squad on each side of the road, along with machine-gun crews. He also had another rifle squad and two tanks just to the rear in support. The lieutenant himself joined the left-hand group, Assault Team Three.

"I was able to get into the first field past the le Parc farm, through a gate in the hedge. No resistance at first, except small-arms fire going over our heads," he recalled.[95]

Staff Sergeant Paul Ragan's tank also passed through the gate. His vehicle had the number "3" crudely painted on its rear, identifying its assault-team assignment. The tank rolled up to the first hedgerow, and its crew pumped shells and machine-gun bullets into an enemy-held hedgerow across a pasture. Artillery fire ripped at the far hedgerow, as did a light machine-gun crew. The noncom, who led the rifle squad, signaled his men forward, and they advanced in a half-crouch, rushing along the pasture sides. Two engineers ran with them. Curley rolled back the artillery fire as the soldiers pushed forward. All the while, Sergeant Bernard J. Cook from the 165th Signal Photo Company captured scenes on motion-picture film.

After the squad reached the next hedgerow, two engineers, who had remained behind, pulled satchel charges from the tank and detonated them on the first hedgerow. The eruption sent earth, smoke, and shredded vegetation rocketing skyward. Curley and everyone still at the first hedgerow rushed up the pasture sides, while Ragan's tank powered through the freshly blown hole. Enemy shells soon fell behind them. The smoke had attracted German attention.

Lieutenant Curley decided against using satchel charges again. When he reached the next hedgerow, German "egg grenades" sailed over its top and down on his men, who, in return, lobbed fragmentation grenades in the opposite direction. Curley saw his squad leader climb the hedge and, fully exposed to the enemy, fire his Thompson submachine gun down the other side.

Ragan's crew tried blasting open the hedge with HE shells, but they only shredded foliage and kicked up clods of earth. Frustrated, the tank commander and Curley hashed over the dilemma on the vehicle's field telephone. Ragan said their support tank had a bulldozer blade. "Bring him up!" Curley replied.

Sergeant Elmer Middleton's tankdozer rumbled forward, "3S" painted on its sides and rear. Curley rushed over, grabbed its telephone receiver, and spoke with Middleton. "I wanted him to hit the hedge low, and, once he broke through, turn and run along the hedge with his blade stuck in the side of the hedge. He said, 'No problem' and took off."[96]

The driver, Private John R. Brewer, lowered the big blade and accelerated. He hit the mound with a jolt, driving deep into its packed clay and tangled roots. He backed up and smashed forward again, breaking through the ancient hedge. Brewer whipped the dozer to the left and began shearing off the mound from its backside, plowing through enemy entrenchments.

About ten Fallschirmjägers wearing camouflaged smocks flew out of their holes, leaped through the hedge, and ran toward the American rear. "Halt!" Curley and his men shouted, before realizing the Germans were running on their toes, as if playing hopscotch.[97] They had chosen to dance through a minefield rather than face the dozer blade. The artillery bombardment had uprooted many mines, making them visible. Ragan and Middleton's machines had somehow rolled across the pasture without detonating a single one. The infantry had merely skirted the area, thus avoiding the mines.

While the enemy fled in panic, Brewer continued to gouge open the hedgerow, reportedly killing three paratroopers by burying them alive.

On the right, Assault Team Two battered its way through enemy-fortified hedgerows.

After more than an hour, Curley's platoon sideslipped Kraut Corner, unhinging its defenses and freeing up a sister platoon bogged down at the strongpoint. The battle immediately turned toward Cloville, ten farmhouses and outbuildings already torn apart by artillery fire.

Lieutenant Curley stood in front of Ragan's tank at the last hedgerow. He gesticulated with his hands, pointing out the new attack route to the sergeant and his driver.

"German tanks!" someone shouted.

The fearful warning seized everybody's attention. Ragan's gunner reacted straightaway, traversing his cannon right to cut loose with a shell. "The blast felt like it tore my head off," Curley recalled. "My helmet flew over the hedge and, semiconscious, I raised up, and they fired again."[98]

Curley woke up groggy with his ears ringing after his platoon messenger poured water on his face. "Are you all right, lieutenant?" asked the soldier.[99] Curley first thought the tank had sustained a devastating hit, but it was Ragan's expert gunner, Corporal Frank Lenhardt, who had caught two enemy assault guns.

Downrange, a Sturmgeschütz burned in an orchard, while a second one sat dead on a dirt road, many of their occupants killed outside, machine gunned by Ragan's crew.[100]

It took Company E another ninety minutes to seize Cloville, permitting the Second Battalion to press its attack over the western slope of Hill 192.

The drizzle and rain showers quit by late morning, and sunbeams occasionally cut through the clouds as the attack continued into the afternoon.

At 5:00 P.M., the tanks and infantry reached the Bayeux-Saint-Lô highway near le Calvaire, where a tall, wooden crucifix stood sentinel along the road. The Second Battalion, 23rd had reached its objective, an achievement unmatched elsewhere by the 2nd Division that day.

Company F prepared to cross the highway and seize the nearest hedgerow on the other side. The foot soldiers asked for help from their armored escort—Sergeant Millard Case, who commanded the lead tank of Assault Team Four, and Sergeant Wilbert C. Mortzfield in the support tank.

An infantryman used the telephone on Mortzfield's tank to request that he move across the highway. Mortzfield obliged. As his driver rumbled toward the first hedgerow, Mortzfield, nicknamed "Stubby" due to his slight stature, poked his head through the commander's hatch and scanned the terrain for German activity. To the left, he spied freshly turned earth, possibly an enemy entrenchment. He shouted fire commands into his microphone. The tank halted, and its turret spun left. His gunner blasted the spot, then suddenly yelled, "Traverse right!"[101]

In that instant, an enemy rifleman hit Stubby. The bullet struck the turret, splashing him with fragments that ripped into his cheek and lower lip and punctured his neck.

Mortzfield blacked out from shock and dropped to the crew-compartment floor. He awoke to see his men staring at their bloodied commander.

The wounded sergeant pulled himself up and scrambled back into the turret. "Put it in reverse, back out of here!" he screamed, fearful that enemy tank hunters were maneuvering to destroy them.[102]

The driver's left foot hit the clutch, and his right hand grabbed the gearshift lever. He steered back across the highway and parked alongside a stone building. With his tank safe, medics evacuated Mortzfield for a hospital stay that would last thirty-seven days.

The tanks from Baker Company took defensive positions along the highway, while its commander dispatched a relief force—the Assault Gun Platoon from 741st headquarters. The platoon numbered six machines, an ad hoc assemblage of every assault gun in the battalion.

Company B casualties for the day totaled four men wounded and one tank driver killed, plus two tanks disabled, one by friendly artillery. Technical Sergeant Marvin M. Helmberger and his company maintenance section arrived on scene to retrieve both disabled machines using their T2 tank-recovery vehicle.

German shells, meanwhile, struck the area, each salvo screaming down on its target. Curley had also crossed the highway, and he spotted a likely perch for the enemy artillery spotter: a distant bell tower at Saint-Jean-des-Baisants. He tracked down Lieutenant John Bruck, who commanded the Assault Gun Platoon, and asked him to hit the tower with 105 mm HE rounds. After two shots, the enemy unfurled a Red Cross flag from a shell hole in the tower, and Curley halted the firing.

The lieutenant found another use for the nearest assault gun when his messenger yelled, "What the hell is that?"[103] What looked like two squat beetles covered with brush emerged onto the highway from a sunken road.

Curley caught the gun commander's attention by jumping up and down and pointing toward the camouflaged targets. The commander dropped down into his vehicle, and seconds later, his howitzer cut loose. One Sturmgeschütz returned fire but missed. The Germans focused on the single American gun they could see, unaware that others lurked nearby. In minutes, both enemy machines sat flaming on the highway, turned into steel cookers by 105 mm antitank projectiles.

★★★

The next morning, German corpses littered Hill 192, some grotesquely contorted in death, others curled up, as if asleep. Abandoned rifles, discarded helmets, and dropped ammunition lay strewn in the orchards and pastures. All across the hill, splintered trees and thousands of shell craters accentuated the scene of havoc.

Among the detritus, the victorious GIs discovered several 8.8 cm recoilless antitank guns. These rocket-firing weapons required a three-man crew and had an effective range of 230 meters. An enemy prisoner revealed that his antitank company had eight such guns. Nicknamed the Puppchen, or Baby Doll, the weapon was toy-like in appearance but packed a lethal wallop.

General Robertson ordered the attack to resume until the 2nd Division reached all its objectives. The First Battalion, 23rd and the First Battalion, 38th jumped off at 11:00 A.M. without artillery preparation and gained their objectives.

American troops at last controlled the hill.

CHAPTER 10 LITTON'S RHINO

July 12, 7:00 P.M.: An enemy position hidden south of the highway fired on the Second Battalion, 38th. The battalion commander ordered Lieutenant Bruck to have an assault gun eliminate the devilishly camouflaged position.

Bruck tapped Sergeant Frank Kurkowski for the job. He and his crew maneuvered behind a house along the highway and aimed straight through its front and back doorways. The gunner fired multiple rounds and hit the position three hundred yards away. Infantrymen later mounted a patrol and counted two victims: an antitank gun and a Sturmgeschütz.

The next day, Baker Company armor from Lieutenant Wilson's platoon replaced Bruck's Assault Gun Platoon. Bruck's unit disbanded, and the vehicles returned to their original companies, all now performing maintenance and receiving replacement personnel.

At the battalion CP near Cerisy-la-Forêt, the officers and enlisted men packed up their equipment and moved Vitamin Rear south of the town. They selected a location where the main road (D34) entered the Cerisy Forest.

Three new second lieutenants joined the 741st: Walter W. Parr, Basil J. Raymond, and Robert L. Dudley. They arrived just as a new series of exercises began. The attack on Hill 192 had exposed a need for more training and revised tactics.

Hedgerows were the greatest factor holding back the tanks and infantry. Satchel charges proved a cumbersome, if not risky, tool. Tankdozer blades worked, but dozers were too few in number, and their presence told the Germans where to expect an assault. The Americans needed a better way to beat the bocage. Rear-echelon officers solicited ideas from those who knew best: soldiers at the front.

Men in the 102nd Cavalry Reconnaissance Squadron (attached to the 2nd Division) mulled over the problem, and a soldier from the light tank company voiced a suggestion. Sergeant Curtis G. Culin III recommended attaching a "snowplow" to each tank.[104] The squadron motor officer, Captain Stephen M. Litton, refined the idea. He proposed welding steel prongs to the front of each machine, a device resembling fork tines. These, he believed, would permit tanks to punch through hedgerows without explosives.[105]

The idea took shape when Litton's men built a prototype from captured German beach obstacles. The angle iron made ideal fork tines. These enabled any Sherman to exploit its weight advantage—thirty-five tons, versus eight to ten tons of hedgerow earth. Stuart light tanks also possessed a weight advantage.

After trial runs, the 102nd Cavalry demonstrated its hedge-busting innovation to the commanding generals of the 2nd and 5th Infantry Divisions, as well as the V Corps commander, and the First Army commander. The device, dubbed the "Rhino," so clever in its simplicity, arrived as a godsend. The army classified it "Secret" and restricted it from combat until the next big push.

The Rhino made possible new tactics. Skaggs and Robertson developed a method known as a "sortie." It called for tanks to mass behind the departure line while infantry hunkered down in slit trenches several hundred yards farther back.

The artillery would deliver fire along the departure line and up to five hundred yards into enemy territory, while the tanks sped forward and slashed through hedgerows, protected all the while by artillery shells bursting in midair. For psychological effect, tank commanders had instructions to switch on their sirens while advancing.

If a hedgerow resisted the first blow from a Rhino, and it became necessary to strike two or more times, the crew needed to judge when the hedge was about to give way, pull back, and burst through at attack speed.

Every lead tank had a support tank for rear protection. Speed was the key to success and the best way to avoid enemy tank hunters. All firing was to occur on the move.

After penetrating the enemy's defenses, the new modus operandi called for the tanks to double back, join the waiting infantry, and advance together through the breach.

The sortie method would make its debut during an offensive being crafted at First Army headquarters. Code-named "Cobra," its planners aimed to break open the Normandy front and bid farewell to the hedgerows. Headquarters set the target date for July 19, 1944, with the 2nd Division to seize high ground between Mouffet and Saint-Jean-des-Baisants. The attackers would pursue additional objectives as fast as enemy resistance permitted.

While the 741st and 2nd Division developed the sortie method, enemy paratroopers on the other side hardened their defenses. They laid down three fortified lines in echelon, a scheme intended to digest an American attack, leaving only wasted remnants.

Well behind the frontline, the hospitalized Lieutenant Sledge wrote to his father, a physician in Mobile, Alabama:

"I am just about recuperated now from my first contact with German steel. I got a little shrapnel in the back of the neck, but the medics dug it out, and I'm good as new. I guess I'll eventually get the Purple Heart for it."[106]

Another decoration also awaited the injured officer. While still hospitalized, he received a summons to Ivanhoe headquarters on July 14, where

General Robertson presented Sledge and Leonard Trimpe with Silver Star Medals for "gallantry" on D-Day.

The lieutenant mailed his medal home before returning to Able Company three days later, after which he immediately penned a V-Mail letter to his dad:

"I just got back to the company a few minutes ago. And it sure is good to get back with the boys again. They're a great bunch, and I'd sure hate to have to be transferred. I'll mail the Purple Heart home tomorrow. Keep it along with the Silver Star. The wee small bar with each medal is to wear in the lapel of civilian clothes. Nice idea isn't it?"[107]

The next day, Sledge also sent home a photograph showing his battle-damaged helmet. He explained that he received the picture from Colonel MacLaughlin, the group commander.

Unknown to Sledge, an Associated Press correspondent named Hal Boyle also received the photo and wrote a story about it. Newspapers across the United States published the account, including *The Mobile Press*. Its editorial staff titled the piece, "Mobilian Has Close Call When Hit by Nazi Shell."[108]

Other papers ran an extended version of Boyle's article, which presented a description of John Brewer's tankdozer maneuver, wherein he buried three Fallschirmjägers.

For security reasons, Boyle omitted any unit identification.

Second Lieutenant George R. Coleman, a platoon leader with Company D, also wrote to his father, a real-estate broker in Elgin, Illinois, and a First World War veteran.

"Pop, I can picture very vividly what you went through in the last fracas. Very useless way for human beings to act. I've been lucky so far. All my lights are still running, and all my boys are O.K. Damn lucky."[109]

Before joining the army, Coleman studied engineering on a football scholarship at Iowa State College. He quit school in 1942 to enlist and seven months later earned a commission. For luck, the young lieutenant wore a gold ring that his dad had worn during the Great War.

In a three-page letter to his son, the elder Coleman reflected on that war, saying that once a man has experienced combat, "he has some things imprinted so deeply in his memory that forgetfulness is impossible." He continued. "You are going through the same things today that I went through twenty-five years ago. Your methods are more modernized and streamlined. You are doing things today that in War #1 would have been impractical, but the overall principle is the same."

George's father went on to write about being the parent of a soldier. "This much I will tell you, old-timer, and speaking as a man who has seen and experienced both sides of this question, it should carry some weight. I make the following assertion in all sincerity. The mental strain is one thousand times greater for those who are left at home.

"Let no man ever boast again of our vaunted so-called civilization, as I am firmly convinced that man of today is more savage, more brutal than any other time in history. Spiritually, it looks as though we are a total flop."[110]

The gray-haired veteran never posted the letter. He decided to save it for his son to read after the war.

Like everyone else in the 741st, George Coleman waited for the coming offensive, but foul weather on July 19 delayed Operation Cobra. Low-hanging clouds grounded Allied aircraft, and July 24 became the new date for the 2nd Division's attack.

Tanks from the 741st began infiltrating to the frontline area each night. On July 21, a field order from battalion headquarters dissected the attack plan.

Preparations continued, especially Rhino installations. Ordnance troops fabricated and distributed over five hundred hedge cutters throughout First Army. Given the lead role in Cobra, armored units in VII Corps received top priority. Demand outpaced supply, and units like the 102nd Cavalry—where the Rhino originated—received none, so the cavalrymen built their own.

The 741st also fended for itself. Service Company trucks ventured to Omaha Beach, where soldiers hoisted rusty beach obstacles onto the vehicles. With acetylene torches showering sparks, 2nd Division ordnance men and battalion maintenance troops cut down the obstacles. After welders mounted the devices on tanks, their drivers began Rhino schooling. For best results, the tankers learned to hit a hedgerow in third gear at fifteen miles per hour.

To toughen their vehicles, crews heaped on sandbags for added protection against Panzerfaust warheads and Panzerschreck rockets. These shaped-charge projectiles could penetrate six inches of armor plate. The bags served as a standoff, detonating the charge away from the armor and absorbing or deflecting the blast. This stratagem was unproven in combat.

By July 23, all 741st tanks had reached the front and sat poised for H-hour the next morning, but low clouds intervened again, foiling an aerial bombardment needed for the VII Corps attack.

With no offensive to wage, the tank crews used their spare time for more training, sandbag piling, and vehicle maintenance. Inactivity also filled many hours, but these were hardly carefree hours of smiles and laughter. The men had too much time for thought. Fear and tension grated on nerves, and many soldiers drank to ease the strain. Normandy was the only region in France that produced no wine, but farms everywhere distilled calvados, a potent apple brandy. Corporal Jack Boardman from Company B recollected that Technician Fifth Grade Joseph Feckete, a tank driver in the company, siphoned gasoline and traded it to farmers for bottles of their apple lightning.

Boardman never forgot those anxious, idle hours. "In the few days preceding the attack we did a bit of sweating," he said. "I was trying to work myself into a state of mind where I could go into battle cheerfully with courage and complete confidence. I never succeeded."

Boardman frittered away time inside his tank. It was his home address and his sleeping quarters. Digging a foxhole outside was too much bother. Private Raymond A. Tuttle Jr., the driver, and Private First Class Enrico Valente, the assistant gunner, had shoveled out a sleeping hole but gave it up to join Boardman. He explained the reason why:

"All they wanted was to keep me company, they said. I found out later that a mortar shell had landed about six inches away from their foxhole and tore their shelter half to shreds. We spent most of the time in our tank or with some other tank crew. It just wasn't healthy to wander around above ground. The Jerries dropped mortar shells in at the most inconvenient times, particularly when we walked down to chow or went to relieve ourselves."[111]

Forward observers on both sides viewed nothing as inviolable, even a soldier squatting to empty his bowels. They considered everything a fair target, although they usually exempted persons, vehicles, and buildings bearing a Red Cross emblem.

CHAPTER 11 **COBRA BEGINS**

On July 25, the clouds dispersed, and the Cobra aerial bombardment began in the VII Corps sector. Over twenty-two hundred Army Air Force planes drubbed the Germans in an epic onslaught. The tank soldiers turned their faces skyward to view the flagrant display of Allied air superiority. The aviators, unconstrained by hedgerows, leveraged their numerical advantage to thrash the enemy.

As the VII Corps attack began, V Corps alerted its units for action. A teletype printer at 2nd Division headquarters clacked out a message with the new H-hour and D-day: 6:00 A.M., July 26. Signalmen flashed the information to 741st headquarters, and Colonel Skaggs drove to la Giffardière to reopen his forward CP. Everything was set. No more delay.

Before H-hour the next morning, Jack Boardman found himself battling diarrhea. He used an ammo can as an improvised toilet and caught hell from his fellow crewmen every time he did. The contents nauseated the tank commander, who gagged whenever Boardman handed him the can to empty.

Outside the stench-filled tank, the sky was cloudless, but, as Boardman later stated, "There was a heavy ground mist, which made things damp and miserable."[112]

The tank crews commenced their start-up procedure by hand rotating the Wright-Continental "Whirlwind" engines in their machines. After sitting idle for hours, crankcase oil in these radial motors sometimes seeped into the lower cylinders, causing hydraulic lock, a condition that could break a connecting rod and cause engine failure. The rotating process allowed crewmen to feel a locked cylinder, remove a spark plug, and drain the oil.

With all cylinders clear, the drivers started their motors and moved out.

Ahead, on the southern horizon, the sky exploded with artillery fire. Shells flashed overhead and detonated in the air above the enemy. "There was a steady, cracking roar from the high-explosive shells landing one hundred yards from us," Boardman remembered.[113]

Baker Company tanks crossed the highway, ready to start their sorties at H-hour.

Sergeant Maddock, Boardman's tank commander, told Rafael Landrove, the driver, to squeeze through a cattle gate and navigate a grassy meadow pockmarked with shell craters. Once across, the crew stopped at the first hedgerow, its leafy mass interspersed with tall, spindly elm trees that looked like pipe cleaners, ubiquitous features of the Norman countryside. Over the decades, the locals had stripped away branch after branch for kindling.

Maddock and his crew waited for their support tank to fall in behind, but it never arrived. Its commander, Sergeant Clyde W. Mercer, missed his mark and entered an adjacent field. The clock, meanwhile, continued ticking, each sortie allotted only twenty minutes—ten in and ten out. Maddock had no time to spare as he held up his microphone and shouted above the motor roar, "Okay Landrove, back 'er up about ten feet, then take off and bust through."[114]

Boardman took a firm grip and braced himself. The machine crashed hard and hesitated a moment before the hedge broke open, and the tank burst through.

The impact knocked out Ray Tuttle, who manned the bow machine gun. Boardman opened fire alone, blasting HE rounds from the cannon and pounding out bullets from the coaxial machine gun. Landrove roared across another beat-up pasture, and Boardman's fusillade continued for about a minute. Time was up. They had to turn back. Landrove threw the vehicle into reverse gear and swung around to exit the field.

After withdrawing to join troops from the First Battalion, 38th Infantry, Landrove whipped the machine around and pointed its bow toward the enemy. Boardman looked through his gunsight and spied an enemy soldier. "He poked his head over the hedge across the field in front of us. Ah ha, I thought, I'll teach you! I hit the top of the hedge with a 75 shell. He poked his head up once more. I fired another shell. Again, his head cautiously rose. Again, I shot."[115]

The elusive Fallschirmjäger never reappeared.

Thirty minutes after H-hour, Jack Boardman watched the advance resume through the narrow perspective of his gunsight. Guns blazing, his Sherman roared back through the first hedgerow and over the meadow. The accompanying infantry squad sprinted across the open ground and gained limited cover against the next hedgerow.

Ferocious opposition erupted. The squad leader scaled the hedge and gunned down the paratrooper who had eluded Boardman, bringing a smile to the tank gunner's face.

He described what happened next:

"Suddenly the doughboy I could see in front of me dropped to the ground. I strained my eyes through the gunsight and traversed the turret back and forth trying to pick up the gun that was firing. The dough's squad leader began yelling over the telephone on the back of the tank, 'We're getting heavy fire from the left corner! Hit 'em, hit 'em!' I snapped the turret to the left and put an HE on delay into the corner of the hedge. It passed through the bank and exploded in the middle of a machine-gun nest."

After Boardman's star performance, the infantry squad prepared to dig out the remaining defenders. "The doughs wanted to know if we had any Tommy guns we could lend them," Boardman remembered. "We immediately handed out three and all the ammo we had for them. Valente began handing out grenades through the pistol port to one of the doughs. Val said his face looked like a kid's receiving Christmas presents."[116]

The infantrymen winkled out the enemy, and, together with Maddock's tank, captured the next pasture. By noontime, the tank sat concealed in a sunken road. Feeling safe, the crew enjoyed a breather. Boardman tore open a dinner K ration and pulled out a puck-shaped can filled with cheese. When he idly glanced through his gunsight, he saw someone's head poke around a curve in the road.

"My mind was elsewhere, on something not even remotely connected with warfare," Boardman recalled. "But when I saw him grin and beckon to another Jerry who stepped out and leveled a bazooka at us, I realized somebody better do something . . . fast!"[117]

Boardman always kept an HE shell in the firing chamber. Without taking aim, he stomped on the foot trigger. The bow gunner also opened up, but both men missed.

The Panzerschreck team ducked behind the curve.

The assistant gunner, Valente, slammed another HE round in the breech, and Boardman aimed the cannon just beyond the curve, hoping to nail the two paratroopers with shell fragments.

He banged out two shells. The infantry later found both of the Germans dead.

Enemy tank hunters also stalked other Baker Company tanks that day.

When a Panzerschreck rocket struck Sergeant Walter R. Jutkins's vehicle, three crewmen sprang out and ran to safety. Private First Class Merle A. Newell stayed behind with the wounded driver, Technician Fifth Grade Ferrell P. Bishop. Newell pulled him into an empty German foxhole and bandaged his bleeding arm. Enemy mortar and small-arms fire thwarted all rescue attempts until evening, when Staff Sergeant Paul Ragan drove his tank directly over the foxhole and stopped. Newell crawled in through the escape hatch and helped lift Bishop inside. Nerve damage and paralysis in Bishop's injured arm earned him a Purple Heart and a medical discharge. Ragan received a Bronze Star Medal.

Another rocket hit Sergeant Donald P. Lucey's machine, killing his gunner, Corporal Robert W. Fry. The other men escaped and recovered their damaged tank later that day.

The enemy wrecked Sergeant Willis D. Warren's tank with an 88 mm antiaircraft gun deployed in an antitank role. The first shot smashed through a track, followed by a shot through the frontal armor between the driver and assistant driver. The last round penetrated the turret just below the gun mantlet and spewed white-hot steel into an ammunition-stowage rack.

The driver and assistant driver escaped through their hatches. Warren opened his hatch to bail out, with his gunner, Corporal Alfonso Rolland Jr., right behind. Rolland glanced back to look for the assistant gunner and saw Private First Class Valjean H. Rich with his head under the cannon, ready to follow. At that instant, fire exploded in the crew compartment.

Rolland dashed fifty yards to a foxhole, his face and thighs burned. He turned, hoping to see that Rich had also survived the flames but instead saw a smoldering form alongside the tank, a body burned beyond recognition. Medics later evacuated Rolland, and he remained hospitalized until December before returning to duty.

Valjean Rich, a twenty-three-year-old auto mechanic from Tucson, Arizona, died that day. The army never positively identified his burned remains.

As for Sergeant Warren, machine-gun bullets chased him while he ran from the tank. He reached safety but never felt secure again. His narrow escape triggered an awareness that he was vulnerable, dreadfully so. "I never felt they could hurt me until the day they shot my tank," he recounted. Week after week, the tank commander had told himself that he was unique, that his ability to survive was superior, that others would be killed but not him. He now felt helpless to prevent his own death, that his time was surely coming.

Warren spent twenty days hospitalized for minor facial burns. When he rejoined Company B and received a replacement tank, climbing into it for the first time "was the hardest thing in the world."[118] It took every ounce of his courage, his hands trembling the whole time.

★★★

In the early morning darkness, Charley Company faltered prior to the first round of sorties when smoke from exploding shells blinded its tank crews. Only five of them launched sorties. After the smoke drifted away, the company began its second round of sorties with two platoons alongside the Second Battalion, 23rd Infantry.

Accompanied by foot troops, Staff Sergeant Malcolm Reynolds's tank rumbled across the highway and downhill, with the sergeant's longtime buddy Phil Fitts in the driver's seat.

The two men became friends in the National Guard before the war. Fitts later joined the 29th Infantry Division as a mechanic in its ordnance company. While stationed in England, he learned that his friend and the 741st were also there. Fitts finagled a transfer to the battalion, a reassignment that meant combat duty, something he initially kept from his wife, Polly.

After crossing the highway, Fitts turned left and plowed through a large hedgerow, leaving a wide hole. Ahead stood a wooded area, the Bois de la Roche. Fitts stopped the tank, and the crew swept the trees with cannon shells and machine-gun bullets.

The motionless tank attracted a lone Fallschirmjäger hiding in the green shadows and clutching a Panzerfaust. The paratrooper had his target. He squeezed the trigger mechanism, and the warhead sailed through the air in a high arc. Its journey ended in an explosion that pierced the vehicle's left sponson and ruptured the five-gallon fuel tank for the auxiliary generator.

Fire and steel fragments blew through the crew compartment. The blast ejected Reynolds from the turret hatch like a jack-in-the-box. The gunner and loader scrambled out through the open hatch, the turret filling with flames behind them. The assistant driver, a recent replacement, screamed in horror, lacking the knowledge to open the escape hatch on the floor behind his seat.

The calamity also trapped Fitts. He tried to fling open his driver's hatch, but it struck the cannon barrel. The gunner had failed to elevate the barrel before exiting, standard practice when abandoning a tank. Fitts's mind raced to find another way out.

He rolled over the transmission, jerked up the escape hatch, and dove out, the assistant driver following him. Fitts paused for a split second and decided to crawl forward, afraid the Panzerfaust man was to the rear and also fearing the tank could explode at any moment.

He emerged and darted around to the left side, searching for Reynolds. Fitts looked up and saw his friend on the tank's frontal armor, bleeding profusely and tugging at the cannon barrel, trying to free up the driver's hatch. Fitts grabbed his buddy's waist belt and yanked him down. After dragging Reynolds into a foxhole, a safe haven also occupied by a dead German paratrooper, Fitts demanded an answer.

"What the hell were ya tryin' to do? Get yourself killed?"

Reynolds fought back the tears as he spoke.

"I couldn't go home and tell Polly I left you in a burning tank."[119]

All the crewmen survived. Reynolds was the most seriously injured—his left leg, backside, and buttocks loaded with steel fragments. He remained hospitalized until October and walked with a limp for the rest of his life. By comparison, Fitts had trifling burns and shrapnel wounds and stayed on duty after receiving treatment from an aidman. For more than a year afterward, Fitts combed tiny steel shards from his hair as they loosened from his scalp. Besides those mementos, he had another souvenir from that day: a P-38 pistol snatched from the dead Fallschirmjäger in the foxhole. Fitts carried that sidearm throughout the war.

As the flames consumed Reynolds's tank, German machine guns whittled away the foot soldiers accompanying the other Charley Company vehicles supporting the Second Battalion. Charging through two or three fields, the tanks carried the attack forward alone. Suddenly, voices from their radios interrupted and directed the tankers to fall back and assist the infantry.

Meanwhile, Mister Swan and his platoon operated in another sector, having teamed up with a rifle company from Third Battalion, 23rd. Together, they moved partway across the first pasture, until the infantry drew automatic-weapons fire and scampered for cover. The tanks crept ahead and began flushing out the enemy, but new orders sent the armor elsewhere. Swan's platoon drove to reinforce another group of soldiers from the Third Battalion, troops pinned to the earth by German bullets. The tanks relieved pressure on these infantrymen and stayed until late in the day when ordered into reserve.

Both tank platoons with the Second Battalion remained at the frontline.

★★★

Able Company completed its initial sorties that morning, encountering only sporadic rifle shots and machine-gun bursts. The tankers fell back and joined the Third Battalion, 38th Infantry. The combined force quickly overcame all resistance but halted to permit lagging units on the flanks to catch up.

Just as the advance resumed, 88 mm armor-piercing (AP) shells interdicted the tanks.

Lieutenant Calvin Meyers's tank rumbled through a hedgerow opening and into the crosshairs of a German gunner. The lieutenant, a George Washington University graduate and former dance studio manager, had joined Company A on June 17 as a replacement. He died when steel fragments struck him in the head. His loader and his assistant driver also perished—Private Robert L. Freed and Private First Class Samuel Popovic.

Shermans from Company A reached les Épinets by early afternoon, an advance of nearly a mile. They remained there with troops from the Third Battalion until late evening.

By 10:00 P.M., all the Able Company tanks had returned to the rear for servicing. Their crews filled empty fuel tanks, snaked in belts of machine-gun ammunition, and conveyed 75 mm shells onboard one after another by human chain. The tankers also swapped out bullet-worn machine-gun barrels and mended battle damage.

First Lieutenant Roger McDonough replaced Meyers, McDonough having rejoined the company after convalescing from his D-Day injuries.

★★★

Company D spent most of the morning on July 26 in division reserve, but at 11:00 A.M. that changed. The light tanks relocated to the 38th Infantry's forward CP. Two platoons stood ready there, while another raced off to help pinned-down soldiers from the First Battalion, 38th.

The tanks pelted enemy emplacements with canister shot, high explosives, and machine-gun bullets. Tank fire knocked out a mortar and a Panzerschreck, which weakened the opposition, but the beleaguered soldiers remained stuck.

At 3:00 P.M., Lieutenant Coleman left the 38th CP and assembled near Cloville, tasked with assisting the Second Battalion, 38th. As the platoon moved out, artillery fire hit the lieutenant's tank, leaving him with a foot injury and lacerating the abdomen of Staff Sergeant John Cahill, the platoon sergeant. Toe and hip wounds hobbled Private William J. Russell.

The company commander, Captain John Sicks, took over the platoon, and, along with the Third Platoon, joined the infantry in clearing a wooded area northwest of les Épinets. Tank cannons sent enfilade fire into the trees ahead of the advancing infantry. The foot soldiers broke the enemy's hold on the little forest, and the tanks withdrew for the night.

At the 2nd Division CP near Cerisy-la-Forêt, General Robertson hosted the First Army commander and Soviet military officers, all interested in the attack and its progress.

By day's end, the 23rd Infantry had gained one thousand yards, as had the 38th Infantry—results far shy of expectations. The 9th Infantry battled spotty resistance and forged ahead twenty-seven hundred yards, supported by armor from the 744th Light Tank Battalion. Saint-Germain-d'Elle fell to the 9th, and, by nightfall, its soldiers held ground near their objective, Mouffet. The toughest fight, however, occurred on the extreme right, where the 38th Cavalry Reconnaissance Squadron advanced a scant five hundred yards. Enemy armor and paratroopers counterattacked, a rarity for the Germans in this sector. The American cavalrymen fell back but later recovered their lost yardage.

Although the 2nd Division reached none of its first-day objectives, it built momentum toward ousting the 3.Fallschirmjäger-Division from its echeloned defenses. Countless paratroopers died in the battle, and precisely 235 now sat inside the Ivanhoe PW cage.

The 741st recorded six fatalities and nine tanks lost, although only two or three of those machines met with total destruction.[120]

That night, the 2nd Division readied plans to punch through the enemy division.

CHAPTER 12 BREAKTHROUGH

July 27, 8:00 A.M.: Company B began its work, along with infantrymen from the Second Battalion, 23rd. The company fielded six medium tanks, three 105 mm assault guns, and a single tankdozer. The assault force swept forward into enemy territory.

Trouble exploded around 9:00 A.M., when a German antitank gun disabled one of the assault guns, but, luckily, no casualties resulted.

Lieutenant Patrick O'Shaughnessy's tank platoon advanced with riflemen, who detected a well-camouflaged position. One infantryman guided O'Shaughnessy's tank into firing position behind a hedgerow. The lieutenant's gunner, Corporal Hopper, struggled for a clear view through thick foliage in front of the tank.

To everyone's astonishment, the infantryman climbed the hedgerow and hacked down branches, all the while exposed to every German gun in the area.

Thinking a Panzer may reside under the camouflage, O'Shaughnessy ordered his loader to shove an AP shell into the firing chamber. Hopper sighted and fired, but nothing happened. The round passed through the target with no sign of damage or movement. O'Shaughnessy called for an HE shell, and Hopper fired again. This time Germans poured out, like bees from a hive. Fleeing a command post, several fell to American machine-gun bullets.

Enemy rifle grenades later struck O'Shaughnessy's tank, with one penetrating its final drive. Everybody escaped unharmed, and maintenance men eventually recovered the vehicle.

More Germans fell that evening along a sunken road near les Maréchaux. The assault gun commanded by Sergeant James W. Johnson pounded them with HE shells. Some escaped, but most died or surrendered. Staff Sergeant Ragan spotted an enemy tank, possibly a Panther. His gunner fired six rounds, but the machine vanished.

By 7:30 P.M., Baker Company had won the ground it failed to secure the previous day, an accomplishment that cost no dead or injured tankers.

Company A resumed attacking that day at 9:00 A.M. with eight tanks and three assault guns. Five vehicles developed mechanical difficulties, and retrieving them for repairs meant dodging incessant shell fire from long-range German artillery.

Sergeant Henry Kult's tank was among those that functioned unerringly. The sergeant and his crew blasted a house near les Maréchaux but also drew antitank fire. The first enemy round landed short, allowing Kult's crew to return fire. Then a mishap occurred: Kult fractured an arm when the recoiling 75 mm cannon whacked him. Medics evacuated the tank commander along with his gunner, an exhausted soldier whose mettle had deserted him.

Company A retired from the frontline at 9:00 P.M., all 38th Infantry objectives having been secured. The company lost no tanks, and all crewmen survived the day.

Company C hoped for equivalent results when it moved out at 8:00 A.M. with the 23rd Infantry. One platoon supported Company B, 23rd and another joined Company G, 23rd.

Near a crossroads called Planquais, Sergeant Clifton B. Markin's tank fell prey to a Panzerfaust. The assistant driver, Private George W. Lee Jr., died when the projectile detonated. Metal fragments hit Markin in the butt and tore open a hand, but he escaped the tank.

Fragments also struck the driver, Technician Fourth Grade Duane E. Tarnow. The flying steel penetrated one of his eyeballs, and he attempted to flee, but the cannon barrel was above his hatch, making it impossible to open all the way. The half-blind soldier tried to squeeze through the narrow opening and became stuck. Staff Sergeant William M. Johnston saw the predicament, dismounted his own tank, and charged over. He climbed inside, elevated the cannon, and pulled out Tarnow as flames filled the tank. Johnston hauled the injured soldier to safety, but the fire had taken too high a toll. The twenty-year-old from La Porte, Indiana, succumbed to burns the next day at a field hospital.

The loss of Markin's tank was just the beginning.

Sergeant Earnest O. Padgett's tank, "Traveling Pillbox," met misfortune when a Panzerfaust slammed into its frontal armor. The men closest to the impact point, the driver and assistant driver, never stood a chance: Privates Pearl Ross and Woodrow W. Kitchen died in an instant. The explosion wounded Padgett in the leg and hospitalized him for sixty-three days.

Lieutenant Edwin Jacobs also lost his tank. As it started to burn, he hesitated to bail out because of machine-gun fire. The gunner, Corporal William T. Brown Jr., elbowed past him, climbed outside, and hit the ground as the tank exploded. The fiery blast killed the young officer, a former real-estate agent from Jersey City, New Jersey. The loader, Private John McKnight, also escaped, but enemy bullets cut him down as he ran away. Corporal Gilberto B. Vidal rushed to aid the mortally wounded McKnight, but Vidal himself became a casualty, collecting shell splinters in his hip and shoulder blade.

Losses continued to mount, with Company C losing six tanks by noontime. Personnel casualties amounted to six dead and fifteen wounded, the worst day for any 741st company since Omaha.

The setback necessitated Company D's transfer to the 23rd Infantry. The regimental commander needed more tanks at once before his attack stalled.

Captain Sicks reported to the Second Battalion, 23rd for instructions. At 4:10 P.M., two light tank platoons attacked with the infantry toward Planquais and seized three hundred yards from the enemy. Due to a Puppchen rocket, they lost one tank, the vehicle commanded by Sergeant Russell L. Burris. He and his crew came through without harm.

The remaining armor temporarily pulled back until the attack reopened at 9:00 P.M. The tankers ran riot, their 37 mm guns spewing canister shot—each round packed with 122 steel balls. The battle blew past Planquais and into Notre-Dame-d'Elle. The tanks and foot soldiers swept the town clean and rolled on for another one hundred yards before darkness fell.

The light tanks demonstrated their value during Cobra. The battalion after-action report for July cited three reasons for their success:

"Primarily, the light tanks proved excellent in situations where the main resistance had been broken, and the situation was essentially a mop-up operation. Two, the light tanks, with greater speed and mobility, were able to keep up with rapidly advancing infantry, even through difficult terrain. Three, 37 mm canister proved very effective against enemy infantry."[121]

★★★

The morning of July 28 found only one tank company in action, the others in division reserve. Company D attacked at 10:00 A.M. with the Second Battalion, 23rd Infantry. They departed near Notre-Dame-d'Elle, aiming for Saint-Jean-des-Baisants.

Dog Company and the foot soldiers encountered only token resistance: snipers, minefields, and booby traps. The tanks and infantry wheeled into Saint-Jean-des-Baisants, a dusty town battered by artillery shells. Its once magnificent church stood in ruins, a giant bronze bell resting atop a rubble heap. By 5:30 P.M., this pitiable place belonged to the 2nd Division.

With the objective in hand, the 741st halted operations for the day. Colonel Skaggs closed his forward CP, and all personnel returned to Vitamin Rear.

Colonel John H. Chiles, operations officer for the 2nd Division, busily prepared for the next day's action. He advised Skaggs to have one medium-tank company relieve the 744th Light Tank Battalion and launch an attack with the 9th Infantry.

Company B motored to the Rouxeville area, arriving just after eight that night. The tank crews exchanged places with the 744th and learned the attack would start that afternoon. Baker Company divvied its three platoons between the 9th Infantry's three battalions.

At 3:30 P.M., the 9th moved out. Many soldiers rode on the tanks until German artillery shells forced them to dismount. The enemy also planted mines and chattered away with their machine guns, but nothing blunted the US attack, which stole roughly six miles of territory from the Germans. Company B left one platoon with the infantry, and the rest pulled back.

The 3.Fallschirmjäger-Division had retreated overnight, leaving only rearguard units. Reconnaissance aircraft reported enemy vehicles streaming away from the battlefront.

Companies A and C spent the day resting in division reserve, where men gathered around a Red Cross Clubmobile while three ladies brewed coffee, fried doughnuts, and played music on a Victrola. Company D stayed with the 23rd Infantry and also received a Clubmobile visit.

The break from combat came "after what seemed like an eternity," remembered Technician Fourth Grade Peter J. Fardella, a radio repairman in Charley Company. When the company kitchen arrived, he ate his first hot meal in four days. Fardella found the mood somber: "There was no excitement or noise, just groups of us sitting around, dirty and exhausted. The gray skies matched our emotions."[122]

During the early evening, Colonel Skaggs and his headquarters staff left Vitamin Rear at the edge of the Cerisy Forest and established a new forward CP near Notre-Dame-d'Elle. Less than an hour passed when Major Browder arrived with news from the 3rd Armored Group. The 741st was now assigned to corps reserve, except for the Baker Company platoon still attached to the 9th Infantry. The new CP closed and everyone returned to Vitamin Rear.

The four tank companies pulled out on July 30 and headed for new bivouac areas in pastures near Cantilly. One Company A tank detonated a mine during the move. The crew climbed out, and a tank retriever towed the disabled vehicle. In the bivouac areas, maintenance work began, but the time for upkeep and overhaul disappeared all too fast.

The enemy was on the run, and experience had shown the need for swift pursuit. The breakout from Omaha Beach had ground to a cautious creep in June, allowing the Germans to dig into the hedgerows and build a robust defensive line. After the 2nd Division pierced that line on July 11, the attack halted at the Bayeux-Saint-Lô road, and the enemy entrenched again.

Hedgerows and the American response to them enabled the German paratroopers to thwart a disproportionately stronger enemy for more than a month. The single Normandy battleground envisioned by Allied planners turned into a thousand hedge-bound battlefields. Major attacks delivered minor gains within this labyrinth.

The use of tanks and infantry as autonomous elements failed. Their individual limitations left them vulnerable to German weaponry nested in the hedgerows. General Robertson grasped this problem and ordered training to meld them into an integrated force, the strength of one balancing the other's weakness. Though they trained at length, their first efforts led to disappointment. The tanks and infantry achieved gains but no crushing victory.

The Rhino device became a catalyst, which, combined with combat experience, yielded success. The 3.Fallschirmjäger-Division would dig no more defensive lines. Its survivors took to their boots in ragged retreat.

The Normandy front cracked open.

CHAPTER 13 CROSSING THE VIRE

On July 30, three hours before midnight, the battalion CP hummed with conversation. Skaggs met with his staff officers and company commanders in preparation for an attack on Vire, a city twenty miles beyond the current frontline. Company B would remain behind in reserve. The other three tank companies needed to move their machines forward under cover of darkness to assembly areas near Mouffet and stand ready for action at 11:00 A.M.

An hour after the meeting, Captain King drove to meet with Colonel Lovless, whose 23rd Infantry would conduct the attack. No definite plan for tank support existed yet.

Not far from the colonel's CP, Captain McDaniel from Headquarters Company established a new forward CP for the 741st near le Perron. Farther south, near Mouffet, Skaggs waited up front for Companies A, C, and D to arrive.

The tankers left their bivouac areas near Cantilly and trundled south to their assembly areas. Once there, Lovless paired Company A with his Third Battalion, Company C with the Second Battalion, and Company D with the First.

By daylight, the three companies lay coiled to strike. Skaggs visited Lovless and the 2nd Division CP, but no plans developed.

The tankers stayed alert, yet an attack order never came. The enemy had pulled out, and the infantry required no armored assistance. The 741st tanks sat idle until evening, when they received word to move south through territory vacated by the retreating Germans. The armor pressed five miles south over vehicle-jammed roads to new bivouac areas.

★★★

Company A soon broke camp, with fifteen Shermans joining the Third Battalion, 23rd for a six-mile foray to the intersection of two national highways (N174 and N175) at le Poteau. The infantry rode on the tanks, eight men per vehicle, and disappeared into the dark at 10:00 P.M.

Captain Thomas was at the helm of his command tank when machine-gun fire cut through the darkness. The infantry dismounted and returned fire, but the lack of moonlight rendered the original plan unwise. The infantry moved ahead alone, while the tank crews halted near the Guilberville train station at la Couaille.

Company D also stopped at the station, having followed in echelon behind Company A and the Third Battalion. The light tanks supported First Battalion, 23rd.

On a nearby siding, the tankers confronted a railway gun, its monstrous 238 mm cannon able to hurl shells twenty kilometers. The Germans had abandoned the leviathan after destroying its breechblock the previous day. It sat derelict in the dark as the crews waited for daylight.[123]

Le Poteau crossroads lay just three miles ahead.

When light came on August 1, the Company A tanks took to the road again and discovered that, during the night, soldiers from Grenadier-Regiment 752 had stalled the infantry short of the crossroads. These rearguard troops continued to lash out with small-arms fire.

Meanwhile, Company D ripped through fields until a Panzerschreck projectile streaked into the lead tank. Its crew abandoned their disabled machine and escaped.

The Americans dropped back to rethink their plan.

After reconnaissance work, Company A and its infantry embarked on an assault at noon. The Germans fought back at a dusty road junction on N175 called la Croix Étêtée.[124] They laced it with mines and greeted the attackers with small-arms fire, along with something hitherto unseen. Enemy soldiers launched mortar rounds in flat trajectory as antitank weapons, or so the Americans claimed, after finding tail fins. In actuality, the fins probably came from an antitank projectile known as the Stielgranate 41, which resembled a mortar shell.

At the road junction, Sergeant Hecox's tank struck a mine and caught fire. Everyone inside escaped the blazing machine.

The Company A tankdozer rolled forward to replace it, but a large projectile slammed into the dozer. Its commander, Bill Smulik, died along with three others: Private Warren H. Price, Private Charles R. Reeves, and Private First Class James O. Thurston. All four lives ended in death by immolation. Only the assistant driver, Technician Fifth Grade Emory T. Hopper, escaped the inferno, though badly burned.[125]

The enemy delaying force pulled back, and la Croix Étêtée fell into American hands.

The light tanks and infantry had also picked up the attack at noon, deploying to the left of N175. They forged ahead against enemy rifles, mortars, and machine guns. The attackers lost one tank to a mine but pushed across country to la Mazière, where the battle died down by nightfall.

Company D and the First Battalion stood a mile short of le Poteau.

While Able and Dog Companies battled down N175, the Second Battalion, 23rd and two platoons from Charley Company headed for N174, but first had to liberate Guilberville.

Second Lieutenant John H. Covington Jr. led the Second Platoon and its six tanks, which jumped off at 11:00 A.M. with Company F, 23rd. The lieutenant later described the expedition:

"Our briefing was all too short, and, before we crossed the line of departure, the plans changed several times, as the 38th Infantry was pushing up on our right and front, while the 9th Infantry was on our left. Light resistance was met most of the way until we came into Guilberville. Several machine guns and antitank guns were in the village. We used quite a bit of HE on the buildings, trying to dislodge the Jerries, but when the infantry moved in there was still a lot of enemy to be found. The antitank guns placed a lot of fire in the area, but no damage was done to any tank. Three or four bazookas were found in the area, plus their dead operators."[126]

Recalcitrant German paratroopers—many from Fallschirm-Aufklärungs-Abteilung 12—held Guilberville and forced the GIs to capture the town three times. Company C recorded no losses in men or machines before the paratroopers fell back to positions on Hill 263.

The Germans had prepared a defense in depth, but, after Guilberville, the Americans built momentum and drove down N174, bashing aside resistance, including an antitank gun.

Before day's end, the Second Battalion, 23rd had driven past the hill and held le Poteau, the first US troops to reach the crossroads.

The second day of August was two hours old when Colonel Lovless sent new attack orders to 741st headquarters near Guilberville. They set 8:00 A.M. as H-hour, but, as the start time approached, it was apparent the Germans had withdrawn.

The attack turned to pursuit.

Stuarts from Company D, with their speed and mobility, outpaced the heavier Shermans from Company C, which fell behind on roads clotted with military traffic.

Nine light tanks under Second Lieutenant Cecil J. Payne avoided the snarled traffic and raced south on twisty dirt roads. Loaded with men from the First Battalion, 23rd, the tanks reached the Vire River and crossed on a makeshift bridge erected by 2nd Division engineers. After crossing, they advanced to la Haie de Bourdière and remained there through the night.

In the morning, the far-flung troops pulled back across the river, with Payne's tanks returning to the Company D assembly area. The lieutenant's force had rushed ahead faster than anybody else could follow.

Forty-five minutes after Payne's return on August 3, Colonel Lovless flashed a message to 741st headquarters: "Colonel Skaggs, it is directed that you have tanks join their infantry units at once. First Battalion has no tanks. Company C has not joined Third Battalion. I have no communication with you or any of your companies."[127]

Skaggs charged out of his CP to direct his armor, and he squared the situation within an hour. The advance was on again but at a sluggish pace due to congested roads.

Under a warm August sun, Company C passed through Campeaux, the last town before the river, and pulled off the road to park in adjacent fields. Within seconds, a shell struck the road behind them, injuring two tankers and several infantrymen. It also wrecked a 1½-ton truck and set fire to a barn and farmhouse. More shells landed as troops scrambled to fill bottles and canteens from a huge calvados cask in the burning house. The property owner stood outside, wringing her hands as ravenous flames gutted her home.

Shortly after 10:00 A.M., Company C crossed the river on a Bailey bridge at Courbe-Fosse, the heavy machines rattling the wooden planks. The other tank companies followed.

Company D crossed last but stayed only a short time. Its vehicles motored back across the bridge and rallied at Campeaux in the early evening.

The medium tank companies bivouacked on the opposite side of the river. Able and Baker Companies briefly took up defensive positions in response to reports of an approaching enemy armored column, but the threat never appeared.

The Germans had no strength to challenge the bridgehead.

CHAPTER 14 **NORMANDY FINALE**

August 4, 5:00 A.M.: Quentin Swan from Charley Company led his platoon to a rendezvous with the 23rd Infantry. "The Third Platoon was supposed to support Company K," he recalled. "We were directed to the wrong field, and the company pulled out without us."[128]

Company K dispatched a guide to fetch the wayward tanks. In column formation, the vehicles followed the lone infantryman, until reaching a creek that looked impassable. Swan radioed his company commander, who said to try fording the stream. The tanks crossed without sinking in the mud and met an infantry officer, who guided them to the frontline.

Thunderous bedlam descended as shells burst all around. The infantrymen never gathered their wits enough to direct return fire by using the telephones on the tanks.

As the ruckus heightened, one vehicle rolled into a creek and bogged down in its silty bed. The tank commander, Sergeant George W. Ganger, poked his head out, and a sharpshooter pegged him with a bullet. Ganger dropped back inside, bloody and with a fractured jaw, an injury that would keep him hospitalized until November.

Swan ordered his tanks to seek cover and wait until the infantry sorted out the tenuous situation. The platoon sat stationary amid an unending barrage of mortar bombs and artillery shells. The hours seemed like eternities. Around 9:00 P.M., Swan received word to withdraw.

The crew of the bogged-down tank abandoned their vehicle in the darkness.

Meanwhile that day, Lieutenant Covington's platoon also found events fraught with disorder when his tanks moved out with Company I, 23rd.

"Our briefing was never received," he later wrote, "as the infantry company commander just didn't seem to know the time of day. We moved southward, meeting no resistance. The infantry ran into personnel mines, and various tank mines were found along the shoulders and at crossroads."[129]

The armor and foot soldiers reached Étouvy, where the tank crews obeyed instructions to hold up. The infantry continued forward, but a guide later returned to lead the tanks into action. The route passed over wide, deep-sunken roads, insurmountable obstacles for tanks. Covington's company commander instructed him to stay put. The crews spent the rest of the day waiting, while American artillery rounds whistled overhead. Several projectiles dropped short and sent fragments into partially open turret hatches, causing clatter but no casualties.

Company A, operating with the Second Battalion, 23rd, encountered no initial opposition on August 4, but that changed after passing through Étouvy. Enemy rearguard troops showered the tanks and infantry with bullets and artillery shells, but the Americans strove ahead.

Panzerschreck rockets hit the tanks commanded by Lieutenant Sledge and Sergeant Adolf C. Rodriguez. Both crews escaped, though one man died after exiting his tank—Corporal Clyde Rohrbaugh, the gunner in Rodriguez's vehicle. Enemy bullets reportedly ended his life, although a story circulated that Rohrbaugh was feigning death when a German bayoneted him in the back. The casualty roster that day also included eleven injured crewmen, with the two tank commanders among them. Sledge refused evacuation, his minor wound being of trivial concern compared to the wire-frame glasses he lost in the scramble. The lieutenant nevertheless earned his second Purple Heart in less than a month.

The remaining Company A tanks exhausted their gasoline by 2:00 P.M., and Skaggs called on Company B to relieve them for refueling. Baker Company complied and resumed the attack in their place. Captain Thornton immediately conferred with the infantry, their joint mission being to vanquish an enemy strongpoint in a wooded area surrounding the Château de la Ruaudière.

The assault commenced at about 3:30 P.M. with two tank platoons abreast and one in support. The crews plastered the enemy with over three hundred rounds of HE and ten thousand rounds of caliber .30 ammunition. The enemy raised white flags but failed to emerge, so the tanks continued hammering away. It was good they did. The surrender flags were a ruse, masking a German withdrawal.

Another ploy unfolded when an enemy motorcyclist flying a white flag drove along a sunken lane toward Sergeant Geddes's tank. The German suddenly ditched his bike and darted away, his dust cloud having concealed a self-propelled gun that fired two rounds. One struck the turret roof of Geddes's tank, wounding the sergeant. The enemy machine immediately fled. Medics evacuated Geddes to England with a fractured skull.

The attack petered out by evening, after the enemy pulled out from the strongpoint. To cover their retreat, the Germans dealt out continual shell fire.

The next day, the Second Battalion commander, Lieutenant Colonel Lewis F. Hamele, opted to hold back on armor. It was plain to Thornton that many infantry officers feared using tanks because they tended to draw artillery fire. Attitudes among the footsloggers had changed since facing shrewdly entrenched Fallschirmjägers the previous month. Against a retreating foe, the infantrymen preferred to advance alone until meeting opposition, and then to call for tanks. Hamele intended to use just such a procedure, but his men encountered no Germans. They had slipped away since their bold stand the previous day.

Thornton's tankers returned to their bivouac that evening, rejoining the other 741st companies, which had never moved during the day. The

741st ceased combat operations on August 6, relieved from its attachment to the 23rd Infantry and consigned to division reserve.

The next day, Captain Sicks stepped aside as the Company D commander, sidelined by appendicitis. Skaggs selected Lieutenant Sledge to fill the vacancy, and the new company commander wrote home saying, "I pray and hope that I can do a good job. And secondly, it's a chance for promotion, if I do all alright. So, wish me luck."[130]

The battalion remained in reserve for four days, servicing tanks, repairing battle damage, and resting its tankers. Despite the chance to unwind, few men could relax, the recuperative power of rest having dwindled under the continual strain of battle.

"Combat exhaustion" was the US Army term for their condition. It usually struck without the suddenness of a bullet or shell fragment. It encroached day by weary day, and in some cases, a soldier might descend into hysteria or depression.

★★★

While the 741st sat in reserve, the 2nd Division pressed south, with the 9th Infantry entering Vire to relieve the 29th Infantry Division. The bomb-blighted city fell to the 29th after the recoiling 3.Fallschirmjäger-Division abandoned it.

The enemy paratroopers and other remnants from the German Seventh Army retreated into what became known as the Falaise Pocket. The 2nd Division received orders to chase the enemy southeast toward the pocket, with Tinchebray as its final objective.

The 9th Infantry attacked at midafternoon on August 10 and drove southeast with two battalions in column formation. Only desultory opposition rose to meet them, but corps headquarters feared a counterattack and ordered the troops to burrow into the dirt and defend the village of Maisoncelles-la-Jourdan.

Sixteen tanks from Company A returned to action, joining the infantry there. Lieutenant McDonough's platoon settled into positions one thousand yards south of Maisoncelles-la-Jourdan, while the remaining tanks stayed closer to the village. The enemy shelled the Americans throughout the night and all the next day, but the Germans sprang no counterattack.

Late on August 11, Skaggs and his operations officer attended a conference at the 9th Infantry CP to glean details about a renewed attack to be led by the First Battalion, with tank support from Company A. Everything was set to kick off at 10:00 A.M.

Less than an hour before the attack, the 9th Infantry's intelligence officer warned that German troops were forming for a counterattack in the area chosen for the First Battalion's assault. The enemy force reportedly included four assault guns and four Tiger tanks with their powerful 88 mm guns.

The 9th Infantry quickly launched its operation, rather than wait for the enemy punch. Able Company sortied alone, while the infantry waited for the armor to return. Sunken roads and sodden earth hampered the way, allowing only four machines to pierce enemy lines.

Staff Sergeant Robert Nicol's tank reached the infantry's first objective, a road one thousand yards from the jump-off point. Nicol directed his driver to halt the vehicle on high ground, and they looked down in surprise at bedraggled German units in retreat. The crew hurled fire upon the withdrawing enemy, with deadly results. Bewildered soldiers ran helplessly in circles, while chattering machine guns cut them down in a sickening spectacle.

The sergeant spotted stone buildings and a stand of trees to the left, places that concealed more Germans. The crew turned its guns in that direction, sending bullets and HE shells at the targets. White-phosphorus rounds finished the job, setting the buildings alight, along with everyone inside. Time now expired for the tanks.

Nicol rallied the four vehicles, and the group hastened back to the departure line. Once there, the infantry pushed off, meeting little opposition after the carnage already dealt. The earlier report about an impending German counterattack proved false.

During the sortie, Companies B and C stayed on alert but saw no combat. In the early evening, Baker Company joined the 38th Infantry in reserve. The previous day, Charley Company had joined the 23rd Infantry on line and abreast with the 9th Infantry. Both companies prepared to push toward Tinchebray.

The 23rd Infantry attacked at 11:00 A.M. on August 13, with Company C launching a sortie into an enemy position southeast of Truttemer-le-Grand. The tanks rushed forward behind protective fire from the 38th Field Artillery Battalion and smashed ahead eight hundred yards.

Several tanks fell into a sunken road near the turnaround point. One machine belonged to Phil Fitts, now a sergeant and tank commander.

His vehicle tipped nose first into the sunken road, and the cannon barrel struck a tree trunk, stripping the pinion that geared the turret to the traversing motor. Fitts saw no alternative but to leave the tank. He disregarded a standing order to scuttle any vehicle abandoned behind enemy lines. In its present state, the tank offered nothing for the enemy, and, once the sortie ended, infantrymen from the 23rd would move forward and regain the vehicle. It could then be winched from the sunken road and the damaged gear replaced.

Fitts and the other four men in his crew dismounted their stricken machine and fled toward a nearby farmhouse. Fitts sneaked up to the stone-built structure and froze at a sudden noise. Germans, he thought. With nowhere to run, the sergeant pulled out his captured P-38 and chucked the weapon aside, before walking forward with his hands raised. Nobody was there. Relieved, he wiped his brow and retrieved the pistol.

As Fitts and his crew backtracked on foot, the other tanks returned and paired up with infantry companies. Together, they pushed off and consolidated the ground covered by the sortie. After reaching a staunchly defended stream near le Houx, the tanks and infantry bypassed it to the east but stopped after friendly troops appeared to the front. They raised alarm about approaching enemy armor. The men wondered if the threat was real this time.

Captain Young followed events from the CP of the First Battalion, 23rd. At the frontline, Lieutenant Covington's tank platoon operated with Company A, 23rd, and Mister Swan's platoon supported that company and Company B, 23rd.

Communications faltered, and uncertainty grew about the tactical situation. The 23rd had no contact with its right-hand neighbor, the 38th Infantry, or with its neighbor on the left, the British 3rd Infantry Division—actually fifteen hundred yards to the rear.

The British hurried to remedy the situation by attacking to bring their line abreast, but they first laid down a seventy-five-minute rolling barrage to prepare the ground. The tanks from Company C and infantrymen from the 23rd endured a ruthless bombardment. No shells hit the vehicles but only by the grace of God.

Once the British shifted their savage pounding elsewhere, the Germans took their turn. They lobbed in shells and mounted a counterattack at nightfall with troops from Grenadier-Regiment 1051 and assault guns from Sturmgeschütz-Brigade 394. The enemy cut off forty men from Company B, 23rd along with two tanks from Swan's platoon. German troops also cut off Captain Young and the First Battalion commander.

Staff Sergeant Norman K. Crisler (pronounced Chrysler) and Sergeant Leonard J. Henkelman commanded the tanks. Ghostly fog rose in the cool nighttime air as the two men and their crews kept silent in their vehicles, hidden but fearing discovery at any moment. They nervously listened to German voices outside. The enemy soldiers seemed confused and unsure of their progress in the misty darkness. The isolated Americans lost all communication with regimental headquarters.

By 11:00 P.M., the infantrymen reestablished communication, but an uneasy night lay ahead for the tankers and foot soldiers caught in the German noose. After daylight, troops of the 23rd Infantry launched a counterthrust that straightened the line and restored tactical control.

On August 13 at 9:00 A.M., Baker Company returned to combat when its tanks and the 38th Infantry pushed ahead through morning fog to cut the Sourdeval-Tinchebray highway at a crossroads named les Croix. One tank platoon worked with the Second Battalion, 38th, and two platoons joined the Third Battalion, 38th.

The platoon with the Second Battalion encountered no enemy at first, but one Sherman churned up to its belly in a muddy apple orchard.

Two other tanks tugged it free, but the ground was too soaked for the armor to keep up with the infantry.

Trouble bolted down the highway when a German car approached the Americans and stopped long enough to unload a mortar crew and their weapon. The vehicle roared away before anyone could react and thump it with bullets.

Trees screened the German mortar team from view while they set up their tube. In minutes, rounds exploded among the Americans in the orchard.

Staff Sergeant Ragan's driver backed under a tree. As the sergeant ducked inside his machine, his spring-loaded hatch caught a tree branch. The steel lid slammed shut on his right hand, amputating part of his index finger and crushing two others.[131]

Sergeant Elmer Middleton's tankdozer was also in the orchard. "Happy" Middleton was a Regular Army soldier from the Appalachian Mountains and a prewar veteran of the 38th Infantry Regiment. The lighthearted tank commander always wore his M1 helmet cocked at a rakish angle. He emerged from his hatch just before the mortar rounds landed. One projectile hit the dozer with a roaring bang, and fragments smashed into his skull. The driver, John Brewer, looked up from his open hatch to see a mangled head drooping over the turret edge. He took over as commander, and the bow gunner drove the vehicle. They and their crewmates retreated along with the other tanks, carrying Middleton's body with them.

Once darkness fell, light tanks from Company D arrived and escorted the infantry forward over the mushy terrain. They consolidated at the les Croix crossroads without incident, the German mortar team having disappeared.

The night passed uneventfully, except for an odd occurrence. Sometime after midnight, a sizable contingent of soldiers came strolling down the highway. An infantry sergeant who spoke German shouted, "Halt!"

"Is this Company E?" a voice called out in English.

"Yah," the sergeant replied.[132]

As the soldiers walked closer, their uniforms grew visible in the moonlight.

Germans!

The German-speaking American sergeant hastened to defuse the dangerous situation before shooting erupted. He spoke with the enemy leader and learned he was their Spiess, or first sergeant. The group included no officers. In minutes, the senior German noncom agreed to surrender his men, and their captors quickly herded them into a field with instructions to bed down until morning.

Later that night, the infantry leader, Lieutenant Charles Curley, noticed light tanks parked in the field with their crews asleep outside. "As it started getting light," he recalled, "we had to grab the Tommy guns from the tankers sleeping on the ground before they woke up and found themselves surrounded by sleeping Germans. It was funny, but it could have been a disaster."[133]

Early on August 14, Colonel Lovless instructed Company C to link up with the First Battalion and attack at 9:00 A.M. to seize Tinchebray, but the attack hour rolled past without action. The infantry needed more time to reorganize after the enemy's nighttime counterattack.

More than two hours elapsed before Lovless committed his soldiers—but without tanks. The infantry surged forward against brittle opposition. The tank crews sat in the lurch until midafternoon, when they pushed off on a sortie, protected by a time-fire barrage.

The sortie included two new second lieutenants, Joseph H. Dew and Victor L. Miller Jr., both fresh from the 3rd Replacement Depot. They had joined Charley Company earlier that morning. Miller led the Third Platoon, and Dew took over the First Platoon, which had lacked an officer since Lieutenant Jacobs's death three weeks prior.

The tank crews hammered out shells and bullets before returning to the departure line, with one platoon assigned to the First Battalion and another to the Second Battalion. The armor and infantry attacked south and east, but around 6:30 P.M. the tanks rumbled to the rear, temporarily withdrawing. Tinchebray remained in enemy hands, its liberation postponed another day.

The 741st stood ready for another call to battle on August 15, but none came. The infantry trudged alone toward Tinchebray. The advance ended there when divisions on either flank pinched out the Indianhead Division from the shrinking Falaise perimeter.

The next morning, Skaggs learned that corps headquarters had detached his battalion from the 2nd Division, the 741st reverting to 3rd Armored Group control.

The battle for Normandy was over.

CHAPTER 15 **EXHAUSTION CENTERS**

Operations in Normandy ended none too quickly. Combat exhaustion and related neuroses clawed at men throughout the battalion. They experienced anxiety, hysteria, reactive depression, and post-traumatic disorder. Men who spent weeks suppressing the instinct to avoid danger could no longer contain their fear. It spilled out. Some wept uncontrollably or flew into stark, screaming madness. Others turned catatonic, unable to speak or move. They sat huddled and silent, inaccessible, like a nautilus in its shell. Whatever the manifestation, combat exhaustion siphoned away each man's value as a soldier.

All the tank companies sent men to exhaustion centers after the Normandy campaign. Company C, for example, transferred eight men on August 14, including Quentin Swan, the warrant officer from Roundup, Montana, who led the Third Platoon. He departed that morning, followed by Jack Ray, Victor F. Brookbank, Carson Connell, Raymond S. Parker, William E. Patton, Joseph A. Speckman, and Virgil B. Stansberry.

These soldiers had suffered a progressive deterioration but endured until the campaign ended. They could now leave without self-reproach. Their comrades harbored no antipathy, their exit marking no breach of faith.

Swan and the others became patients at the 622nd Medical Clearing Company, a treatment center for combat exhaustion cases located two miles west of Saint-Sever-Calvados. An identical unit, the 618th Medical Clearing Company, shared the caseload. Both centers offered standardized care and provided facilities for special procedures unnecessary when treating physical injuries. The centers also reduced congestion at evacuation hospitals and lessened the potential for "infecting" other patients with neuropsychiatric symptoms.

Many soldiers admitted to exhaustion centers suffered from anxiety in its most extreme state. Their treatment moved through three phases.

The first phase entailed psychiatric assessment and, in some cases, hypnotic exploration with the aid of sodium pentothal. The appraisal process ended after twenty-four hours.

Patients started the second phase by swallowing sodium amytal in blue tablet form. The drug produced a trance-like sleep intended to have a cathartic effect, but the slumber often spawned nightmares. Terrible shrieks reverberated through the wards as wounded minds revisited the battlefield. The process lasted forty-eight hours, with patients only awakening to eat and use the latrine. They emerged from the second phase numb and barely able to walk, but the clinical staff immediately pushed them onto the final phase—rehabilitation.

This phase occurred in another ward and emphasized each man's identity as a soldier rather than as a patient. The regime included team sports, close-order drill, and calisthenics. Group and individual psychotherapy helped create a sense of normalcy. Each man then underwent another psychiatric assessment. Those deemed suitable for battle—the most common judgment—returned to their original units with new clothes and equipment.

The unfit received non-combat duty but no cure for the emptiness deep inside.

CHAPTER 16 **HOLIDAY IN PARIS**

On August 18, the 3rd Armored Group issued an order calling for the 741st to make an "administrative march" to a town called Sées. It lay on the Orne River, one hundred miles from Paris.[134]

Headquarters Company led the column now that further combat seemed unlikely. After an initial delay, the 741st started moving during the early hours of August 20. The tanks drove through Lonlay-l'Abbaye, Domfront, and la Ferté-Macé. Disarray in the darkness required them to halt three times to tighten their overstretched length. The confusion evaporated when the sun broke the horizon, and the column reached its destination at 8:45 A.M.

The battalion intelligence officer, having gone ahead to Sées with an advance party from the 3rd Armored Group, greeted the leading vehicles. Seated in his jeep, he guided the men outside the town to bivouac areas around a hamlet named Fréneaux. All the companies encamped there before noon.

Company D received the only mission during the layover at Fréneaux—escort prisoners. The light tanks departed before sunrise on August 21 to join the 428th Escort Guard Company in confining twelve thousand enemy captives, sweepings from the Falaise Pocket. The job took two days, during which time the rest of the 741st moved to la

Monniere. Company D rejoined the battalion at the new location after the guard assignment ended.

The tanks took to the road again on August 25. Wooden signboards marked their route, each sign painted with an arrow and the V Corps code name—Victor. The column chased one sign after another, traveling 124 miles. It was the farthest single-day advance for the battalion since landing on D-Day.

The advance continued for a small party from Company B, which dashed ahead in a jeep. First Lieutenant Vincent A. Cinquina rode in the vehicle, and he later wrote about the adventure:

"Supreme Headquarters had ordained that the French 2nd Armored Division was to have the honor of liberating the famous French capital. However, the undying curiosity of Vitamin Baker yearned to be satisfied, and consequently a representative group composed of Lieutenant O'Shaughnessy, Lieutenant Dudley, Staff Sergeant Grossi and I proceeded, entirely without orders, to assist in the accomplishment of this memorable task.

"Upon our approach to the city we encountered long columns of French armor making their triumphant march. It was a gala holiday. The Parisians lined the streets and welcomed the troops with unparalleled enthusiasm. We joined the French columns and participated wholeheartedly in the vigorous hand waving and cheering. The French are a frank and demonstrative people, as we were to learn. Upon being recognized as Americans, we were greeted with calls of 'Vive l'Amérique,' showered with flowers and practically drowned in champagne. This was the type of battle we truly enjoyed, but, like most good things, it was short-lived. After riding along with the French forces for an hour or so, we decided to start a parade of our own and proceeded into unexplored sections of the city.

"We stopped our vehicle for a short rest, and immediately the celebrating crowds swarmed about us. An exchange of cigarettes, hand signs, and crippled French soon completed our introduction. The crack of a rifle and bullets whistling overhead dispersed the crowd in three seconds. We sharply awakened to the fact that the war was still going on. An excited Frenchman rushed forward to inform us that a group of Germans was still holding out in this sector of the city. Without further ado, Sergeant Grossi wheeled the peep around and cruised away at about sixty miles an hour. We decided that our curiosity had been sufficiently satisfied and resolved that all future premature sightseeing would be done with the comfortable coat of our tank's armor about us."[135]

★★★

As the 741st approached the "City of Lights," Lieutenant Coleman of Company D longed to be with his men after spending twenty-two days hospitalized, mostly in England. "I want to get back to the company in time to march into Paris," he wrote home.[136]

"I am in good shape again. Not raring to go, but a little job has to be finished," he said in another letter. "Boy, if there is ever any talk of another war after this one, here is one guy who is going to get into the Services of Supply or the Quartermaster Corps."[137]

The lieutenant would spend a month at the 3rd Replacement Depot waiting for his former unit to requisition a platoon leader. Special rules applied to hospital returnees, which permitted him to wait and avoid reassignment to another unit.

★★★

On August 26, Coleman still sat in limbo as the battalion edged forward to Janvry, less than ten miles from Paris. The next day, the 741st motored on and reached Sceaux, on the French capital's outskirts. Along the way, crowds spilled into the streets to greet their liberators.

The tankmen bathed in the spontaneous generosity of citizens eager to share what little they had saved through the German occupation. The French plied the GIs with cider, wine, champagne, and assorted liquors. They also doled out apples and homegrown vegetables.

On August 27, Colonel Skaggs drove to meet Brigadier General Norman D. Cota. The general had become the 28th Infantry Division commander two weeks earlier, replacing a commanding general mortally wounded by a sniper's bullet on his first day with the division. The dead general's predecessor had departed in disgrace, cashiered for poor leadership. The 28th (code-named "Holiday") now belonged to Cota, and the 741st was his, too.

Colonel Skaggs conferred with Brigadier General George A. Davis, the assistant division commander. Their discussion sent Company B to the 112th Infantry.

The next morning, Captain Thornton visited that regiment's CP to coordinate future operations. When the 28th Division received an unexpected interim assignment, word swept through the ranks. The 28th and 741st would parade on the Champs-Élysées in Paris.

Eisenhower ordered the event in response to a request made by the self-appointed leader of the French Republic, General Charles de Gaulle, who had asked that a US division parade through the city to show that Allied support rested behind him. American military muscle on display would help him rein in all Resistance factions, especially the Communists, as he maneuvered to install his government in the French capital.

The 28th Division got the nod to fulfill de Gaulle's request. The choice was easy because the division planned to attack on August 30 from assembly areas in northeast Paris, and its troops had to pass through the city center to reach those areas.

The tank crews hustled to spiff up both their vehicles and themselves for the show. They removed sandbags from tanks and scrubbed grimy hulls. Shoes and uniforms came under close inspection for the first time since England. One enlisted man from Company C recorded his thoughts on the tidying work:

"All day Monday, August 28th, we were polishing up. We had learned that we were to parade through Paris the next day. Frankly, we didn't like the idea of cleaning up. We'd fought through mud, and the tanks looked just as they had at Saint-Lô. We didn't see why they should be cleaned up to give the Parisians the idea that war was a pretty business, waged in a condition of spit and polish. But our opinions had no weight."[138]

Besides cleaning tanks and uniforms, a truck convoy from the 741st took battalion members on a rolling tour of "Paree." The men craned their necks to see historic landmarks as the vehicles twisted through the cityscape.

That evening, the warm August sunshine turned to rain. With the atmosphere dampened, the tank crews snoozed in their machines, hoping that morning would bring back the sun. It did.

By 10:00 A.M., the battalion had reached the Bois de Boulogne, a sizable wooded park in the city. The parade organizers had set 2:00 P.M. as the start time.

As noon approached, the tankers began to assemble on the parade route, four vehicles abreast. While they fell into formation, deliriously happy Parisians surged at them and romped around the tanks. The shouting celebrants thrust wine and cognac bottles toward the tankers. Women hugged and kissed the men.

While the populace swarmed the 741st, a solemn ceremony transpired at the Arc de Triomphe. Generals De Gaulle, Gerow, Hodges, and Bradley paid homage to France's Unknown Soldier. Bradley laid a gladioli wreath on the stone slab that capped the hallowed grave. Across the wreath, a ribbon bore the words: "From the American Army to the Heroes of France." The generals saluted as a band played the national anthems of France and the United States. The group then drove down the Champs-Élysées to the Place de la Concorde, where engineers had constructed a reviewing stand from Bailey bridge components.

The parade commenced on schedule.

General Cota led the procession in a jeep. He first passed the reviewing stand, then his driver pulled over, and the general climbed out and took a spot on the platform.

Armored cars from the 28th Cavalry Reconnaissance Troop followed Cota's jeep, and behind them, Division Headquarters. Next came two infantry regiments, the 112th and 110th. They marched in battalions, twenty-four men abreast. The regimental cannon and antitank companies rode in trucks, their guns in tow.

More motorized troops followed. The lineup included the 630th Tank Destroyer Battalion, the 447th Antiaircraft Automatic Weapons Battalion, the 81st Chemical Mortar Battalion, the division's four field artillery battalions, its engineer battalion, and its medical battalion.

The 109th Infantry appeared next. Like the previous regiments, the foot soldiers marched en masse, an olive-drab ocean that flooded the street, curb to curb.

After a fifteen-minute break in the procession, more artillerymen followed, and then came the 741st. Peter Fardella from Company C offered his recollections:

"We formed for the parade on the Avenue de la Grande-Armée and spent most of the morning trading cigarettes and rations for drinks. By the time the parade started, some of the boys were alcoholically enthusiastic. We paraded around the Arc de Triomphe and down the Champs-Élysées to the Place de la Concorde, past the reviewing stand. It was too far away to see who was in it. The crowds were wild with enthusiasm."[139]

The revelry ran contrary to all that the tankers had experienced in France. Fate inflicted upon them the highest losses of any tank unit on D-Day. Hedgerow warfare plunged them into a grinding hell. Now the men stood as honored guests in a dreamland, a world away from blazing tanks and incinerated comrades. The rapturous euphoria of a liberated people faded the fearful memories, at least for the moment.

"As the tanks moved slowly ahead," another Company C soldier remembered, "people fought to get to us and to shake our hands. Others cried openly at the joy of seeing Americans after four long years of German domination. Perhaps the liquid celebration accounted for a bit of the happiness, but most of it was honest."[140]

Howitzers from two V Corps field artillery battalions, the 108th and 187th, rolled behind the 741st. M10 tank destroyers from the 629th TD Battalion followed next. The 953rd and 200th Field Artillery Battalions brought up the rear.

The parading forces increased speed and split into two columns after passing the reviewing stand. They moved by separate routes through narrow Parisian streets toward rallying points at the city limits from where they would launch a northward attack.

The screaming tide of humanity swirled about the clattering 741st tanks as they drove for the outskirts of Paris. Dusk found the tankers in varying states of sobriety and camped near le Bourget airfield. That night was lost to alcohol-laden slumber.

★★★

The attack began in the morning, and the enemy offered no fight. Company B accompanied the 112th as previously planned, and Company A now paired with the 110th Infantry. Division headquarters formed an ad hoc force to mop up behind fast striding tanks and infantry. Two platoons from Company D, two assault guns from Headquarters Company, and the battalion reconnaissance platoon joined the force.

On the final day of August, two combat commands from the 5th Armored Division passed through the 28th and raced ahead, until stopped by German holdouts in the Forêt de la Compiègne. Foot troops from the 28th cut down all opposition in the forest and helped seize the city of Compiègne. The mop-up force arrived there on September 1 to subdue any lingering resistance but found nothing.

The Wehrmacht was streaming back to the Fatherland in mad retreat, the exodus so rapid that infantrymen from the 28th mounted trucks to give pursuit. The titanic invaders of 1940 abandoned their conquest, leaving only blown bridges in their wake.

Victory in 1944 seemed palpable, but a sputtering commotion on high pointed toward another conclusion. As they lay in their bedrolls that first September morning, the tankers heard a single V-1 flying bomb overhead, a sound new to their ears.

It portended more fire and death ahead.

CHAPTER 17 **DASH TO THE OUR**

Place names from a bygone nightmare punctuated the geography of northern France. The First World War ran its murderous course there a quarter century earlier. The butchery between 1914 and 1918 saw mighty armies ground to corpses along the Aisne and Somme rivers.

Now, in 1944, a fleeing enemy and pontoon bridges allowed easy passage. American armor and infantry swept across the French countryside in long dusty columns. Horrific slaughter grounds from the past slipped by with little notice.

Ahead, on the Belgian frontier, lay the Meuse, another river line from the previous war. This winding waterway and the Ardennes Forest farther east stood as the last natural obstacles before Germany. The US forces intended to deny the enemy any chance to defend them.

The 741st barreled east with the 28th Division. One tankman jotted down a paragraph describing the relentless march:

"The steady clank of the track, the roar of the engine, the smell of gasoline and hot rubber becomes just a part of the tanker's life. Then there are things to which he never becomes accustomed: the dust kicked up by the tank ahead, coating his face, filling his nostrils and eyes, and choking him; the endless night marches with nothing but two winking cat eyes on the tank ahead for guidance. Eyes straining, nerves tensed, one's imagination soon has dark shapes flitting across the road. Imaginary walls seem to pop up and block the way. Sleepiness fogs the brain."[141]

The 28th followed the 5th Armored Division toward the Meuse, and their supply lines grew thinner with each passing mile.

At this late date, most supplies still arrived from the invasion beaches and Cherbourg harbor, the only deep-water port in Allied hands. Ample material reached France, but Quartermaster personnel had too few trucks to haul it all forward. Distances grew ever greater, as did the number of units needing supplies.

Ammunition remained plentiful because no pitched battles developed in August, but gasoline grew short in supply.

The US Army called its vehicle fuel MT80 (Motor Transport gasoline, eighty octane). Allied forces coped with the deficit by allocating MT80 only where most needed. Those units leading the charge had priority. Despite this rationing, the breakneck advance promised to exhaust supplies.

Having received a fuel ration, tanks from the 5th Armored lurched ahead, and, on September 5, seized bridges over the Meuse.

That day, the 28th Division responded to an order that it form a provisional task force. The new force included a truck-borne battalion from the 109th Infantry, a 4.2 inch mortar platoon, a light tank company, a field artillery battalion, a combat engineer platoon, and a tank destroyer company with towed antitank guns.

Task Force "S" was the name assigned to this hodgepodge. The letter "S" referred to Lieutenant Colonel Daniel B. Strickler, the infantry officer chosen as its commander. His unit had a threefold mission: (1) clear out resistance pockets bypassed by the leading elements, (2) reinforce the 5th Armored in "seizing and holding vital terrain features," and (3) relieve 5th Armored units whenever necessary for deployment elsewhere.[142]

Company D, 741st provided Strickler with light tanks. First Lieutenant Sledge commanded it. His crews and their machines left bivouac at 6:00 A.M. on September 6 and rolled toward Launois-sur-Vence, a riverside town near the Belgian border where Strickler assembled his force. Dog Company was the only 741st element with its gas tanks full. The other companies halted and remained idle, all now detached from the 28th Division.

On September 7, the 28th Division coasted through the 5th Armored and poked into Belgium. Company D and Task Force "S" followed the 112th Infantry Regiment and encountered no enemy interference. The 112th did uncover German defenders but rooted them out and raced beyond their regimental objective for the day.

Strickler's men overtook the 112th late on September 8 and struck through the darkness toward a new target—Etalle, in southernmost Belgium. Ahead of this town lay Pin and Tintigny. Stuarts from Lieutenant Sledge's company spearheaded the drive.

Sergeant Randall Goodman commanded the lead vehicle and was the first to encounter the Germans. An enemy self-propelled gun fired at his tank, penetrating its thin armor. Goodman sustained a fractured tibia and flesh wounds to the leg. The bow gunner, Private William R. Hough, suffered a cracked skull and a fractured arm. He later died from his injuries.

Sergeant Gerald E. Wiltrout's vehicle followed Goodman. Wiltrout's driver jockeyed for a better firing position in the darkness, but a truck loaded with infantrymen slammed into his tank's rear. At that instant, an enemy shell scored a hit on Wiltrout's tank. The sergeant ordered his crew out, while he remained behind, protecting their withdrawal with machine-gun fire. He killed two Germans and saved his fleeing men.

After Wiltrout's fusillade, German resistance vanished, and the self-propelled gun escaped, but the Americans ventured no farther in the darkness. Task Force "S" had failed to reach Pin, the first town on its path to Etalle.

Daylight on September 9 brought resumed movement by Task Force "S," but roadblocks hampered the way. Enemy infantry stood and fought at Pin. Light tanks from Company D and infantrymen from the 109th broke into the town and eliminated the resistance. The tanks soon pulled back, while two platoons from Company C, 109th Infantry mopped up and captured four hundred Germans. Movement toward Etalle resumed by midafternoon, but road obstacles again shackled progress. Strickler's force ended the day at Tintigny, still short of Etalle.

During the afternoon, word arrived at 741st headquarters that the battalion had rejoined the 28th Division. Captain King, battalion operations officer, departed for division headquarters and returned after dark with instructions for the 741st to move into northern Luxembourg near Ulflingen. This town was only four villages from the German border.

For the first time, the 741st had orders to enter a place with a German name, although it was in Luxembourg and bore no allegiance to the enemy. Like other towns in the region, it also had a French name—Troisvierges—and its inhabitants quickly pointed to that fact.

In the morning, the battalion reconnaissance officer left for Ulflingen to locate a bivouac spot. An order from corps headquarters set midnight, September 11, as the deadline for reaching the new area. The battalion had to make itself ready for an attack the next day if called upon.

While the 741st prepared to rejoin the 28th, Task Force "S" drove through Etalle and pressed into the Belgian city of Arlon. The task force set up defensive positions, and one platoon from Company D provided roadblock protection.

A day later, the other 741st companies converged on the new bivouac area. Peter Fardella, the Company C radio repairman, recalled the journey:

"The 11th saw the battalion move 103 miles into Belgium and Luxembourg to Ulflingen. During this march, we moved with infantry of the 28th Division. ... Our route led through Sedan, Neufchâteau, and Bastogne. At the latter city, a woman rushed out to our halftrack with a decanter and bottle and poured a bubbly, colorless liquid for us. Tom Young tasted it, and his face fell as he found it was soda water."[143]

Fardella and his buddies continued east and entered Luxembourg after dark. They finally halted near Ulflingen, their eyes red and burning from dust and wind.

Earlier that day, troops from the 28th Division had reached the town and celebrated its liberation with townsfolk, who rewarded them with cold beer, the one refreshment missing from most liberation scenes in France. For all their prowess in producing libations, the French fell flat in every effort to produce palatable beer.

Sunrise found two 741st members back in Belgium at Bastogne. The battalion commander and one of his men stood before General Gerow at V Corps headquarters. The general pinned a Distinguished Service

Cross to Colonel Skaggs and another to Sergeant George Habib. Both men earned their decorations for deeds on Omaha Beach.

Task Force "S" disbanded, and Strickler's men returned to their parent units. Company D rejoined the 741st at Ulflingen.

Another ad hoc unit now formed. Verbal orders from Bastogne called for an ad hoc police detachment controlled by corps headquarters. Within hours, two officers and forty-one enlisted men readied for the assignment. Captain Duke, battalion intelligence officer, headed the detachment. The next day, his group traveled by truck to Bastogne to join several others as part of a Provisional Military Government Police Force Battalion. Lieutenant Colonel Church M. Matthews from the 3rd Tank Destroyer Group became its commander.

The colonel's policemen would enforce curfew and blackout rules, keep civilians from obstructing military traffic, guard captured enemy material, safeguard financial institutions and monuments, organize searches for hidden enemy soldiers and arms caches, report suspected collaborators, and assist in arresting them as directed by Counterintelligence Corps personnel. The police battalion had instructions to cooperate with the civilian police in Belgium and Luxembourg. Together, they had to disarm Resistance groups, which represented a threat to public safety. Some groups had been responsible for looting, indiscriminate private-property appropriation, and mass roundups of alleged turncoats, who were shot or imprisoned without investigation or trial. Civilian police agencies had too few men and weapons to exert control alone. They needed additional muscle and firepower.

The battalion quartered at Heintz Barracks in Bastogne. The men spent the next day at the former Belgian Army caserne and attended lectures on the Military Government Police (MGP) and its duties. Each enlisted man received a Thompson submachine gun and fired one twenty-round magazine at an indoor range. Close-order drill added to the day's agenda.

The next morning, it was off to Malscheid, Luxembourg. Commanded by Captain Duke, the police detachments from the 741st and the 941st Field Artillery Battalion journeyed onward to the city of Wiltz. Duke reported to Major Ralph W. Yuill, who headed up Civil Affairs for the city and its environs. The major allotted billets to four officers and eighty-seven enlisted men. He left authority for employing the police force with Duke. Men with "MGP" stenciled on their helmets soon patrolled Wiltz.

Private First Class Earl J. Robinson, a bow gunner from Company B, was among them. He felt the army had finally given him a suitable occupation, a job in civil service, something he would pursue as a career after the war.

On September 11, the day tanks from the 741st first rallied at Ulflingen, foot patrols from the 5th Armored Division and 28th Infantry Division set their boots upon the Fatherland. Just after 6:00 P.M., five cavalrymen from the 85th Reconnaissance Squadron waded across the Our River and climbed a hill. They became the first Allied soldiers to enter Germany.

Farther north and several hours later, a patrol from Company B, 109th Infantry also infiltrated the enemy homeland. The patrol crossed the Our and entered Germany eight miles east of the 741st bivouac. The patrol passed over the Georges-Wagner Bridge to gain entry but later came back across. Other incursions happened that day, too. Company I, 110th Infantry crossed the Our and penetrated less than a mile to Harspelt. The company occupied the village and became the first American unit to establish a command post inside Germany.

These initial forays over the border detected no resurgent enemy. Hostile patrols appeared on several occasions, but no tenacious resistance materialized. The few Westwall positions discovered stood empty, but these were only outposts. The primary bunker network lay over a mile behind the border. The main German force had apparently retreated there or beyond.

The nagging gasoline shortage and thinning ammunition stocks delayed the planned push toward Koblenz. In the 28th Division sector, the 109th and 110th were the regiments assigned to lead the charge against the Westwall once all attacking units had sufficient quantities of gasoline and ammunition.

Between 1936 and 1940, the Germans had constructed the Westwall, a defensive line stretching from Holland to Switzerland. It included 17,261 fortifications, most built of reinforced concrete. They included troop shelters, command posts, observation posts, and pillboxes, the latter being the most prevalent. Built in many shapes and sizes, pillboxes housed machine guns or antitank guns. Each structure also included living quarters for its gun crew. To link these concrete positions, the Germans buried telephone lines and dug myriad trenches.

As a defensive network, the Westwall depended on mutually supporting fortifications, but these frequently lay in disrepair after years of neglect since 1940. German troops hurried to restore many dilapidated structures, although the neglect produced an unexpected benefit: disuse and climbing vegetation enhanced their camouflage.

Aside from manned fortifications, the builders of the Westwall fabricated concrete antitank obstacles, which the GIs dubbed "dragon's teeth." The teeth stood in rows four to eight deep and zigzagged across the countryside. Each one resembled a pyramid with a flat top. The Germans also utilized other antitank obstacles, such as concrete walls, Czech hedgehogs, water-filled ditches, steel beams set in concrete, and wooden pilings sunk in the ground.

A simple principle guided the placement of antitank obstacles and manned fortifications. They were sparse where a natural barrier made it difficult for armor and infantry to attack. The opposite was true where open terrain created an avenue of approach. Sometimes in those areas, the Germans built a second fortification belt several miles behind the first.

The entire Westwall had a single purpose: to delay and weaken an enemy. Mobile reserves situated behind the concrete defenses wielded the stopping power. These troops would launch counterattacks to defeat the invaders.

In September 1944, however, Allied planners doubted that any German mobile reserves still existed, although few believed the Westwall was an empty shell. Many exhausted survivors from the Normandy nightmare resided there, and this created the potential for thorny opposition. Even the most outnumbered and run-down troops could be heroes behind reinforced concrete.

The first assault on the Westwall in the 28th Division sector came on September 13, when General Cota sent forth two reinforced battalions.

The battalions planned to execute a pincer movement, converging just behind the concrete fortifications at Niederüttfeld and Oberüttfeld, collectively called Üttfeld. The move would breach the Westwall and open a door for the main attack toward Koblenz. Such was the hope.

The attacking battalions quickly encountered problems. German fire pinned down the assault force, whose soldiers lacked flamethrowers and satchel charges to subdue pillboxes. For the first time since Normandy, the 28th Division faced Germans intent on holding ground.

Around noon, the reconnaissance platoon of the 741st departed from battalion headquarters. Led by First Lieutenant Legnini, the platoon aimed to scout three spots where tanks could ford the Our River. The recon troops drove south in jeeps and turned east toward Dasburg, a riverside community on the German side. Enemy bullets and mortar shells halted the men at a roadblock. The patrol returned fire and withdrew, unable to accomplish its mission. Later that afternoon, the recon troops tried again. They found two crossing points in the 28th Division sector: a bridge on the north flank and a ford in the center near Harspelt.

That night, Colonel Skaggs joined General Cota in a conference to discuss Üttfeld, which was again the objective. The upcoming attack would pick up where the September 13 effort had fizzled. Cota directed Skaggs to furnish two medium tank companies. Company B would operate with the 109th Infantry, and Company C would work with the 110th. Companies A and D would have no role in the attack. Cota held the former in reserve and sent the latter to the southern flank to conduct security patrols.

Captain Young's Charley Company began moving to an assembly area from where it would enter the coming clash. Its tank crews steered their vehicles to a point near the 110th CP. Their sixteen Shermans and the company assault gun joined a halftrack and two assault guns from Headquarters Company. Captain Thornton's Baker Company moved later, during the predawn hours, to an assembly point at Heinerscheid. Both companies were yet in Luxembourg.

The night sky poured rain as the men waited in their tanks for word to cross the border.

CHAPTER 18 INTO THE DRAGON'S TEETH

September 14. Dragon's teeth cut across the muddy road that connected enemy-held Kesfeld and American-occupied Grosskampenberg.

The Second Battalion, 110th Infantry planned to use the road as its axis of advance: one rifle company to the right and another to the left, each having a tank platoon from Company C for support. Three assault guns—one from company headquarters and two from battalion headquarters—would anchor the center.

Captain Young's men began their day before sunup. Rain beat down on the tanks, and condensation fogged their periscopes and gunsights. The crews left Luxembourg and entered Germany for the first time, cautiously feeling their way to Grosskampenberg, where they joined the Second Battalion.

Fog smothered Grosskampenberg and the surrounding countryside. Everything in the town looked dour and gray—the sky, the streets, the buildings.

The two lead platoons and assault guns exited the little community and rumbled up to the dragon's teeth at 8:40 A.M. They commenced firing at pillboxes on the other side, at least those visible through the murk. The crewmen had instructions to continue this bombardment until demolition teams arrived and exploded gaps through the teeth.

Lieutenant Joseph Dew led the First Platoon and its four tanks through pastures to the right of the road leading to Kesfeld. He had joined the platoon a month earlier, and the soldiers under his command already felt lucky to have him. None would ever forget his ingenuity, raw courage, and piercing blue eyes.

The twenty-four-year-old officer grew up in Redfield, Iowa, his paternal grandmother a descendant of Charlemagne, the warrior king and Holy Roman Emperor. Tall and blond like the king, Dew played halfback for his high school football team and won election as class president. Lack of tuition money later cut short his college studies, so he alternated between farming, meat cutting, and trapping small game. He also hitchhiked and bummed rides on freight trains to find employment out-of-state, washing cars one summer in Los Angeles.

Newspaper ads enticed Dew to Alaska, where he helped construct roads and buildings at the naval station on Kodiak Island. After Pearl Harbor, he built ammunition bunkers at an ordnance depot in Oregon. The Iowa native ultimately returned home, where he enlisted in the Army Air Force and became a P-40 mechanic until officer candidate school led him to Fort Knox and the Armored Force.

After two-plus years in uniform, Dew entered combat on September 14 with less than a day's experience in battle. He waited until 9:15 A.M. for the demolition teams to arrive, and, when enemy fire thwarted them, he ordered his driver to approach the teeth.

The lieutenant decided to create his own gap.

As if impervious to enemy shot and shell, he climbed from his turret and squatted on the frontal armor by the cannon barrel to help his gunner aim AP rounds at close range. Private (soon-to-be corporal) Albert R. Wallace shattered the dragon's teeth, one by one. He cleared a needle-thin gap just wide enough for a Sherman.[144]

Dew and his platoon scooted through the breach at 9:25 A.M. and blasted pillboxes that had pinned down the infantry.

The lieutenant's ears perked up when he heard an AP round whistle past. He saw an antitank gun nestled alongside a building eight hundred yards ahead. His crew destroyed the target and swung left toward Kesfeld.

Dew spied another antitank gun alongside yet another building, but gunner Wallace failed to score a hit before the enemy gun crew towed their weapon away.

Captain Young radioed the lieutenant, telling him that an antitank gun had knocked out a tank from his platoon. Dew and his crew pulled back to hunt for the culprit. They quickly located the gun hidden among distant trees and put it to ruin with three HE rounds. Turning to help stranded comrades, Dew's crew provided cover fire for the men from the knocked-out tank as they scurried for shelter behind a pillbox, dragging an injured man with them, Private First Class William M. Polyak. He had steel fragments in his back and buttocks.

Meanwhile, another tank from Dew's platoon retreated through the dragon's teeth, its cannon rendered inoperative by a jammed shell.

Dew and his last remaining tank held tight near two pillboxes. The crews waited for the infantry to advance and secure those fortifications, which still held enemy defenders.

After Dew's platoon penetrated the dragon's teeth, three tanks from the Second Platoon followed to help exploit the breakthrough. Lieutenant Covington led the platoon and commanded the first tank through the gap. He sighted the tankless crew that Dew had covered and used his own vehicle to shield the men as they withdrew to the rear, carrying the injured Polyak.

Covington returned to lead his men.

Staff Sergeant George J. Bernstein, nicknamed "Fearless Fosdick" after a comic-book character, commanded a tank in Covington's platoon until an AP projectile pierced its turret. The bull's-eye decapitated Corporal Kleophas E. Kovar, the gunner. Fragments penetrated Bernstein's thigh and abdomen and fractured the skull of Corporal John M. Pugh, the assistant gunner. The driver, Technician Fourth Grade Cleveland T. Pulley, dragged Bernstein to a captured pillbox.

As the fight raged, Technician Fourth Grade Harold R. Clements's tank received sniper fire, wounding him in the head. The crew removed him for medical care, but he later died.

Only the Third Platoon avoided losses. Lieutenant Victor Miller had four tanks, those commanded by himself, Sergeant Roy Vernon, Sergeant

Leonard Henkelman, and Staff Sergeant Norman Crisler. Like Dew, this was the lieutenant's combat baptism.

The infantry in Miller's area advanced in short rushes. When bullets from five pillboxes held up progress, Miller and Henkelman pulled forward to the dragon's teeth and opened up with cannon shells and caliber .50 machine guns. This "buttoned up" the fortifications and allowed the infantry to subdue them with pole charges, bazookas, and hand grenades.

The foot troops in Dew's area made progress in fits and starts and with heavy losses.

At 1:20 P.M., Captain Young reported the infantry had reached their initial objectives west of Kesfeld. He asked permission to withdraw his company for gasoline and ammunition.

Before the tanks withdrew, Lieutenant Covington dropped back to escort litter bearers he expected were coming to aid Pugh, Pulley, and Bernstein. No litter bearers appeared, and he drove alone to help the men. His gunner, Corporal John R. McGowan, dismounted and repeatedly dashed across an open field to retrieve his comrades, all the while dodging German fire. He loaded the injured men onto the tank's engine deck.

McGowan's work no doubt saved Bernstein, the most critical of the three. Fearless Fosdick spent 266 days hospitalized but eventually returned home to Brooklyn, New York.

September 14, 8:00 A.M.: Captain Thornton radioed the 741st CP that he had received instructions for the attack from Colonel William L. Blanton, the 109th commander.

The colonel planned to commit two battalions. He attached Thornton's Second Platoon to the First Battalion and his other two platoons to the Second Battalion. The units would launch separate attacks, converging on Roscheid atop high ground within the Westwall. Once that village fell, the tanks and infantry would slash through to Sengerich and meet the 110th at Üttfeld. Blanton's last order to Thornton was for the captain to move his tanks across the river to Harspelt and Sevenig without delay. These were the jump-off points.

In his command tank, Thornton accompanied the two platoons bound for Harspelt, the First and Third led by Lieutenants O'Shaughnessy and Cinquina, respectively.

Their attack route followed a dirt road from Harspelt to Irsen Creek and then up Hill 517 to Roscheid. The Germans had dug two antitank ditches at the foot of the hill. The road offered the only passage over them, but the enemy had obstructed the passageway by detonating explosives, which left a yawning crater. Two tanks tried to circumvent the hole by crossing the ditches, but both machines bogged down. Lieutenant Cinquina later explained how his company commander found a solution:

"Captain Thornton ordered the bulldozer up front with the mission to fill the crater, which had impeded our advance. The dozer succeeded, and the two platoons proceeded across the obstacle and continued up to the high ground and took the objective. In achieving this, the platoons overcame ten pillboxes and destroyed two houses containing enemy troops."[145]

O'Shaughnessy added to the story: "Thornton, with the First and Third Platoons, led the attack in front of the infantry to a hull-defilade position on the high ground northwest of Roscheid. The infantry moved up, still behind the tanks. Heavy artillery and mortar fire fell around the tanks."[146]

That shelling killed Sergeant Fred A. Schild, one of O'Shaughnessy's tank commanders and a former interior decorator from Covington, Kentucky. He was twenty-seven years old.

Thornton and O'Shaughnessy edged forward to peek over the hill and immediately drew fire from an antitank gun. The projectile glanced off O'Shaughnessy's right sponson and left a gouge in the armor plate over the forward ammunition rack. The shot seemed to come from Roscheid. The lieutenant's gunner fired HE at the probable source, while his driver backed into hull-defilade along with Thornton's tank.

Plans changed when regimental headquarters radioed Thornton. The captain and O'Shaughnessy's platoon cut across the hill toward Lieutenant Dudley's Second Platoon. Cinquina's platoon remained in place with Second Battalion troops.

Dudley—or "Uncle Dudley" as the men called him—had approached Hill 517 from Sevenig along with the First Battalion. He lost one tank when it became stuck in Irsen Creek. Enemy resistance soon stymied both armor and infantry. Sergeant Ray A. Dickson's tank detonated a mine, disabling the vehicle and injuring two of its crewmen. Sustained fire from Dudley's remaining tanks exhausted their cannon shell supply. Thornton's tank and O'Shaughnessy's platoon fought their way to the First Battalion area, and Dudley's tanks withdrew to the rear for a reload.

O'Shaughnessy felt alone and exposed. Soldiers from the First Battalion were nowhere in sight. "We sat exposed to observation and direct fire from all directions," he later explained. "The captain told us to turn around and to proceed back in reverse order."[147]

Sergeant Lawrence E. Cerletti's tank led the retreat, until an enemy antitank gunner brought the vehicle to a dead stop with a crashing blow. Cerletti and Corporal Michael Letso died in a split second as their tank caught fire. Three survivors—the driver, assistant driver, and assistant gunner—bailed out.

O'Shaughnessy's tank picked up the driver and assistant driver, Technician Fourth Grade Ernest S. Reed and Private First Class Joseph C. Cheek. Neither man had injuries, but they said their assistant gunner needed help.

Private First Class Ralph R. Garber had a deep laceration in his lower back and right buttock. The wound left him unable to walk, and he lay on his belly near the tank. The lieutenant informed Sergeant Walter Jutkins, and the sergeant's tank pulled close to the wounded man, using the burning machine for cover. The sergeant dropped his escape hatch and rescued Garber.

Jutkins and O'Shaughnessy headed to join Cinquina's platoon, now in a reserve position after drawing too much artillery fire atop the hill. Jutkins's crew transferred Garber to infantry medics.

The nineteen-year-old Garber departed for a hospital bed, ultimately receiving a medical discharge in April 1945. He and his brother Roy had both now earned Purple Hearts while serving with the 741st. The two brothers from Boston had joined the battalion on June 14 as replacements. Roy received his medal for a July wound and had recently returned to duty.

After the enemy destroyed Cerletti's tank, Thornton and Sergeant Maddock told their drivers to divert into a defilade position. They were safe there but separated from the other tanks until Lieutenant Dudley's platoon returned under artillery fire. Dudley had only two machines after losing the remainder in Irsen Creek on the return trip.

The First Battalion had meanwhile fought its way up the hill. An infantryman approached the tanks and asked Thornton to cross a road beyond some captured pillboxes where the foot soldiers hoped to spend the night.

"They wanted us to lay down fire while they finished digging in," Maddock remembered. "As usual, there wasn't supposed to be anything out there. In fact, we had already been to that crossroads earlier in the afternoon. This time, however, Jerry had moved in. We got three-quarters of the way to the intersection when they threw everything they had at us."[148]

An enemy antitank gun blazed away. Its first shot flew on a straight line and hit Thornton's tank. He ordered his driver to back up.

Another shot struck its mark, but the tank kept rolling.

The enemy hammered it again, setting the vehicle on fire.

Corporal Newton R. Leonard was Thornton's gunner. "I believe the captain was knocked out of the turret by the second hit,"[149] he recalled. Leonard pulled himself from the flaming wreck but lost contact with his crewmates during his desperate escape.

Soldiers from Company B, 109th spotted Thornton on his backside near the burning tank. The infantry company commander, Captain Charles R. Thomas Jr., and his orderly rushed to the blazing vehicle. The battered tanker's face startled the infantry officer. It was a face from his past, a face from his school days at The Citadel. There lay his classmate, Jimmie Thornton!

"He had lost his left leg above the knee and had been shot through the left arm," Thomas later stated. "He was not in much pain and greeted me with a grin."

"This is a hell of a mess, isn't it, Charlie?" Thornton said.[150]

His pain-free condition belied the severity of his wounds. Shock caused that reaction. Thomas gave his friend water, clamped a tourniquet around his leg, and called for a medic. There was little time for conversation, and Thomas quickly departed to resume leading his company. He left two soldiers to guard Jimmie Thornton until the medic arrived.

Company B, 109th continued to advance and took its objective. The Germans counterattacked for the next four hours and drove the Americans back. The action took place at night, and confusion reigned.

Several stragglers later reported to Captain Thomas that enemy troops had captured his friend Thornton and rendered first aid.

The following spring, Thomas wrote to Jimmie's father, who had received no further word concerning his son since a War Department telegram reported him missing. Thomas dictated his remembrances from a hospital bed while convalescing from wounds received a few days after encountering his college chum:

"When I saw him last, he had lost much blood and was in severe shock. I did not expect him to live at the time. It was impossible to evacuate him, and he was without medical attention, other than first aid, from four to six hours. I do not see how he could have lived. However, I was told that when he was picked up by the Germans, he was still alive."[151]

Thornton's wife, parents, and siblings prayed he might be in a German hospital, but passing months dashed all hope. In September 1945, the War Department declared him dead in accordance with the Missing Persons Act, and his status changed to Killed in Action. The US Army conducted a field investigation but closed Thornton's case in August 1951, declaring his remains non-recoverable.

Newton Leonard was the sole survivor of Thornton's crew. When Leonard bailed out, he left behind the loader and the driver, as well as the captain. Fortunately, there had been no assistant driver aboard that day.

In March 1945, Graves Registration men lifted two bodies from a derelict Sherman in a meadow near Roscheid, Germany. The charred corpses, scarcely recognizable, eventually reached the US military cemetery at Hamm, Luxembourg. Five years elapsed before forensic examination fixed names to them. They were Private Warren E. Kettles and Technician Fourth Grade Arthur E. Kinkade. The army finally knew the whereabouts of the loader and driver, and their families later elected to have the fallen soldiers permanently interred in the United States.

Captain James Thornton's resting place remains a mystery to this day.

CHAPTER 19 YARD-BY-YARD

While pushing toward Kesfeld, the 110th Infantry netted prisoners from the 2.SS-Panzer-Division „Das Reich," all belonging to SS-Panzergrenadier-Regiment 4 „Der Führer." After crippling losses in Normandy, the regiment rebuilt its infantry strength with technical personnel culled from the Luftwaffe. Reichsmarschall Hermann Göring released these men in July and transferred them to an SS training center for thirty days of infantry instruction.

Two non-SS units reinforced Der Führer—Luftwaffe-Festungs-Bataillon 1 and Feldersatz-Bataillon 78. Before arriving at the Westwall on September 13, the Festungs battalion had existed for just three weeks, during which time its air force personnel received ground-combat training. The ersatz battalion filled its ranks with middle-aged men, including many recently drafted miners from the Ruhr, who had lost their jobs to Soviet prisoners of war.

Elsewhere on the 28th Division front, the 109th Infantry bagged prisoners from the 2.Panzer-Division and from Kampfgruppe Kentner. Numbering about eight hundred men, the Kampfgruppe, or battle group, drew its troops from an officer candidate school at Wiesbaden. The school commandant, Oberst Albert Kentner, led the group to Waxweiler, nine miles from the Westwall. The former officer candidates received bicycles, and Kentner's I.Batallion pedaled to Roscheid, where its men took up frontline positions. On September 14, they faced the 109th and Captain Thornton's tanks.

★★★

Thornton's capture that day left a void in Company B, and Lieutenant Cinquina became the new commanding officer.

Vincent Arnold Cinquina grew up in Chicago, the oldest son of Italian immigrants. He graduated from the University of Notre Dame and later earned a law degree from DePaul University. He suspended his legal career when the war began and joined the 741st after receiving an OCS commission. His wife, Rose, lived with her parents in Pennsylvania while her husband was overseas.

The thirty-three-year-old attorney took command as his company readied itself to continue the drive toward Roscheid. The First and Second Platoons combined their remaining vehicles and joined the Third Battalion, 109th. The attack force started out from Sevenig at 8:00 A.M. on September 15.

The tanks pushed off after engineers cleared a minefield. Two vehicles soon dropped out. Lieutenant O'Shaughnessy's machine bogged down, and Sergeant Jim Johnson's tank detonated a mine previously missed. Lieutenant Dudley led the remaining six tanks to the objective and worked with the infantry to overpower six pillboxes.

One tank, Sergeant Millard Case's vehicle, received a hit to its gun shield. The impact welded the shield to the turret, which rendered the cannon useless.

At around 7:00 P.M., tank destroyers from the 629th TD Battalion relieved Dudley's force. Progress for the day fell short of Roscheid.

★★★

On September 15, Company A received its first combat mission for the month. The men and tanks replaced Company C in Grosskampenberg and provided fire support for the First and Second Battalions of the 110th Infantry.

Concealed by dense fog, the tanks motored into firing position at 8:00 A.M. When visibility improved around noon, one platoon fired on the right flank where German infantry had infiltrated a wooded area.

Company A drew an immediate reaction from enemy artillery. The tank crews relocated to new firing positions, and from there, laid down shells on pillboxes and other targets.

Company C left its bivouac at 8:30 A.M. on September 15 and groped its way through the fog to Heckhuscheid. The day's plan called for the company to join the First Battalion, 110th and seize Hill 553. Rolling forward at 10:15 A.M., two tank platoons advanced abreast—the Third Platoon on the left and the Second on the right. The First Platoon stayed behind in reserve.

The tanks and infantry reached the dragon's teeth only to discover no opening. Enemy fire had driven away engineers tasked with blowing a gap. The teeth in this area were larger and in greater quantity than those Lieutenant Dew pulverized the previous day. No repeat performance was possible, and the infantry beseeched the engineers for help. One rifle company managed to work its way through the teeth, but German fire immediately pinned down the foot soldiers. They desperately needed tank support.

Two tanks attempted to squeeze through spaces between the teeth, but muddy ground sent them sliding into the concrete obstacles. Enemy AP shells added to the difficulties, and Captain Young received permission to withdraw his tanks and wait for the engineers.

The captain and two platoon leaders dismounted their machines and scouted on foot in preparation for another crack at the enemy's defenses. Lying prone, Young, Miller, and Covington studied several enemy pillboxes through binoculars.

At 6:00 P.M., the tanks commenced fire on the pillboxes to keep their occupants from shooting at the engineers. Laden with explosives, the intrepid soldiers aimed to demolish a steel-and-concrete barrier that stood where the dragon's teeth crossed the Üttfeld-Heckhuscheid road. To the rear, waiting foot soldiers and tank crews kept asking, "How far are they? Where are they now?"[152] After creeping up to the barrier, the engineers planted satchel charges and hotfooted it away. The consequent blast ripped open the barrier with a thunderclap.

Captain Young's tanks penetrated the teeth at 6:40 P.M. All three platoons climbed Hill 553, and, along with tank destroyers, fired point-blank at pillboxes. Crouched and at a run, infantrymen stormed up the hill to finish the job, capturing fifty-eight prisoners. The tanks pulled back after dark to reload and refuel. The crews also needed rest.

With the hill captured, the 28th Division had won key high ground within the Westwall, but a depressing truth was all too evident. The war had reverted to a yard-by-yard struggle like Normandy.

Under interrogation, officers among the German prisoners said that their opponents had "made a big mistake to halt on reaching the Westwall." The captives believed that as late as September 11, the American forces "might easily have rolled on to Koblenz."[153]

CHAPTER 20 **KEMPER STEIMERICH**

With his troops entrenched atop Hill 553, Colonel Theodore H. Seely, commander of the 110th Infantry, planned to have his First and Second Battalions converge on the regimental objective, a hill overlooking Üttfeld. It bore the name Kemper Steimerich.

The enemy had other plans.

September 16 had just arrived when a counterattack bolted from the darkness and struck the Second Battalion. The Der Führer Regiment launched a three-pronged assault against six pillboxes occupied by Second Battalion soldiers. As the clash unfolded, SS grenadiers spread terror and confusion with flamethrowers mounted on two halftracks. Fiery tentacles swept back and forth through the black woolen night. In less than two hours, the Germans regained three pillboxes and captured thirty-six GIs, including a company commander.[154]

The Second Battalion, with its position compromised, bowed out from further action that day. Seely now had to commit his reserve battalion to recapture the three pillboxes.

The First Battalion was the only force available for the Kemper Steimerich operation.

At the Charley Company bivouac, tank engines roared to life one by one. Some started in an instant, running smoothly like jeweled watches, while many others resisted with coughs and jarring backfires. The symphony of Shermans rattled the trees and shook the ground as they motored through heavy fog to reach the start line on Hill 553. Their first hurdle was Hill 568, also called Losenseifen. The effort to seize it began at midmorning after the soupy haze lifted. The armor pushed forward with First Battalion infantrymen. Dirty and unkempt, the foot soldiers trooped across the muddy ground, their shoulders hunched with aching tiredness. Enemy artillery held up the advance, but counter-battery fire quieted the artillery, and Losenseifen fell.

The attackers reorganized and prepared to seize the next hill, a dot on the map called Spielmannsholz. The infantry left one rifle company to hold Losenseifen. The tanks and the other rifle companies assaulted Spielmannsholz, securing it before darkness fell.

Kemper Steimerich still awaited.

Among Company C casualties for September 16 was Staff Sergeant William M. Johnston, who took a bullet through his forearm. Medics evacuated him to the 45th Field Hospital, but nerve damage ultimately ended his days in uniform. Three other men were non-battle casualties. Corporal Delbert D. Lambert and Private First Class Dominic L. Ferro both suffered from combat exhaustion. Private First Class Philip J. Zodda injured an arm when a caliber .30 casing exploded in his coaxial machine gun.

Zodda had served in Lieutenant Miller's crew as the loader since the lieutenant arrived in August. The crew frequently operated without a fifth man, the bow gunner. This was the case during the push toward Kemper Steimerich.

As the tank rumbled forward, Zodda shoved a shell into the 75 mm cannon and loaded the coaxial machine gun. The coax fired wherever the gunner aimed the cannon. As the gunner sprayed bullets, Zodda fed belt after belt into the weapon. As he loaded a fresh belt through the feed block, he heard a sharp crack. Tiny metal shards cut his right arm and nicked his right eye.

"We're hit!" Zodda shouted.[155]

He glanced at the machine gun's barrel as it glowed red. Embarrassed, he realized that heat had caused a round to explode.

Miller reported Zodda's injury to the company commander, who responded, "Send him back, and I'll get you another loader."[156] After sprinkling sulfanilamide powder on the injury and wrapping it with a dressing, Miller told Zodda to climb out and hike back to the First Battalion aid station.

With fear in his eyes, Zodda looked straight toward the lieutenant. The tank was in a downhill position facing the enemy, and, if he exited the vehicle, a sniper would likely shoot him. Miller immediately understood. The driver backed over the hill and released Zodda in an area hidden from the enemy.

The injured tanker grabbed a Thompson submachine gun and several hand grenades and started toward the aid station. Along the way, he encountered infantrymen with a group of enemy prisoners. Zodda approached an officer and offered to help march the prisoners to the rear. The officer hollered at two riflemen, who were escorting the captured Germans, "This tanker's gotta go back to the aid station. He'll take the tail."[157]

After delivering the prisoners, Zodda ambled to the aid station where medics took his Thompson and grenades before shoving him into an ambulance. His journey ended in Belgium at the division clearing station operated by the 103rd Medical Battalion. He stayed there four days, seemingly ignored by the staff.

On the last day, someone called him to the building's second floor. Several officers sat around an oblong table, and they invited him to have a seat.

"Stay calm, and tell us how you got wounded," one officer said.

"I don't understand what this is about. I've been here four days, and not one doctor has looked at me," Zodda said.

"Just tell us how you got wounded."

Zodda related the incident inside his tank. He ended by saying, "Call Cap'n Young, and tell him you want verification from Lieutenant Miller." The inquisition ended without explanation, and the private was sent downstairs and loaded into an ambulance, bound for the aid station from where he came.

Zodda reported to Young after reaching Company C. The captain advised his returning subordinate that he would rejoin Miller's platoon.

"You're gonna go out tomorrow with the tanks," Young said.

Zodda hesitated.

"Cap'n, if you want me to go out tomorrow, I'll go out, but I'm as useless as a dog's ass. I can't flex my right arm, and I'm right-handed."

Zodda's job as loader required full dexterity with his right arm because the cannon breech opened from the left.

Young looked puzzled.

"What did they do for you at the hospital?" he said.

"They didn't do shit! See that bandage on my arm. That's the same one Lieutenant Miller put on in the tank. They never even changed it."[158]

The captain decided that Zodda needed convalescence and assigned him to Staff Sergeant Frank W. Blahnik, the company's mess sergeant.

Each day, in the early morning blackness, the Company C tankers stood in line for breakfast with their mess kits open and canteen cups upturned. Zodda endured their sneers and smart-aleck remarks as he ladled out food. He recounted the experience:

"Every time the guys came in for chow the bullshit started. 'Look at that guy, the goof-off.' Every day it would be the same story."[159]

After four days, Zodda appealed to Young for a reprieve. He returned to Miller's platoon.

★★★

September 17. Company C stayed in bivouac all day, taking a breather. Company B took over for them and assisted First Battalion, 110th. Lieutenants Dudley and O'Shaughnessy led eight Shermans from the 109th sector to a holding position near Heckhuscheid. They stood ready to provide fire support, but the tank crews mostly dodged incoming German shells.

Able Company joined Third Battalion, 110th, which had previously regained the ground lost by Second Battalion. Enemy troops had since infiltrated the area and reoccupied several pillboxes, causing Colonel Seely to temporarily halt the drive for Kemper Steimerich. One rifle platoon from Third Battalion partnered with Lieutenant McDonough's tank platoon to recapture the troublesome bunkers, which stood in a row leading toward Kesfeld. The fight was a brutal, deadly business requiring a special tactic:

At each pillbox, the tanks rained bullets and 75 mm shells on its firing embrasure. This kept the position buttoned up and drove inside any Germans in nearby foxholes and slit trenches. After the tank fire ceased, infantrymen crawled close to the box, which was about twenty-five yards away. Three or four soldiers peppered the front of the box with gunfire, while two flanking groups maneuvered behind the structure. They covered the rear entrance, while a single man sneaked forward and tossed hand grenades into the embrasure. Enemy troops who survived the grenades usually waved an improvised white flag. If no surrender occurred, a tank rolled up to blast open the rear door. Once a fortification fell, several infantrymen remained behind to destroy the weapons inside, while the other soldiers and tanks pushed off to assault the next pillbox.

After regaining all the bunkers, the victorious tankers and riflemen stayed in place, while three other tanks and fresh infantrymen pushed ahead and seized Kesfeld.

The tactic used to capture the bunkers was one of several developed by the Americans. On other occasions, foot soldiers climbed atop a pillbox, blew the cover off its ventilation pipe, and slid white-phosphorus grenades down the metal tube. Toxic smoke sent those inside running pell-mell from the structure. Tank crews devised yet another method. The gunner zeroed in on the firing aperture using his coaxial machine gun. After hitting the opening with tracer bullets, he switched to the main gun and could often place a WP shell on the same spot. The first shell through the hole usually killed or expelled the occupants.

The 28th Division received instructions from corps headquarters to limit September 18 operations to local attacks—no push to Kemper Steimerich.

The First Battalion, 110th sought to improve its present position by capturing several pillboxes. Company B rolled into place at 7:00 A.M. with nine tanks led by Lieutenants Dudley and O'Shaughnessy. Fog limited visibility as O'Shaughnessy's platoon and foot soldiers descended on three concrete forts. The first one yielded seven prisoners, but enemy mortar, artillery, and small-arms fire pinned down the soldiers, checkmating their efforts.

O'Shaughnessy's tanks stayed with the infantry until midafternoon, when five Stuarts and two tankdozers from the 747th Tank Battalion relieved his platoon.

An hour after departing, O'Shaughnessy's platoon, along with Dudley's, moved into an attack position behind the infantry. The tanks instantly drew artillery fire. After the fog lifted, the enemy possessed nearly perfect observation, and that precluded any possibility for success. The attack force scrubbed plans to assault more enemy fortifications.

Attention shifted back to the pinned-down infantrymen, abandoned without explanation by the 747th. Dudley and O'Shaughnessy led their tanks to the scene and gave cover fire while the infantry withdrew. The tanks hauled out ten wounded riflemen.

★★★

The day also brought a success story, a new tactic for combating pillboxes.

Headquarters Company sent its tankdozer to join Company A's dozer. Both machines used their blades to bury captured pillboxes. The dozers pushed earth over pillbox entrances and embrasures, thereby foiling any Germans who attempted to reoccupy them at night.

During the earth-moving operation, the dozers fell under enemy eyes and became the target of artillery and small-arms fire. When the area grew too hot, other tanks from Company A pulled forward and engaged the Germans in a firefight. This support allowed the two dozers to continue their work, but other problems interfered. The dozers managed to bury four or five pillboxes, until mechanical difficulties forced them to abort the operation.

Aside from a new tactic, Company C obtained two new machines from the 70th Tank Battalion. These recent arrivals were no ordinary Shermans, both having their bow machine guns replaced with flame guns. The weapon used compressed air to shoot gasoline thickened with napalm, a jellylike concoction made from naphthene and palmitate. The flamethrower had an effective range of sixty feet, which meant that a tank had to approach its target dangerously close to shoot the blazing gasoline on the enemy.

To operate the new tanks, the 70th and 741st had exchanged two full crews. Peter Fardella from Company C felt the 70th had pawned off its riffraff.

"The crews were unrated and probably GFUs (polite translation: Gosh darned Foul-ups). Their morale was bad, and that caused plenty of trouble, refusing to go into combat after awhile. Captain Young was going to recommend court-martial, but finally we got rid of them."[160]

All that took several weeks. The malcontents were with Company C at 7:00 A.M. on September 18, when its armor rendezvoused with the Third Battalion, 110th.

Dense fog prevailed throughout the morning and delayed plans. Many infantrymen were still snoozing when the tanks arrived. With no fight to wage, the tanks sat idle until 1:00 P.M., when the fog thinned, and an attack finally rolled forward.

Lieutenant Covington's tanks led the way with their 75 mm cannons belching shells at seven pillboxes. The pounding suppressed fire from them, and the infantrymen snatched four boxes. Captain Young accompanied Covington's platoon during the action. He watched the infantry fall back after taking the fortifications, leaving them to the enemy.

"Heavy artillery caused the infantry to retreat," Young recalled. "We waited about an hour for their return. During the wait, three prisoners surrendered to Sergeant Suydam's tank from a pillbox on our far left. There being no infantry in the vicinity, it was necessary for him to herd them back to the reserve platoon with his tank."[161]

After several calls, the infantry finally returned at about 4:00 P.M. and cleaned out the four pillboxes again. Darkness halted the operation, and the tanks returned to their bivouac area.

During the day's action, a flamethrower tank engaged one pillbox, but the exercise proved disappointing. The flame gun failed to burn its mark. The weapon needed greater range and increased fuel capacity for sustained use.

Company C also implemented the new bunker-burying tactic. Its tankdozer crew located eight empty pillboxes and covered their entrances with earth before retiring for the day.

Charley Company reported for duty at 7:00 A.M. on September 19 and found the Third Battalion troops still asleep in their holes, as had been the case the day before. Today they would remain there again, their ranks too casualty-ridden to mount an assault.

The tanks stayed on call should the battered Third Battalion be thrown into action or attacked. Only the tankdozer found work. Under Lieutenant Miller's direction, the dozer crew plowed dirt over pillboxes seized two days earlier. The crewmen entombed eighteen bunkers but left one only partially buried after dud artillery shells impacted around the dozer. The crew thought it wise to sidestep the unexploded ordnance.

The next day, September 20, Company A participated in another try for the regimental objective, Kemper Steimerich. The tanks drove south from Heckhuscheid and passed through the infantry's frontline positions. The armor pushed forward alone.

Three tanks and a tankdozer motored through pastures to the left of the main road, led by Second Lieutenant Alfred W. Steineker Jr. The Louisville, Kentucky, native had joined the company as a replacement on July 27 but two weeks later became a casualty.[162] He spent the next month hospitalized before returning to his post.

To the right of the road, Lieutenant McDonough led four tanks supported by three assault guns. Corporal Joseph Basile, the gunner in McDonough's tank, surveyed the landscape through his periscope and watched the enemy react with shells and bullets. American artillery replied with terrific air bursts. Hell swept over the countryside like a tide.

Amid the furor, two tanks churned to a halt in deep mud. They had no way out under their own power and stood helplessly like tin ducks in an arcade game.

Basile and his crew mounted a rescue effort. With shells exploding all around, the corporal leaped from his tank and dashed about attaching tow cables. The steel lines permitted two successful extractions. Another problem then arose—enemy armor.

Two or three German tanks appeared, possibly Panzer IVs. Several tankers reported fire originating from nearby woods. Staff Sergeant Lloyd Ball, McDonough's platoon sergeant, saw high-velocity shells coming from another direction, where a lone Panzer sat concealed behind a house. One of its projectiles hit Steineker's turret.

The lieutenant, like most tank commanders, had raised his head through the turret hatch to better observe the battlefield. The violence of the impact and explosion launched him into the air like a champagne cork. Both of his comrades inside the turret died: Corporal Harold C. O'Riley, the gunner, and Sergeant Merwin C. Beck, the loader and assistant tank commander.[163]

In the lower hull, the jarring blast slammed Private Frank F. Frimel into unconsciousness. He awoke moments later with his head hanging on the bow machine gun. His eyes focused on a nightmare behind his seat. Flames swirled in the turret basket. Another enemy shell had ignited a fire. He saw the bodies of Beck and O'Riley draped over one another on the turret floor.

Frimel had yet to budge from his seat, his mind swimming in a fog. The hatch above his head suddenly opened and two arms reached in, hauling him up and out. The driver, Private Clarence Olson, saved Frimel from the flames.

Once outside, he heard a painful plea.

"Frimel, help me. I've been hit!"[164]

It was the lieutenant. He had landed near the tank, his lower left leg severed.

Frimel jumped down to help, but a stinging pain racked his left side. He grabbed his chest and pulled back his own bloody hand. Fragments had torn into his back, chest, left hip, and left shoulder. Steineker again begged for assistance.

"I can't, lieutenant. I've been hit myself!" Frimel said.[165]

Olson pulled Frimel to his feet, and the two spoke with Steineker, promising to send a medic and a litter team for him. The duo staggered away as bullets whizzed over their heads.

Lloyd Ball witnessed the destruction of Steineker's tank. Ball had partially exposed his own head when a bullet smacked against his turret and pelted him in the face with tiny fragments. While receiving first aid, he told his medic about Steineker's calamity. Ball agreed to lead the way to the injured lieutenant.

Near the ruined machine, they found the lieutenant still conscious. The medic tightened a tourniquet around his left leg and squeezed in the contents of a morphine syrette. Steineker's right leg had two compound fractures. Litter bearers carried him away. He eventually reached an operating table at the 45th Field Hospital in Belgium. His journey ultimately ended at Lawson General Hospital in Atlanta, Georgia, where the army retired the amputee officer from active service in April 1946.[166]

Though wounded, Ball remained on duty until the next day, when he too departed for the 45th, where he spent twenty-two days after having metal removed from a cornea.

After Olson and Frimel left Steineker, they found a stone building sheltering troops from the 110th who urged them inside. Frimel felt his strength drain in a sudden rush when blood loss and overexertion finally toppled him. The Germans cut loose with an artillery barrage that sent stones and debris sailing in all directions. Amid the deafening roar, Frimel made out the sound of Olson saying he was going for help.

After several minutes, the Company A command tank rolled up to the building. Sergeant Mayne B. Youngblood climbed down from the machine. "He was a big guy," Frimel reminisced, "and he picked me up like King Kong picked up Fay Wray."

Youngblood lowered Frimel into the command tank and drove toward the rear, until they encountered Captain Thomas and his jeep. Youngblood hauled out Frimel as the captain rushed over, exclaiming, "My God, Frimel! What happened?" The wounded tanker wheezed a response while pointing a bloody finger south. "Lieutenant Steineker's tank got knocked out down there," Frimel said. "He's in bad shape."

Thomas sent Frimel to a foxhole where he could lie down, safe from shell fragments, but he found no comfort there, gasping and choking all the while. He had a punctured lung, and blood bubbled from a hole in his chest. Thomas came to check on him, and Frimel groaned, "Cap'n, if you don't get somebody to help me, I'm gonna die."

A buck sergeant standing beside Thomas said, "Let me take the jeep." Ten or fifteen minutes later, the sergeant returned with aid men, and one immediately jabbed a morphine syrette into the gasping casualty. Frimel's suffering at last abated, and the medics transported him to their aid station located in a captured pillbox.

Flat on a litter, Frimel soon encountered the surgeon in charge of the station. Triage and treatment began. The doctor scissored off Frimel's pants and shirt and started blood plasma draining into his arm. He then operated to stop the bleeding and stabilize his patient for transport to the 45th Field Hospital. The plasma and morphine improved Frimel's condition.

"What can I do for you now?" the doctor asked.

"Coffee," Frimel said.

That was easy enough, and the doctor began heating water for a cup of Nescafé. Three other casualties were also there, waiting for an ambulance. When it arrived, the doctor announced, "The ambulance is here, and the coffee is ready." Frimel said, "Forget the coffee, let's go!"[167]

The trip to the 45th dragged on and on, and eventually the morphine wore off. Frimel writhed in agony by the time he reached the medical facility. He received more plasma and morphine before undergoing surgery. A ten-month hospital stay followed, ending at Brooke General Hospital in San Antonio, Texas.

CHAPTER 21 **RHINELAND IMPASSE**

Baker Company also engaged the enemy on September 20. Its tanks and men from the Second Battalion, 110th moved against Germans, who had retaken portions of Losenseifen Hill.

Tank shells pummeled pillboxes and helped draw forward the infantry. Once a box fell, the company tankdozer began work to prevent its reoccupation by the enemy. The driver's limited visibility slowed the job, as did the inability of the dozer blade to bite deep into the earth like an excavating shovel.

Sergeant Donald P. Lucey and his crew operated the Baker dozer, taking over from John Brewer's crew, who had earned a rest. Lucey's men buried five pillboxes on September 20. The bow gunner, Private Kenneth V. Powell, directed the operation from outside to compensate for the driver's limited visibility. Powell avoided harm, despite artillery fire throughout the day. After the infantry retook the entire hill, the dozer and the Company B tanks returned home.

When morning arrived on September 21, the tanks and dozer returned to the battle area and helped the infantry assault a ridge and six pillboxes built along its crest. Lieutenant Dudley's platoon took the lead and drove east from Heckhuscheid toward the ridge, while O'Shaughnessy's platoon stayed behind and bombarded the bunkers. Lucey's driver cautiously approached the first one.

"We spotted heads popping up out of the box," Lucey remembered, "and we fired on them until they buttoned up. We received in return two bazooka shots and mortar fire, but our continuous fire soon had old Jerry waving the white flag."[168]

The driver, Tony Morgando, advanced to the first bunker's rear entrance and lowered the dozer blade. Germans in a neighboring position detonated a charge buried near that entrance. Lucey later described the explosion:

"Helmets flew one way, my crew flew another. I found myself wedged tightly in the turret with my gunner, Corporal Louis Robles, and loader, Homer Cupit, urging me to abandon the tank as I was blocking their exit."

The crew sprang out as flames consumed their vehicle. While running through black smoke, Lucey saw something unexpected.

"I looked to my left to find a brawny Kraut running in the same direction. I looked at him, and he looked at me. Each, trying to outdo the other, ran that much faster. As a result, it was almost a photo finish as we reached our lines and safety."[169]

The German had run in the wrong direction, but his compatriots on the ridge remained in possession of their ground. The American tanks and infantry retreated.

Lucey's crew escaped with only bruises, singed hair, and scorched clothing.

The Germans had learned to defend against the new tactic. Besides remote-detonated charges, they planted antitank mines—sometimes three combined—near pillbox entrances. The enemy also deployed remote-guided demolition vehicles, each resembling a dwarf tank. Known as the Goliath, one version packed over two hundred pounds of explosives. Steered by wire, the Goliath seldom achieved success, often falling victim to small-arms fire or mechanical failure.

★★★

The most serious casualties on September 21 occurred behind the frontline, near Grosskampenberg, when mortar fragments struck two radio operators from Headquarters Company.

Technician Fifth Grade Donald Momany was one of the victims. He had previously earned a Purple Heart for a thigh wound inflicted by a shell fragment on D-Day. He spent forty days hospitalized in England during the summer.

After his return to Headquarters Company, Momany became friendly with Technical Sergeant Nelson Hall. Both men worked on a newsletter produced by enlisted men in the battalion and then distributed to all 741st members. After the newsletter lampooned several battalion officers, its publication ceased, expressly forbidden.

Owing to their newfound friendship, Momany asked Hall about joining the headquarters section as a crewman in the command halftrack. Momany belonged to the tank section and disliked his posting. He was a qualified radio operator and required no training to join the halftrack crew. Hall saw an opportunity.

"We had acquired a fellow in our halftrack crew, who did not get along with us, primarily because of his womanish personality and a penchant to challenge most orders. He simply did not fit in well. We welcomed his request when he asked to be considered for a different place in the organization."

The misfit left, and Momany joined the crew. He grew up in Michigan as the son of a schoolteacher. Hall fondly recalled Momany. "He was a handsome, blond fellow, very outgoing, and well suited to the military. He quickly melded into our little group and made us all glad he was one of us."[170]

September 21 found Hall busy with paperwork inside the wheeled trailer that had served as his portable office since Headquarters Company acquired it in Normandy. He busied himself updating the unit journal, an event log required by regulation. Hall derived most entries from radio messages received or intercepted in the command halftrack at the forward CP. After taking the messages in pencil, he typed them into the journal.

The command halftrack often sat at the rear CP when no fighting was underway. With combat looming on the morning of September 21, the halftrack crewmen readied themselves to leave for the forward CP. Hall usually accompanied the crew, but he begged off this time because he had fallen behind typing journal entries. Momany willingly took his place. The halftrack rumbled off to the forward CP, which was nestled in an apple orchard behind Grosskampenberg. Everyone considered the location safe.

Hall later recounted Momany's tragic fate: "That afternoon, as he sat in the driver's seat reading a comic book, with the metal window shield in place, with only the vision slit open, a German mortar squad slipped past our infantry and fired one round at the halftrack.

"How a piece of shrapnel from that burst managed to bounce from the metal motor cover of the halftrack and travel through the vision slit and through Don's metal helmet remains a damnable mystery. How a piece of Don's brain ended up on the motor cover is even more damnable."

Technician Fifth Grade Martin W. Coyne fell victim to the same mortar round. He had an injured hand and lacerated posterior.

The halftrack returned to the rear CP, where Hall witnessed its arrival.

"I still can see the look of pain and hurt on the face of Ike Everts, our driver, as he pulled up to the rear CP and yelled at me, 'Momany's been hit, bad. That's part of his brain there on the halftrack.' Don was on his way to a field hospital, he told me, but the medics said that if he lived, he wouldn't be right. There was too much damage.

"All the details came pouring out as I stood there in silence, realizing it was I who had put him in that place of danger. And I suddenly blamed myself for having let him transfer into our crew. I don't know how other people handle these things. But I have never found the way."[171]

Coyne and Momany reached the 45th Field Hospital. Momany survived surgery there and transport to a hospital in England, where he lay partially paralyzed. He developed non-infectious meningitis before succumbing to his injury on October 19.

The 28th Division had a provisional unit called Task Force M. The letter "M" denoted its commander, Colonel MacLaughlin, from the 3rd Armored Group.

The colonel planned to use Company C for an attack on September 22 to capture objective "Zebra," code name for the ridge where Sergeant Lucey lost his dozer. The tanks would neutralize the pillboxes, while the engineers blasted them to bits.

MacLaughlin soon amended the plan. Because no engineers were available, the armor would go alone. MacLaughlin intended to use flame-thrower tanks to subdue the pillboxes and tankdozers to bury them.

At 7:00 A.M. on September 22, Company C assembled for the attack, but a reason for delay beamed brightly. The sun skimmed the horizon dead ahead and blinded the attackers. Headquarters consented to a postponement until the sun climbed higher.

Reports circulated that the Germans had reinforced "Zebra" with Panzerfaust-wielding soldiers. MacLaughlin had no infantry to counter them. Colonel Skaggs advised MacLaughlin about the danger and sought a reprieve: "Request that task be abandoned until supporting infantry or engineers are available."[172]

MacLaughlin called off the attack.

German prisoners captured in recent days seldom belonged to Das Reich, but instead to various Wehrmacht units fighting alongside the SS. One prisoner from Feldersatz-Bataillon 78 said that his battalion commander stayed perpetually drunk at the battalion CP in Habscheid. A political officer gave most orders in the battalion and shot several men who tried to desert.

The chief intelligence officer for the 28th Division summarized the enemy situation: "Although morale of PWs is low and despairing, most of these troops are defending their own country and are well indoctrinated with individual aggressiveness."[173]

While the intelligence chief acknowledged German determination, most reports focused on the shabby replacement troops manning the Westwall: cooks, bakers, butchers, invalids, coal miners, truck drivers, and non-combat personnel from the Luftwaffe and Kriegsmarine.

American replacements amounted to much the same. Since reaching the German border, few replacements arrived as fully trained infantrymen. The army had culled thousands from the artillery, antitank outfits, antiaircraft command, and air force ground crews. The poorly trained men received by the 28th, combined with over one thousand casualties in September, and the absence of reserve troops, left the division unable to expand or exploit the slight gain achieved by the 110th. The situation offered little to justify further offensive action.

So ended the attempt to cross the Rhineland and seize Koblenz on the heels of the crushing victory in France. American commanders turned their eyes north toward another attack route, a picturesque woodland called the Hürtgen Forest.

CHAPTER 22 IVANHOE RETURNS

The Germans exploited the sparse American presence on the 28th Division's southern flank by dispatching patrols across the Our River into Luxembourg. The infiltrators ambushed US patrols and snared prisoners for interrogation. Their first success occurred on September 22.

Fifteen minutes before 7:00 A.M., a lone jeep roared away from the Company D encampment with First Lieutenant John H. Belcher and five enlisted men. The group headed for the Kalborn area, where high ground permitted observation across the river into Germany. Belcher's team hoped to study enemy activity but soon became the target.

German infiltrators fired from Kalborn and caught the men in the open. They were simple prey for the Germans, who swooped in and forced their surrender at gunpoint. Soldiers from the 17th Field Artillery Observation Battalion watched the episode from afar and witnessed five prisoners being marched toward Germany. The hapless bunch included Belcher, two sergeants—Robert F. Stover and Elmer F. Williams—and two privates—Daniel L. Calderaro and James P. Caughey. A sixth man, Private First Class Efford L. Brown, lay dead in front of the jeep, his crumpled remains riddled with German bullets.

At 1:00 P.M., light tanks from Company D set out to locate the missing jeep. The recovery force included the Mortar Platoon from Headquarters Company under First Lieutenant Russel A. Dotson. To screen the operation, the mortarmen fired smoke shells from their three 81 mm tubes mounted in halftracks. The foray ended without incident after less than ninety minutes. The recovery force found the wrecked jeep and collected Brown's body.

Anxious to stop German infiltrators, division headquarters formed an ad hoc unit to bolster the division's porous southern flank. Task Force X consisted of Company D, the 28th Cavalry Reconnaissance Troop, and part of the 630th Tank Destroyer Battalion. This amalgam had instructions to conduct "aggressive reconnaissance and demonstrations" during daylight hours and to man roadblocks and listening posts after dark.[174]

★★★

The 741st refocused on the Grosskampenberg area, where Lieutenant Dew had blasted an opening through the dragon's teeth on September 14. Since that day, the enemy had reclaimed all the ground won by Dew's tanks and put several pillboxes back in use. The Americans now had to recapture the concrete emplacements and render them useless.

Two platoons from Company B drew the assignment, reinforced with a flamethrower tank, two tankdozers, and the five assault guns. The assembled armor obtained support from the 110th Infantry.

When morning arrived on September 23, both tank platoons passed through the dragon's teeth. Enemy bullets, along with mortar and artillery shells, pinned down the accompanying infantrymen. While trying to flank the pillboxes, one tank sustained a hit to its right, rear idler wheel. Crippled but still able to move, the vehicle remained in action until the infantry and tanks pulled back to regroup for another try.

Once more, German fire halted the attackers, except for one tankdozer that crept up to a pillbox and piled earth against it. The operation was otherwise a waste. The tanks pulled back without the foot soldiers, enemy fire pinning them down. Their leaders contemplated an artillery barrage as cover for a retreat but abandoned that scheme—too great the chance for casualties. Another idea seemed better. The Mortar Platoon from Headquarters Company received instructions to lob smoke shells as it had done with success the prior day in Luxembourg.

The platoon's three mortar crews pumped out shells to screen the infantry. The falling rounds mingled with HE from infantry mortars. The combined effect enabled the foot soldiers to exit the battlefield and reach safety by midafternoon.

Late in the day, Colonel Skaggs received word to shift Company C to the Grosskampenberg area. Instead of trying to seize the pillboxes again, the company would batter them from long range with cannon fire. To assist in the effort, Skaggs also moved the Mortar and Assault Gun Platoons from Headquarters Company.

At 8:00 A.M. on September 24, Lieutenant Dew's platoon pulled into position south of Grosskampenberg. The Assault Gun Platoon deployed on an adjacent ridge. The crewmen trained their cannons on targets stretching from Kesfeld to Leidenborn. The Mortar Platoon occupied a draw behind the assault guns and prepared to fire smoke rounds wherever needed.

Abysmal weather delayed firing until well after 10:00 A.M. When Dew's men finished hammering out their fire mission, they cranked up their motors and left the area to fetch more ammunition as smoke concealed their withdrawal. Lieutenant Covington's platoon drove forward to occupy the vacated position and commenced firing.

At 11:20 A.M., a frantic officer from the 110th ran to the 741st forward CP and reported tank shells falling on the infantry. Covington's platoon ceased firing. As the tankers double-checked their targets, bad weather crept back, forcing a halt to the operation. Once more, the mortars launched smoke projectiles to cover a withdrawal.

Clearer weather prevailed at 4 P.M., and Dew's contingent returned. Another fire mission boomed across the rural outland. It ended within thirty minutes, and the tanks retired to the rear. The 110th requested more of the same for the next day.

Lieutenant Miller's platoon pushed off in the morning, arriving in position at 9:30 A.M. Fog limited target selection to houses and a forested area. The platoon opened fire and received a response from German mortar teams. The tanks blasted back and hushed the enemy retort. The operation lasted less than an hour before the platoon rolled home.

★★★

On September 27, Colonel MacLaughlin visited the 741st rear CP and sent Skaggs to the 112th Infantry CP. This regiment had been on loan to the 5th Armored Division but was now back under 28th Division control.

The 112th had a new mission—a limited attack to clear the Westwall in the northern third of the 28th Division zone. The enemy's defenses had remained untouched in that area and formed a salient between the 28th and its neighbor to the north, the 4th Infantry Division. Corps headquarters wanted the menacing bulge erased.

After reaching the 112th CP, Skaggs and regimental staff officers headed out to observe the salient and the planned attack area.

Written orders from division headquarters sketched out the parameters for the operation. Charley Company was the only 741st participant. The news soon reached its officers, and the unit headed to the 112th before midnight.

In column formation, Company C armor drove north to the Belgian city of Saint Vith and turned east. The tanks entered Schönberg, and the column twisted its way across the German border and into Mützenich, where the company made camp. It was 3:00 A.M., September 28.

After sunrise, the tank commanders trekked ahead on foot to view the attack area. They grimaced when they saw it. The rugged terrain was untankable—absolute treachery for armor. All that remained was to wait for the foolishness to begin. Halfway through the afternoon, the company retraced its tracks to the battalion bivouac area. The corps commander had rescinded the attack plan, and rumors abounded that the 28th Division would soon be relieved.

In the Heckhuscheid area, the 110th launched another "limited-objective" attack on September 29. Two platoons from Company A joined two Second Battalion rifle companies for an assault against pillboxes on the ridge stretching from Hill 553 to Hollnich.[175]

Company B had no part in the attack but moved to conduct a diversionary display with help from the Mortar and Assault Gun Platoons, all under the direction of Major Browder, battalion executive officer. The assault guns and mortars started work at 7:30 A.M., with sustained fire on targets south and southeast of Grosskampenberg. Baker Company joined in later with four, thirty-minute fire missions.

The attack kicked off just after 9:00 A.M. Tank cannons buttoned up pillboxes and nearby positions, while infantrymen descended on the fortifications. Three pillboxes fell within thirty minutes, and four more boxes followed despite heavy enemy artillery fire.

The tanks drew attention from a German antitank gun. Its fire scotched one vehicle by striking a drive sprocket. The gun then drilled a second American machine and set it alight. Staff Sergeant Nicol and three other men crawled out from the burning tank. The driver, Technician Fifth Grade Emmett E. Rickman, had burns severe enough for hospitalization until January. The bow gunner, Private Ray I. Sloan, died in the tank.

German fire also struck an observation post used to direct the diversionary display. An artillery shell roared in like a locomotive and exploded overhead. Fragments hit Lieutenant O'Shaughnessy as he operated a range finder. The jagged steel tore muscle from his left arm and ripped through his right, obliterating bone and muscle.

The same explosion injured First Lieutenant Leonard L. Copeland, who pulled two small splinters from his face and one from a finger. He suddenly heard Skaggs and Browder shouting about O'Shaughnessy. Several men hoisted the bleeding lieutenant onto a jeep's hood, where he lay on a litter smoking a cigarette placed between his lips by Skaggs.

The vehicle raced to an aid station, the colonel in the passenger seat holding onto his wounded officer.

An ambulance later transported Copeland to the 67th Evacuation Hospital. His medical treatment spanned multiple hospitals, multiple surgeries, and 602 days. He separated from the army in 1946 at Halloran General Hospital, near his home in Richmond Hill, New York.[76]

O'Shaughnessy never regained full use of his right arm.

His job as leader of First Platoon fell to Lieutenant George Coleman, who had returned to the battalion after a month at the 3rd Replacement

Depot. He was unable to rejoin Dog Company because no opening emerged there for him.

Elsewhere on September 29, Lieutenant Romolo Legnini and a group from battalion headquarters reported to Company D in Luxembourg and laid mines near Heinerscheid. Their digging and planting ended after dark, even though the task remained unfinished.

The next day, nine men from Dog Company planted mines at a roadblock north of town. Upon completing their work, they decided to investigate some machine-gun fire they heard echoing through the hills.

Technician Fifth Grade Vernon C. Wilson and the others were plodding through a field near town, when German riflemen concealed in the houses let loose. One bullet knocked Wilson to the ground, boring a hole in his chest, perforating his right lung, and fracturing his right scapula. He lay hemorrhaging and unable to use his M3 submachine gun or to raise himself.

After the firing stopped, Wilson eyed a lone German walking his way. The enemy soldier was a pubescent conscript, scarcely taller than the Mauser rifle he toted. Surely the boy meant to use his weapon to finish off the downed American. Instead, the kid knelt down, fumbled with a bandage, and dressed his adversary's punctured chest. All the while he jabbered away in German. Wilson understood several words. The kid was on no mercy mission. He had come to claim a trophy: an American caliber .45 pistol. No luck here; Wilson had left his sidearm behind. As the young German turned to leave, Wilson caught a pathetic sight. The kid's wool trousers had a huge tear, and his bare bottom hung naked to the autumn chill. Was the fearsome Wehrmacht sending its legions into battle without underwear?

The boy left Wilson to perish while other Germans took captive four of Wilson's comrades. Three others evaded capture. Sergeant Stanley V. Nation so far was the single fatality, though when help finally arrived, Wilson was near death.

Captain Sledge, the Dog Company commander, rolled up in a jeep. He and another soldier rushed over, administered a morphine syrette, and loaded Wilson into their vehicle. As the jeep approached tents sheltering a field hospital, the gravely wounded enlisted man drifted into unconsciousness. Within minutes, a surgical team began operating on his injuries, though he would require a second operation to reverse an infection caused by wool fibers from his winter combat jacket, which the bullet that felled him had driven into his chest. Wilson's injuries eventually earned him a medical discharge.

During the first hours of October, reports told of enemy forces occupying Heinerscheid. The Mortar and Assault Gun Platoons hurried there to support the 20th Engineer Combat Battalion, now fighting as infantry, to retake the village. The operation expelled the enemy without loss to either 741st platoon.

The rumors of impending relief proved true. The 28th Division had orders to relinquish its present positions to VIII Corps troops arriving by train and motor convoy from France. The 28th and V Corps were pulling out and moving north to an adjacent sector. As the shift unfolded, the 741st remained behind to join VIII Corps.

The 28th Division operations officer informed Skaggs on October 2 that his battalion would remain in place. Ivanhoe had reclaimed the 741st. General Robertson's 2nd Division, in need of rest and replacements after a furious bloodletting in Brittany, prepared to assume a sizable chunk of the 28th Division sector.

Skaggs felt honored by the news. "This reassignment to the 2nd Infantry Division was one of the highest compliments ever paid a military unit. General Robertson had to go all the way to Army Group Headquarters to get the 741st back with his division."[177]

Hours before the 2nd Division arrived, the 3rd Armored Group sent both its chaplains on a final visit to the 741st, where the two clergymen had often held worship services.

Chaplain William C. Poepperling, a priest from Saint Louis, and his assistant, Technician Fifth Grade Joseph E. Krywany, conducted Mass as well as heard confession for the Catholic faithful.[178]

Chaplain Richard C. Lipsey, a Presbyterian minister from Mississippi, and his assistant, Corporal Harold C. Sanger, offered Protestant services. They also brought Chaplain Samson M. Goldstein, a Brooklyn rabbi assigned to V Corps Headquarters, who led services for Jewish members of the battalion.

Henceforth, the 741st would call upon VIII Corps Headquarters and the 2nd Infantry Division for chaplain support.

CHAPTER 23 CAFE HENKES

Late on October 3, the 23rd Infantry Regiment trudged forward in the darkness to relieve part of the 28th Division. The other two 2nd Division regiments relieved 4th Division soldiers to the north, in rugged woodland called the Schnee-Eifel.

Orders from 2nd Division headquarters in Saint Vith moved the 741st to provide support across the entire division front. Only Company C remained in place, staying at its bivouac on a wooded hill northwest of Grosskampenberg.[179] Company B shifted north to the Schnee-Eifel, landing in the Belgian town of Schönberg. The Schnee-Eifel also became home to Company A, which encamped at Wischeid, Germany. Company D moved to a stretch of unforested countryside above the Schnee-Eifel, known as the Losheim Gap. The shift north concluded on October 5, with Headquarters Company and Service Company taking up quarters in Schönberg.

Colonel Skaggs installed his CP in a Schönberg cafe owned by Peter and Barbara Henkes. The building sat at the village center on the main road. No name or signboard adorned its façade, though the locals knew it as the "Cafe Henkes," or simply "bei Henkes."[180]

Inside the structure, its restaurant furnishings had long ago disappeared. The incoming headquarters men filled the empty space with bunk beds and a wooden table that Skaggs used for a workstation. His telephone in its leather case sat on the tabletop.

The Henkes family—mother, father, and daughter—occupied a single bedroom adjoining the command center. The US troops inferred that a son serving in the Wehrmacht lived across the street, recuperating from wounds. Another son and daughter were absent. The family cow lived only a few feet away in a barn integral to the building. Upstairs, Skaggs and Major Browder claimed an unused bedroom as their sleeping area. Far from exquisite, the cafe took on a cozy atmosphere, due in no small measure to the hospitality of its kindhearted owners.

At every contact, the Henkes family greeted Skaggs and his men with warmth and smiles. Peter "Papa" Henkes served in the Kaiser's army during the First World War. Nelson Hall remembered the man and his wife:

"He was medium height, brown hair, blue eyes, probably 140 pounds. His face was wrinkled, and his skin might have been frostbitten at one time. He was outgoing and straightforward. Though he spoke no English, he found ways to communicate with us. He was no doubt a good barkeep, and was, at the same time, well thought of in Schönberg.

"'Mama' Henkes was about the same size and complexion as her husband. She wore high-top shoes, long skirts, a well-worn sweater, and a shawl that often covered her head. She was wonderfully clean and anxious to keep the household and environs clean as well. She was an excellent cook."[181]

Elsewhere in town, the Company B tankers appropriated billets, the first time since England that many of them had lived indoors. The company officers lodged together, and, among the group, Lieutenant Coleman found time to catch up on his correspondence.

"The boys are getting a needed rest," he told his parents. "We're all a bit tired of slinging the old ammunition at these Krauts."[182]

In another letter, he mentioned a recent trophy. "I'm going to send a little sword I happened to run into a few weeks ago. It ought to look good in a den. It's about four feet long, so don't be surprised when they drop it off in a truck."[183]

Coleman could only inform his parents that he was "somewhere in Belgium," although the accommodations in Schönberg offered little to write about. The billets at least provided insulation against the chilly, autumnal temperatures descending on the Continent.

The seasonal change presented the greatest challenge for C, D, and Service Companies, which had no buildings available. Some men burrowed underground, constructing dugouts for warmth. Others built log huts for living quarters. The men yanked wood-burning stoves from abandoned homes to heat their improvised dwellings.

Company A vacated Wischeid on October 8 and relieved Company D. The light tanks then drove south. One platoon joined the 38th Infantry, taking station in Germany at Steinbach. The other platoons also settled in Germany but with the 9th Infantry near Schlausenbach.

At Steinbach and Schlausenbach, the platoons sat just behind the frontline. There they remained on standby should the infantry call for support, something light tanks could do better than medium tanks in the thick woodland.

Besides swapping positions with Company A, the Company D men said goodbye to their commander, Captain Sledge. He took the reins of Able Company on October 16, replacing Captain Thomas, who moved up to battalion headquarters to become operations officer. His predecessor, Major King, departed the 741st three weeks earlier with bronchitis and rheumatoid arthritis.[184] Dog Company, in turn, received a new commander, First Lieutenant Thomas N. Snyder, formerly the maintenance officer with Headquarters Company.

Four days after moving to the Losheim Gap, one Company A platoon found itself assisting six 105 mm howitzers from Cannon Company, 9th Infantry. Telephone wires connected the tanks to the company's fire direction center. The tank crews received fire missions and sent shells arching toward enemy work parties, observation posts, and suspected troop concentrations. The tank platoon expended more ammunition that month than any other 741st unit. The firing concluded on October 21, when Company A relocated to Schönberg.

That same day, Chaplain Poepperling also arrived in Schönberg and held Mass for the battalion's Catholic soldiers. His presence came in response to a telephone call from Major Browder, who reported a total lack of chaplain support since joining VIII Corps. Poepperling also arranged for Protestant services, Chaplain Lipsey being hospitalized. The padre promised better support henceforth.

Two days after Poepperling's visit, seventy-six enlisted men from the battalion reported to the 2nd Division for special duty. They formed an Internee Guard Detachment under the direction of two 741st officers, Captain Duke and Second Lieutenant Charles R. Izor.

Men from the detachment guarded civilian internees in confinement areas at Vielsalm, Ligneuville, and Saint Vith. These Belgian towns held assorted counterintelligence arrestees: enemy agents, line-crossers, collaborationists, civilian enemy nationals, and other citizens suspected of hostility toward Allied forces.

The detachment began its work while the battalion commander, his operations officer, and two company commanders attended a flamethrower demonstration by the 70th Tank Battalion at Hünningen, Belgium. After the display, First Lieutenants Klotz and Dudley and five enlisted men traveled to Hünningen for a three-day course on tank flamethrowers. They returned to Schönberg to establish a similar school for the 741st. The battalion already possessed two Shermans with flame guns and expected to have more guns mounted in other medium tanks.

Across town from the new school, the daily routine at battalion headquarters included a never-ending battle fought by Frau Henkes. Her cafe sat along a major supply route, and each day GI trucks wheeled past, churning up dust clouds or spraying mud. Time and again, the middle-aged woman emerged from her front door with a broom in hand. Her American guests implored her to forget about sweeping away the mess, but she ignored them. This had been her duty for years, and a world war at her doorstep changed nothing. Fixation on trivial matters kept her inner gyroscope spinning level, just as it did for many tankers in combat.

While Frau Henkes swept, her attractive daughter Maria attracted generous attention. One suitor, Private Douglas C. Adair, the colonel's driver, found an opportunity to use a German sentence taught to several soldiers by "Papa" Henkes. "*Willst du spazierengehen mit mir?*" was an invitation to go for a stroll. Adair recited the words, and Maria accepted.[185]

Skaggs, meanwhile, took an interest in Maria's mother and her culinary prowess. He later wrote about her work in the kitchen:

"We always had lunch with the Henkes. We provided the food, and it became a stopping point for the CIC [Counterintelligence Corps] team in the area. Our noonday meal was always pleasant. One day I expressed a desire to have a meal for General Robertson and his Chief of Staff, Colonel John Stokes. The question was, what could we serve? Mama Henkes suggested that the local policeman, who had collaborated with the Germans, had a goose. The CIC people quickly determined that the goose was German and therefore subject to confiscation.

"It was quite an experience to see Mama Henkes prepare that goose. It took several days. General Robertson arrived with his two gleaming stars, but before we could be seated, Colonel Stokes greeted Mama Henkes in the most beautiful German. At that moment, Stokes became the guest of honor, and General Robertson had to take a less-honored seat at the table. It was all in good fun."[186]

Life in Schönberg also brought a painful moment for Skaggs, when unrelenting truck traffic claimed the life of his dog, a mongrel he adopted from the town's streets.

The day before the little animal died, a speeding truck clipped Nelson Hall, though he escaped unharmed. This near calamity precipitated little reaction from Skaggs, but, minutes after the dog's fatal encounter, the colonel was on the phone to division headquarters.

"Yesterday one of your damn trucks nearly hit my operations sergeant," he exclaimed. "Now one has just killed my dog!"[187]

Skaggs was nearly in tears.

The dead canine was the only recent fatality in the battalion, a thankful fact, but the lack of combat led to a problem. Boredom ran rampant.

Peter Fardella, the radio repairman for Charley Company, described the consequences among his buddies at their campsite in Germany:

"The unusual leisure had some bad effects as it gave the boys time to bitch. Morale went down. The captain ordered the company to stand reveille so that everybody would get up and eat breakfast instead of sleeping until dinner. This and a few other things changed his popularity to unpopularity."[188]

CHAPTER 24 BUZZ-BOMB CORRIDOR

As monotony festered among the tankers, *esprit de corps* rose among their adversaries. The Fatherland's defenders looked skyward to witness the manifestation of Herr Doktor Goebbels's bellicose threats about unleashing an arsenal of super-secret vengeance weapons. Day and night, one after another, V-1 flying bombs winged their way toward targets in the Belgian interior. Mobile launch units cast them into the sky just behind the frontline. Each pilotless aircraft carried a ton of explosives and had a pulsejet that created "an earsplitting racket like a badly timed truck engine without a muffler."[189]

In Schönberg, the US Army endeavored to track the V-1s using mobile radar units. Vans sprouting cables and antennae arrived in the village. This cumbersome hardware—looking more like Buck Rogers science fiction than a workable defense system—provided the flight path of each bomb to antiaircraft batteries.

To assist the effort, personnel at the battalion CP telephoned division headquarters whenever a bomb flew over. Besides telephone reporting, Nelson Hall noted in the unit journal the exact time each V-1 buzzed over the area.

The bombs occasionally displayed erratic behavior. Some of them nosedived from their course and crashed in the 2nd Division sector.

Others fell before reaching American lines. Several reversed direction after passing over the division sector and sputtered back toward Germany. Still others wobbled and spun into aerobatics, causing suspenseful seconds for onlooking GIs.

The first near miss for the 741st occurred on November 24, when one of fifteen V-1s passing overhead careened toward earth and impacted in the Service Company area. The detonation left a giant crater—and one unsightly soldier.

Private Fred A. Torres lay asleep in a dugout when the bomb exploded. On a shelf above him, an uncorked ink bottle overturned and bathed his face and neck in a purple wash. For weeks to come, Torres scoured his face with every cleanser he could find, but he only managed to change the color to a yellowish green.

The day after the ink calamity, fourteen more V-1s flew over, one dropping from flight above Schönberg. One bomb struck just five hundred yards behind the Cafe Henkes, shattering windows and knocking down ceiling plaster. There were no injuries.

★★★

While V-1s droned over the 2nd Division sector, the three medium tank companies hurled cannon shells in the opposite direction. Two Shermans—and later three—from each company conducted fire missions to augment the 2nd Division's artillery battalions, which were contending with an ammunition shortage. The arrangement began on November 2 and continued through month's end. The crews rotated, and those off duty resided in dugouts. The tanks stayed in defilade, parked away from enemy view.

Truck drivers hauling shells confronted the greatest peril. Technician Fifth Grade Harlann J. Darlage drove a truck that schlepped ammunition to the Company B tanks, and he found himself under German observation during his first trip. While steering his vehicle, shell after shell slammed into the earth behind his tailgate. He narrowly evaded enemy artillerymen as they fiddled to adjust fire on his speeding truck. On future trips, Darlage used a less visible route.

Back in Schönberg, the 741st flamethrower school began classes on November 7. All tank commanders and bow gunners from Companies A and B attended a two-day course. Fifty pupils completed the program, including four men from Company C. Everyone thought that flamethrowers would see extensive use during the next thrust into Germany.

As they waited to attack, the first snow fell on November 9, and it quickly covered the countryside in deep drifts.

George Coleman, now a first lieutenant, noted the weather change in a letter. "It's getting a little chilly out," he wrote, "and the snow is just a trifle deep." He asked his mother to send wool socks, heavy mittens, and a woolen face mask.

His letter shifted to news from home, his mother having informed him that a friend named Chick Bennorth had burned his face and hands when enemy fire knocked out his tank. Another friend, Bob Connell, was missing, shot down while piloting a B-24 over Borneo.

"I'm sorry to hear about Chick. I hope he's not permanently disabled. Those burn jobs are usually pretty bad. Not trying to be discouraging."

The lieutenant continued: "Boy, this old world is really kicking hell out of the guys I used to hang around with. Don't mind me; I'm just a little bit down. One of those moods. I hope to God that Bob is O.K. Another one to sweat out."[190]

At battalion headquarters, news arrived that War Department officials had curtailed Officer Candidate Schools in the United States. This move promised to sharply reduce, or perhaps halt, the shipment of replacement officers to Europe in early 1945, but the replacement pool had already begun to evaporate. The shortage necessitated an increase in direct appointments to qualified warrant officers and enlisted men.

Battlefield appointments drew heavily on experienced combat leaders from the enlisted ranks. The army recognized that one such veteran was worth a half-dozen shavetail lieutenants from OCS. Rank without experience was often folly in combat.

The 741st had three battlefield appointments. Staff Sergeants George Habib and Turner Sheppard, from Companies A and B, respectively, received honorable discharges from the Army of the United States on November 12 and became second lieutenants the following day. Staff Sergeant Malcolm Reynolds, from Company C, became a lieutenant on November 29. He had rejoined the company in October after convalescing from injuries suffered on Hill 192. After his promotion, he transferred to 741st headquarters and became the battalion motor officer. Habib and Sheppard remained with their companies, each man becoming a platoon leader.

The battalion also had its black sheep. Joseph Feckete from Company B distinguished himself as the most incorrigible, a soldier who simply could not submit to discipline. After an AWOL conviction in Normandy, the former New York City dockworker spent most of September under arrest because of two more offenses in August. His antics brought him before a special court-martial in October. The court reduced him to private and confined him in the battalion stockade at Schönberg. He remained there until hospitalized due to illness in mid-November.

Private Francis F. Archer from Company C also resided in the stockade, after he pleaded guilty to disobeying an order from Captain Young. The court sentenced him to three months confinement at hard labor and forfeiture of twenty-eight dollars per month during that period. He had no prior convictions, and the court later remitted his sentence after he served nine days.

Private Thomas E. Young and Technician Fifth Grade James M. Sollars, both from Company C, faced AWOL charges. The court restricted each man to the company area for three months and busted Sollars to private. Private James T. Bumbalough, from Headquarters Company, sat in the stockade for nine days after an AWOL conviction. Private Roy C. Inman, from Service Company, became well-acquainted with the stockade after an identical conviction. While finishing his sentence at the 2nd Division stockade, he fell down some steps and suffered a lower back injury. He spent 166 days hospitalized, before receiving a medical discharge from the army.

Special courts-martial took place at 741st headquarters, and various battalion officers took turns serving as court members. The supply officer, Captain Albert O.G. Joven, usually acted as defense counsel. The Brooklyn native was an attorney.

General courts-martial involved more serious charges and also occurred at battalion headquarters. Technicians Fifth Grade David Neill and Robert H. Meriweather from Company D stood before a general court-martial in October, both men charged with being absent without leave and for violating the 64th Article of War, which involved willfully disobeying a superior officer. The court convicted the soldiers, reducing them to privates and confining them to the stockade until year's end. They ultimately finished their sentences in the 2nd Division stockade.

Throughout the war, numerous 741st soldiers faced AWOL charges, but only one man faced a desertion charge—a capital offense. Private Donald A. Kepplin, from Chicago, turned up missing at sunrise on October 6, 1944. The Able Company first sergeant searched the company area and tank park without finding the twenty-four-year-old bow gunner. He had wandered off toward Saint Vith, according to Captain Sledge, who saw the Chicagoan and another soldier walking in that direction. Both men had saluted the captain as he drove past. Kepplin traveled much farther than Saint Vith, absconding all the way to Paris, where he lived a clandestine life. After its liberation in August, the French capital became a mecca for Allied deserters, who formed gangs and thrived on profits from stolen goods, especially US Army gasoline. Life there resembled the Al Capone era in Kepplin's hometown.[191]

While the Chicagoan hid out, numerous battalion members received passes to recreation centers. These junkets started when the static battlefront permitted the army to relax its "all towns off limits" policy. Saint Vith and Verviers became favorite destinations for 741st tankers, as did the VIII Corps rest camp at Arlon. The 2nd Division camp at Vielsalm was another gathering spot. GIs there bunked in brick barracks and patronized a movie theater dubbed the "Paramount." There was also a

Red Cross club, an indoor auditorium for USO shows, and, most glorious of all, showers with hot water.

Back at the battalion, Thanksgiving Day brought traditional feasts. The cooks in each 741st company received frozen turkeys from the Quartermaster Corps. Unknown to the tankers, who wolfed down the birds, they contained bacteria left alive by undercooking. At battalion headquarters, a putrid trail of half-digested turkey and dressing stretched from the cafe to the latrine out back. That scene repeated itself in each company area. Tainted turkey became an enduring autumn memory for many men.

Lieutenant Coleman avoided the subject in his letters, not wishing to panic his mother. He instead brought up his war trophy: "The sword may not be sent for awhile yet. It is so big that it won't fit into the mailbags. I may have to save it and bring it home with me."

The platoon leader also thanked his parents for snapshots they had sent, especially a picture of his mother and another of his teenaged fiancée, Bettie Phelps. There was also a photo of his twenty-year-old sister, Jane, posing in a skirt and black pumps. He mentioned that one in a letter:

"Say, Jane, I showed that picture, the one taken at Lords Park, to a friend of mine, and he said he would sure like a letter from you. He is tall and blond and a hell of a nice guy. Comes from Mobile, Alabama, and is a graduate of The Citadel. His name and address is Captain Eddie Sledge, Co. A 74lst Tank Bn, APO 230. I wish you would slip him a line."

As usual, Coleman signed off with the words, "Love, Joe." The name was something of a handle, referring to his reputation as a "Regular Joe."[192]

The first of December found the medium tank companies still supporting the 2nd Division artillery. Each day, the Shermans continued to pump shells at enemy targets.

The commotion aloft faded as V-1 sightings decreased. The Germans shifted many of their mobile launch units north, in an attempt to circumvent American antiaircraft gunners. The radar vans departed as the antiaircraft defenses moved to counter the enemy change.

In Schönberg, the locals celebrated Saint Nicholas Eve. The "saint" appeared on December 5, clad in a white robe, bishop's miter, and holding a shepherd's staff. He and angel assistants paraded from house to house, bestowing treats and candy on the children, the first for many since the war began.

Also in December, the army inaugurated a home-furlough program. Before this time, emergency leave due to a family death was the only way home, other than serious injury.

The First Army filled a quota of 670 rest-and-relaxation furloughs in December. Priorities for eligibility were: (1) twice hospitalized for wounds; (2) twice decorated for valor; (3) six months in combat; (4) compassionate grounds with at least six months overseas.[193]

Two Company A men received furloughs. George Habib departed for his hometown, Fond du Lac, Wisconsin. Staff Sergeant Lloyd Ball traveled home to Monroeville, Indiana. Both men had Purple Hearts. Ball had also earned two Bronze Star Medals and a Silver Star Medal, while Habib held the Distinguished Service Cross. One Company D soldier also received a furlough. Frank A. Rosino returned to the Bronx. All three men had ninety-days leave, beginning December 8.

Ball, Habib, and Rosino made their exits none too soon. The day before their departure, Colonel Skaggs, his intelligence officer, operations officer, and supply officer attended a conference at 2nd Division headquarters in Saint Vith. They learned about a plan to seize two dams on the Roer River, targets requiring capture or destruction.

The Royal Air Force had attacked the dams four separate times in early December, but their bombs only made ugly dents. One option remained, the costliest option—ground assault.

For its part in the operation, the 2nd Division had to assemble at Camp Elsenborn, a former Belgian Army post. The plan set December 13 as the date when the division and an attached combat command from the 9th Armored Division would pass through the frontline and seize six objectives in rapid succession, the last being the Urfttalsperre Dam.

The second dam lay downstream, and the 78th Infantry Division had responsibility for its capture.

On December 9, the 741st marshaled in and around Schönberg.

Early the next afternoon, billeting parties from each company left for Ovifat and Robertville, two villages just a few minutes west from Camp Elsenborn.

Meanwhile, fresh-faced soldiers from the 106th Infantry Division arrived to take over the Schnee-Eifel. Clad in spotless field jackets and neckties, they looked thoroughly green to the 741st men.[194]

Snow flew as the tank crews departed for Ovifat and Robertville.

CHAPTER 25 **WAHLERSCHEID**

December 11. The iron clatter of tank tracks filled the air along snow-packed streets as the 741st slid into Ovifat and Robertville.

Men from the billeting parties darted back and forth, guiding the platoons here and there. Many tank crews boarded with civilians. The battalion had yet to settle in when Skaggs and Captain Sledge drove to the 9th Infantry forward CP, where they met the regimental commander, Colonel Chester Hirschfelder.

In preparation for the upcoming offensive, the colonel's unit had become a regimental combat team with the attachment of a tank company (that belonging to Sledge), a self-propelled TD company, and a company of 4.2-inch mortars. Additional strength came from a field artillery battalion, an engineer company, and a company of towed antitank guns. Officers from these units and the infantry battalion commanders gathered at the CP.

Hirschfelder possessed a hard, weathered face, like a great Indian chief, and was the most decorated soldier in the 2nd Division at that time. The colonel had three times received the Distinguished Service Cross. He also had four Silver Star Medals and two Purple Hearts. Since first donning a uniform in 1909, he had worn nearly every rank in the US Army between private and colonel, a fact well known to his subordinates. His roughneck reputation also drew attention. It brought him demotions while an enlisted man but favored him in combat. During the First World War, he became a second lieutenant on the battlefield and received his first Distinguished Service Cross. To earn that award, he single-handedly destroyed an enemy machine-gun nest and shot down two aircraft from the famous Richtofen Flying Circus. After becoming an officer, his demotions ended, though his reputation marched on, especially in the flow of libation at 9th Infantry banquets.

On December 11, while presiding over the officers assembled at his CP, Hirschfelder battled pleurisy, every breath radiating pain through his chest. He had refused advice from his regimental surgeon to seek a hospital bed. The colonel carried on and outlined the attack plan.

Two infantry battalions would advance abreast to secure the first objective, a road junction nestled within the Westwall and situated deep inside the Monschau Forest. The junction, named Wahlerscheid, was a customs station on the Belgian-German border. The axis of advance would be the single-lane road that ran north to the station. To preserve the element of surprise, there would be no preliminary bombardment by mortars or artillery. After capturing Wahlerscheid, the combat team would seize two more objectives and be prepared to advance on a third. Everything was set to kick off at 8:30 A.M. on December 13.

Hirschfelder gave instructions that all attacking units move to an assembly area northwest of Krinkelt-Rocherath, a pair of twin farming villages below the forest. Captain Sledge would start moving his company to the area at 2:00 P.M. on December 12.

Company A took to the frozen roads at the prescribed time and was in position by 6:00 P.M. The company stood ready with thirteen medium tanks, a tankdozer, and a 105 mm assault gun.

About the time Able Company rolled into position, Captains Thomas and Sterrett, along with five enlisted men, established a forward CP for the 741st. They chose the home of Mathias Hönen, Nr. 100 Rocherath. His home was a combination house and barn, and it stood near the stone-built church, which dominated the twin villages.[195]

Hönen was one of several village residents who remained in Rocherath after it became a frontline village in September. That month, he lost his only daughter to a land mine set by an infiltrating German patrol. Two other locals died under similar circumstances before the population migrated west to Malmédy, far from the frontline. Hönen and a few others stayed behind to tend dairy cattle. He had one son in the Luftwaffe and two others in the German Army, but he opened his home to Skaggs and his men.

Nelson Hall was among the five enlisted men who helped establish the forward CP, and he later described the place:

"We chose the kitchen for our work area, a modest-sized room that contained a wood stove, a table, and a couple of chairs. Our field desk was placed on the table, with its green-covered face opened to accept the typewriter we always carried. On the wall adjacent to the desk, we mounted our operations map, which showed the position of our battalion and its elements. It also displayed regimental and divisional positions, so that we had firm knowledge of who was where and what was going on.

"There were a couple of bedrooms adjacent to the kitchen, which were used as sleeping quarters during our stay. Surrounding the kitchen was a large, screened porch, which must have been a fine place to be during the summer months but was now cold and drafty in December.

"Parked behind the building was our halftrack, where our radio operators communicated with our company commanders and monitored the chatter that went on between the tank crews."[196]

December 13. George Coleman dashed off a pencil-written note to his family:

"Well, today is my birthday. It doesn't seem like any other day except for past memories. I am starting on my third year now. Think I will take a little drink tonight just for the hell of it, just to keep up the old courage."[197]

On his twenty-second birthday, the big attack opened as planned, but obstacles on the road to Wahlerscheid thwarted armor. The Germans had downed trees over the road, interlacing their trunks and rigging them with mines and booby traps. Shell craters added to the impasse. Only Sledge's tankdozer participated in the attack, helping 2nd Division engineers clear obstacles. The Company B dozer joined the effort later that morning.

Ahead of the two dozers, infantrymen struggled through the woodland on either side of the road. Knee-deep snow and intertwined branches slowed the soldiers. Daytime temperatures edged above freezing, which sent globs of slush dropping from spruce boughs. The melted snow soaked Hirschfelder's men from field jacket to skin. The water also penetrated boot leather.

At noon, the soggy vanguard reached the Westwall fortifications. Through the fog, scouts spied wisps of smoke drifting skyward from stoves burning inside enemy pillboxes. The Germans seemed unaware of the American presence. Word quietly passed among the leading rifle companies: drop packs and overcoats and prepare to attack.

In front of their fortifications, the enemy had chopped away all the trees and filled the open space with a cobweb of barbed-wire entanglements. These included a profusion of antipersonnel mines attached to an invisible network of trip wires.

When the leading companies reached the scene, enemy machine guns rattled, and rifles cracked. Hope for surprise died as bullets ripped over the entanglements. In the first shooting frenzy, one platoon on the right charged through five aprons of wire before being pinned to the snow. Efforts to circumvent the entanglements ended with the jolting pop of mines. The snowscape became a tableau of bleeding casualties.

Somehow, a rifle squad found its way across the cleared area, slithering beneath two hundred yards of wire. The soldiers reached the trench network connecting the enemy pillboxes. Part of another squad followed, snipping a narrow gap with wire cutters. The two groups spent five hours fending off furious attempts to kick them back before finally relinquishing their gain. News of their breach was slow to ascend the chain of command.

The early darkness of wintertime fell across the battleground by midafternoon, and freezing temperatures came with it. Boots, gloves, and uniforms stiffened and froze. Hirschfelder ordered his troops to break off the attack, tuck in for the night, and make ready to renew the assault in the morning.

German shells rained down all night. The men huddled beneath icy shelter-halves, clutched frosty rifles, and waited for dawn.

★ ★ ★

Before sunrise on December 14, four tanks and an assault gun from Company A prepared to join the infantry companies, but enemy obstacles on the road continued to prevent passage. German shells and the short winter days made clearing the obstacles a daunting task for the engineers despite tankdozer help. The forest was impassable for armor on either side of the road. There was nothing to do but wait for the engineers to finish. The infantry would have no tank support.

Hirschfelder's combat team stormed the German defenses head-on and also attempted to flank the position. The enemy reaction was again violent. Treetops flashed with mortar and artillery explosions. Machine-gun fire beat down the attackers.

Throughout the day, the defenders claimed more and more American casualties, some groaning and crying, some swearing, others silent.

The thunderous retort of American artillery diminished the enemy's ability to inflict losses but did nothing to push forward the attack. In many cases, the artillery failed to register on targets; impossible terrain and inadequate observation were to blame. The troops needed direct fire from tank guns.

In the failing light of December 14, freezing temperatures overtook the forest. Dazed and half-frozen after their failure to break the enemy's hold on Wahlerscheid, the soldiers of the 9th Infantry could do nothing but hunker under blankets and shelter-halves for another night of cold and terror.

After dawn on December 15, the first pallid traces of light invaded the forest, but a dense curtain of fog limited the sunlight to a murky glow. Men strained to see a few feet from their holes. American fighter-bombers zoomed overhead, but their pilots saw only a blanket of impenetrable whiteness. With nothing visible to bomb or strafe, the planes roared away. If the fog cleared, they would return, but as morning turned to afternoon, there was no change. Hirschfelder held back his infantry, opting to forgo a direct attack until after an air strike.

No tanks from Company A rolled forward during the delay, their pathway still blocked. The company's T2 recovery vehicle did, however, work its way ahead and helped remove road obstacles with its crane and winch.

At 741st headquarters in the twin villages, word had come from division headquarters that the corps commander wanted a tank platoon made available to the 395th Infantry Regiment of the 99th Division. Troops of the 395th had launched an assault against Westwall fortifications in the rugged woodland to the right of Wahlerscheid. Like the 9th Infantry, the 395th had encountered enemy pillboxes and entrenchments. If tanks could somehow negotiate the severe topography, their cannons would lend a mighty hand to the infantry.

Accordingly, Captain Young and Lieutenant Covington from Company C reported to 99th headquarters. The division operations officer gave a synopsis of the 395th situation and used a map to pinpoint the spot where the tanks should rendezvous in the forest. The two tank officers then left to make contact with the 395th. Soon afterward, Covington's platoon moved forward in the failing light of midafternoon.

Back at Wahlerscheid, Hirschfelder's troops were on the march after sundown, unlike the two previous nights. The lieutenant colonel commanding the Second Battalion saw an unexpected opportunity in the darkness. During the initial surge of the attack on December 13, soldiers had cut a gap all the way through the wire, but the battalion commander had just received news of it. His awareness of that breach changed things. It was time for another try.

The Second Battalion dispatched a patrol to locate the gap and cross to the German side. Once across, they reported the situation. No enemy activity. The defenders were snoozing.

One rifle company after another, the Second Battalion crept through the gap and infiltrated the enemy's network of trenches and pillboxes. At one point, the Germans detected movement and blindly fired into the darkness but to no effect. Soon, the battalion commander himself was through the gap and atop a pillbox. On the heels of his battalion, the Third Battalion infiltrated through the opening. Caught unawares, enemy soldiers quickly surrendered or succumbed to hand grenades and bunker-busting munitions called "beehive" charges (so named because of their hive-like shape). By midnight, a string of fortifications near the breakthrough point belonged to the 9th Infantry.

The Third Battalion now had orders to seize the Wahlerscheid road junction several hundred yards ahead.

During the early hours of December 16, the 9th Infantry commenced an assault to stomp out all opposition at Wahlerscheid. The GIs killed, injured, and captured scores of the enemy. The fighting ran bunker to bunker, with several surges needed to vanquish the defenders.

By 11:00 A.M., the 2nd Division controlled Wahlerscheid. The 38th Infantry now moved to exploit the penetration. Its troops, fresh to the fight, would pass through the 9th.

Two tank platoons from Able Company joined the Second Battalion, 38th on the road to Wahlerscheid. Ahead, the Company A tankdozer labored with 2nd Division engineers. The dozer rumbled along the roadway, bashing aside the last German obstacles, clearing the way for the 38th and the pair of tank platoons.

The push for the Urfttalsperre Dam had at last gained momentum, but after all the misery and death, there would be no further progress. The men of the 741st and 2nd Division would soon face an enemy offensive so immense it would dwarf all other battles in US Army history.

The American frontline on the German border with Belgium and Luxembourg exploded with artillery in the predawn hours of December 16.

Hitler's legions in the West were on the march.

CHAPTER 26 TURNING THE TABLES

At the 741st CP in the twin villages, Nelson Hall at first discounted the enemy's artillery.

"Our first hint the Germans were suddenly active came when a jeep driver arrived from the rear area to report that artillery and rocket fire had been coming in near the division CP. We had caught these sounds but guessed they were in another sector. Since we were on the attack, we had every reason to believe we were not about to be targets."

He and others at battalion headquarters had seen intelligence reports that mentioned an offensive in the works. German prisoners had related such stories, but everyone discounted them. As Hall explained, "We had seen the destruction of the enemy in France, and their weakness along the line seemed to assure us these stories were untrue."[198]

The fallacy of that belief and the magnitude of events had yet to unfold in most American minds, especially at Wahlerscheid, where the enemy offered only minor counterattacks aimed at regaining their lost bunkers.

In their grand scheme, the Germans had chosen to bypass the Monschau Forest and its vast wooded acreage, instead moving to the north and south. Ignored by the Germans, the 2nd Division consolidated its gains at Wahlerscheid and prepared to continue toward the Urfttalsperre Dam. The division called for more tanks.

In Ovifat, Company B responded. Its commander instructed his men to rally north of Rocherath in a patch of woods. The tanks rumbled out of town in midafternoon, spending more than an hour winding along country roads before reaching their destination.

At the rallying point, Private First Class George E. Beeley of Providence, Rhode Island, shivered in the bow gunner's seat of Lieutenant Dudley's tank.

"We sat in our tank, frozen stiff, hoping to get the word that we were having a dry run. The metal of a tank holds in the cold and steals the heat from your body. I had never been so miserable in my life. There was no way to get warm. My teeth chattered so badly my whole body shook. When I urinated in my ammunition box, it turned to yellow ice."

Having hastily departed Ovifat, none of Dudley's crewmen grabbed rations. By afternoon, cinching hunger pains multiplied their anguish, but the driver, Technician Fourth Grade Ivan E. Schmidt from Dieterich, Illinois, saved the hour according to Beeley.

"Schmidty brought out a fruitcake he had been saving to celebrate the end of the war. He felt that we were all so hungry it was a good time to bring out the cake. I have never enjoyed anything as much as I did that crumbled fruitcake."[199]

★★★

While the 2nd Division focused on the dam, the corps commander pondered the meaning of German activity along the 99th Division front. He and the commanding general of the 99th presumed the activity was a response to Wahlerscheid. They had anticipated an enemy reaction, but if a German penetration occurred, the 99th had only one rifle company in reserve. The corps commander instructed his deputy chief of staff to help assemble reinforcements. General Robertson received a telephone call.

Twenty minutes after 12:00 P.M., Major Daniel Webster, operations officer of the 2nd Division, telephoned 741st headquarters. He told Captain Thomas, his counterpart in the 741st, that Company C—minus Covington's platoon already on loan to the 395th Infantry—should prepare for deployment with the 23rd Infantry as an additional reserve for the 99th.

Webster soon called back and notified Skaggs that Charley Company would be attached to the 23rd Infantry at 1:00 P.M. He then asked Skaggs to "please have the company commander report to 23rd headquarters right away."[200]

At the 23rd CP, the regimental commander spelled out the plan. Sometime in the afternoon or early evening, Company C would follow a truck convoy carrying the Third Battalion to a reserve position in the Krinkelter Forest northeast of Rocherath. The tankers would bivouac nearby in the Rocherather Forest.

At 3:10 P.M., Company C moved out from Robertville and reached its bivouac area after dark.

Stuarts from Company D and a company of towed antitank guns from the 612th TD Battalion also received orders to join the 23rd. Just after 8:10 P.M., the Company D commander departed for 23rd headquarters to obtain instructions. His tanks reached Camp Elsenborn around midnight and joined the tank destroyer company.

With each passing hour, the decision to reinforce the 99th looked more and more prudent. News arrived that the 28th Division was under assault in Luxembourg. Sketchy reports abounded of attacks on the 106th Division. All three 99th Division regiments were in hard combat with German infantry.

General Robertson and the corps commander grew worried by day's end. They requested that the 2nd Division halt at Wahlerscheid and withdraw to defensive positions behind the 99th.

At the Company B rallying point, Sergeant Ray Dickson recognized that events had turned for the worse. "We were gassing up our tanks when all hell broke loose. Buzz bombs came over every few minutes, sometimes two at a time, and artillery hit all around us."[201]

The fireworks continued all night.

December 17, early A.M.: G-2 maps at command posts throughout First Army precincts charted the progress of German thrusts against the 28th,

99th, and 106th Divisions. To even the most nearsighted intelligence analyst, it was plain that the Germans had offered more than an answer to the Roer dams attack. Even so, at First Army headquarters—located in the comfy confines of the Hôtel Britannique in Spa—there was no indication that enemy troops had pierced 99th Division lines and, thus, no reason to alter the stance of the 2nd Division at Wahlerscheid.

5:30 A.M.: Word rippled through the communications network that German tanks and infantry had penetrated the 99th front and turned the division's right flank. The attackers blitzed through Honsfeld and turned toward Büllingen. The latter town lay a mile from the 741st CP in the twin villages and the same distance from General Robertson's forward CP in Wirtzfeld.

Little stood between the two CPs and the enemy. Expecting corps headquarters to cancel the Roer dams attack, Robertson phoned the 9th Infantry and ordered the regiment's antitank company to Wirtzfeld. He also called for help from the 644th TD Battalion and the single battalion of the 23rd Infantry still under his command. The infantry battalion was to divide its strength between Wirtzfeld and Krinkelt-Rocherath, two places that needed to remain under American control to prevent the enemy from pinching off the 2nd and 99th from behind.

German tanks and infantry now occupied Büllingen, a major supply dump for the 2nd Division. Kampfgruppe Peiper from the 1.SS-Panzer-Division had captured it.

8:00 A.M.: M10 TDs from the 644th had just settled in the Wirtzfeld area when Peiper's armored vanguard crawled into view on a ridge just south of town. Artillerymen from the 2nd Division took a crack at the approaching threat, but their shells dropped short. With the Germans in full view, two TDs unleashed AP rounds. The projectiles hit four Panzers and a halftrack, which sent the Germans lurching backward toward Büllingen. Riflemen from the 23rd Infantry, who arrived during the fight, ascended the ridge and found destroyed tanks wreathed in smoke and flame. The Americans collected ten SS prisoners, including four tankers.

★ ★ ★

News of the SS breakthrough reached the 741st CP in Rocherath while the Wirtzfeld clash was underway: "0819 Enemy has thirty tanks and one infantry battalion in vicinity Büllingen."[202]

Within minutes, Companies A and B received urgent orders to protect the twin villages. Headquarters personnel lent two machine-gun squads to the defense, one armed with a caliber .30 light machine gun and the other with a caliber .50 weapon. Twenty-five tanks assembled to guard the villages, nine from Company A and sixteen from Company B.

The Baker Company contingent included Sergeant Dickson's tank.

"The company commander came down, giving us news that twenty-five or thirty Kraut tanks had broken through the 99th Division lines. We moved out at once to meet the attack. We moved into the village of Rocherath, about a half-mile away, and drew what we thought was an easy assignment—the Second Platoon was to remain in company reserve."[203]

With the Second Platoon in reserve, the First and Third Platoons motored to the edge of Krinkelt and watched the road from Büllingen. One platoon from Company A joined them, while another Company A platoon drove to Krinkelt but guarded the road from Mürringen.

First Lieutenant Roger McDonough led the Able Company platoon observing the Büllingen road. His tanks carried troops from the 23rd Infantry, along with their battalion operations officer, Captain Herbert C. Byrd.

While moving into position, the tankers and infantrymen spied the SS vanguard previously engaged by the 644th TDs. McDonough's crew opened fire on a halftrack and claimed its destruction.[204]

This was likely the same machine credited to the 644th. Such was the way with armored warfare where enemy targets often fell in the sights of multiple gunners. Confused and inflated claims resulted, but such errors never lost a war. On the battlefield, nobody bickered over who shot what. Outlasting the enemy was all that mattered.

The tankers of Companies A and B kept a continuous watch, but their enemies disappeared, having hustled back to Büllingen. Kampfgruppe Peiper departed the town and pressed westward, away from the 2nd and 99th Divisions.

Company D tank crews stationed at Domäne Bütgenbach witnessed the German exodus from Büllingen. The tankers observed the scene along with antitank gunners from a TD platoon as well as combat engineers.[205] The Americans watched enemy armored vehicles stream out of Büllingen onto a secondary road that angled away from Domäne Bütgenbach. German artillerymen spotted the US troops and pounded them. The picket line fell back to safer ground under cover fire from Company D.

Two hours later (3:00 P.M.), soldiers from the 1st Infantry Division began arriving and relieved the engineers. Dog Company stayed put along with the antitank gunners.

CHAPTER 27 MILLER VERSUS MÜLLER

As Kampfgruppe Peiper turned away, greater peril mounted in the forest east of the twin villages. One 99th Division regiment had slugged it out there with enemy infantry for more than twenty-four hours, losing several acres of woodland to the Germans. Aware that his troops needed backup, the 99th commanding general took steps to reinforce them. He controlled two battalions from the 23rd Infantry, and he ordered one to the Krinkelter Forest.

When the Third Battalion, 23rd reached the forest late on December 16, its men unshouldered weapons and dropped equipment. Would they attack or defend?

After entering the forest, the Third Battalion commander, Lieutenant Colonel Paul V. Tuttle, summoned his company commanders and apprised them of the situation. German forces had partially surrounded a battalion of the 393rd Infantry, and the Third Battalion would counterattack in the morning and help drive back the enemy. The company commanders departed to brief their subordinates but soon received an update from Tuttle. German strength was too great for one battalion to launch a successful counterattack. "Hold at all costs" was the new order.[206] Tuttle also said that tank support would arrive at 7:00 A.M. the next day.

GIs began the well-practiced digging ritual throughout the Third Battalion area. After scraping out hasty positions, the men spent the night drifting in and out of whatever sleep was possible in the miserable cold. All the while, artillery rounds whistled over the treetops.

Bivouacked in the Rocherather Forest, Company C tankmen also tried to salvage a few hours sleep from the frigid night. Wrapped in blankets and balled up in sleeping bags, they shivered while others plotted their fate.

Lieutenant Dew's platoon expected the Third Battalion assignment, but his tank had a faulty generator, and Sergeant Neidrich's tank had mechanical difficulties, too. The job passed to Lieutenant Miller's platoon. Dew's men would remain on standby at the bivouac area. (Covington's platoon was still with the 395th Infantry.)

Miller learned about his assignment during a powwow with his company commander. Miller and his men then broke camp and headed out with the lieutenant in the lead tank. They motored through the morning darkness and watched muddy, bleary-eyed troops retreating in the opposite direction. The sight raised eyebrows among the tankers.

The lieutenant was a natural leader and inspired confidence despite the situation.

Victor Leroy Miller Jr. grew up as an only child in Tulsa, Oklahoma. His father was a dentist, and his maternal step-grandfather, who lived with the family, was an oil driller and promoter. Victor played football for Holy Family High School, where his peers voted him senior class president. He also played softball for his grandfather's oil company. The talented athlete graduated in 1938 and matriculated to Oklahoma A&M College. He played baseball there, joined the ROTC program, and became president of the school's Theta Kappa Phi Honor Society. He also married his wife, Dorothy, while at A&M. The call to service came in April 1943, and he departed for Fort Knox where he received his commission. He transferred to the 16th Armored Division at Camp Chaffee, Arkansas. The need for replacements in Europe necessitated his departure from the 16th. He arrived overseas in June 1944 and joined the 741st on August 14 along with Joe Dew. Both men became first lieutenants the next month.

Miller now faced his stiffest challenge. When his tanks reached the Third Battalion's sector, he received deployment instructions and subsequently split his platoon.

He sent two tanks inside the forest to establish a blocking position. Sergeant Ernest O. Padgett and Staff Sergeant Norman K. Crisler commanded these machines, which entered the woods at a road fork called Ruppenvenn and turned toward the Third Battalion company led by Captain Charles B. MacDonald. The road—an unimproved logging trail—passed through MacDonald's position and tracked farther east to the besieged 393rd battalion.

Catching the sound of American tank engines and squeaking suspension wheels, MacDonald phoned one of his lieutenants and asked him to guide their armored friends into a defilade position. Crisler and Padgett could provide cannon support from there.

Outside the forest, Miller situated his three remaining tanks well back from Ruppenvenn, which was the portal for vehicles entering or leaving the Krinkelter Forest. The three tanks deployed to cover this gateway. Sergeants John W. Welch and Roy G. Vernon commanded the pair of Shermans with Miller. No German vehicle could exit the forest without coming to terms with the threesome, which stood ready on high ground near a remote and lonely inn named Gasthaus zur Waldeslust.[207]

10:30 A.M.: Crisler and Padgett spotted a jeep from the 393rd battalion. Worn and wasted foot soldiers followed the vehicle. One by one and in bunches, withdrawing GIs infiltrated through the trees. A platoon-sized formation hiked through MacDonald's position. A couple of men stayed to fight with the captain's company, while the others pushed toward the rear and uttered words of warning. "Kraut tanks! Tigers!"[208]

Since being placed there, Crisler and Padgett had disliked their position. Their thoughts raced with the news of Panzers on the loose. Would their two machines have a prayer against enemy armor? Was there a more advantageous location? They needed to engage the German tanks with flanking shots fired at close range. Their present disposition would force them to tackle the tanks head-on. Firing shells at the frontal armor of a Tiger or Panther from any distance greater than point-blank was like trying to blast through granite with a water pistol.

Crisler and Padgett relocated. They drove back toward Ruppenvenn and parked at the other end of MacDonald's company. Here they found terrain and trees that afforded them a chance to meet the enemy at close range and from the side. The tank commanders executed this shift without consulting or advising MacDonald. He radioed Tuttle for an explanation. Where the hell were the tanks going? Nobody had given a withdrawal order. The colonel checked with Crisler and Padgett, who now sat within walking distance of his CP. Tuttle informed MacDonald that the tankers had moved to "improve their positions."[209] The captain reluctantly accepted this explanation. The tank commanders knew their job best, and there was no option other than placing trust in their judgment and experience.

As Crisler and Padgett waited, the forest began to ring with the crack of German rifles and the rattle of machine guns. The first wave of attackers lurched toward MacDonald's men, their bullets buzzing wildly between the tree trunks.

The approaching troops wore Feldgrau, or field gray, overcoats and garments with camouflage patterns distinctive of the Waffen-SS. This was Kampfgruppe Müller from the 12.SS-Panzer-Division „Hitlerjugend."

SS-Sturmbannführer Siegfried Müller was in charge of the enemy force. This was his debut as the commander of a battle group, having previously led engineer units.[210] The only armored vehicles organic to his group were two tank destroyer companies fielding ten Jagdpanzer IVs each, but other armor had unexpectedly arrived at Müller's disposal. During the early hours of December 17, he gained a tank battalion from Kampfgruppe Kuhlmann, which sat behind the front, unable to advance along its assigned attack route because of stiff resistance by the 99th Division. The same opposition had also blocked Müller from his original path until his superiors permitted him to use the northernmost route in the Hitlerjugend sector. This was the only route where US defenses had faltered enough to allow access.

Müller's armor and grenadiers joined the fight, and, by 11:00 A.M. on December 17, they had overwhelmed the 393rd battalion that stood in their path. German intelligence showed no American reserves beyond

Summer 1943: Donald G. Adkins (right) of Wartburg, Tennessee, and Douglas L. Arthur of Brighton, Tennessee. In 1943, both men joined the 610th Engineer Light Equipment Company, a subordinate unit of the II Armored Corps. On D-Day, Adkins commanded Tankdozer "10," which supported Gap Assault Team 10.

Gaetano R. Barcellona, Company A,
San Antonio, Texas

Lloyd C. Ball, Company A, Monroeville, Indiana
Receives BSM from Major General Walter M. Robertson

February 9, 1945: T/4 John H. Barner, Company A, receives oakleaf cluster to his Silver Star Medal from General Robertson. Barner was the only member of the 741st twice awarded the SSM. His hometown was Albany, New York. Photo by Leander P. Zwick Jr., 165th SPC.

John E. Barry, USS LCT 549 (Div. 71, "DD" Flotilla 12)
Jackson Heights, New York

Jack B. Boardman, Company B,
Downers Grove, Illinois

ABOVE
Jerry B. Brown,
Companies A and C,
Euclid, Ohio

LEFT
Jack H. Browder,
Battalion Headquarters,
Duncan, Oklahoma

RIGHT
Charles R. Buchanan,
Company A,
Jacksonville, Illinois
Sole survivor of Fair's crew

ABOVE
Thomas D. Chastain, Company A
Highlands, North Carolina

LEFT
Robert C. Buckholtz, Company C
Worthington, Ohio

George R. Coleman, Companies B and D
Elgin, Illinois — Killed December 18, 1944

John H. Covington Jr., Company C
Santa Ana, California
Receives SSM from General Robertson

George K. Cuthbert Jr., Company C
Jackson Heights, New York — Killed April 18, 1945

Joseph H. Dew, Company C
Redfield, Iowa

Thomas R. Fair, Company A
Petrolia, Pennsylvania — Killed July 11, 1944

Valentine Fister Jr., Company A
Akron, Ohio

Philip L. Fitts (left), Danville, Virginia
Malcolm A. Reynolds, Principio Furnace, Maryland
Both men served with Company C
Reynolds, battlefield commission November 29, 1944

Walter J. Fryer (left), Coburn, Pennsylvania
Frederick A. Neidrich, Philipsburg, Pennsylvania
Both men served with Company C
Fryer killed April 13, 1945
Neidrich, battlefield commission May 12, 1945

Ralph R. Garber (left) and Roy A. Garber, Company B
Hyde Park, Massachusetts
Roy killed December 18, 1944

Donald W. Hecox, Company A
Los Angeles, California
Battlefield commission May 5, 1945
Photo made at Starý Klíčov, Czechoslovakia

Florus W. Laird, Company B
Edinburg, Texas
Killed June 6, 1944
Buried at sea, June 8, 1944

Romolo P. Legnini, Headquarters Company
Upper Darby, Pennsylvania

Virgil H. Lohman, Companies A and B
Friedheim, Missouri

Edward R. Lucas, Medical Detachment
Brownstown, Indiana
Marriage to Anna Kadlecova, August 22, 1945

Roger J. McDonough, Company A
Long Island City, New York
Killed April 29, 1945

Victor L. Miller Jr., Company C
Tulsa, Oklahoma
Killed December 17, 1944

Jacob R. Moon, Battalion Headquarters
Goodwater, Alabama
United States Military Academy, Class of 1924

Tom C. Morris,
Regimental Headquarters, 38th Infantry Regiment
Waxahachie, Texas

Patrick J. O'Shaughnessy, Company B
Richmond Hill, New York

Walter M. "Robby" Robertson,
Division Headquarters, 2nd Infantry Division
Arrington (Nelson County), Virginia

RIGHT
Ernest O. Padgett, Company C
Johnston, South Carolina

BELOW
Joseph T. Rutyna,
Company C and Headquarters Company
Detroit, Michigan

BELOW RIGHT
Robert N. Skaggs, Battalion Headquarters
Los Angeles, California
Portrait made in April 1953, Carlisle Barracks, PA

Edward S. Sledge II, Companies A and D
Mobile, Alabama

Richard Smith, Company C
Valley Park, Missouri
Sole survivor of Victor Miller's crew

Bolick "Bill" Smulik, Companies A and C
a.k.a. Bolesco Smulik
New Mine, Pennsylvania
Killed August 1, 1944

Alfred W. Steineker Jr., Company A
Louisville, Kentucky

Patrick Henry Sullivan, USS LCT 600
Torrance, California
Postwar portrait

Lawrence G. Sweeney, Company A
Pittsburgh, Pennsylvania

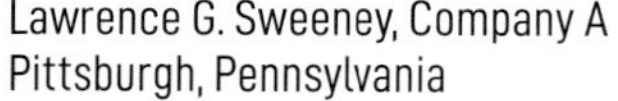

Quentin J. Swan, Company C
Roundup, Montana

Cecil D. Thomas,
Company A and Battalion Headquarters
Reidsville, North Carolina

James G. Thornton Jr., Company B
Wilmington, North Carolina
Mortally wounded September 14, 1944
Died in German captivity; body never recovered or positively identified

Gerald E. Wiltrout, Company D
Connellsville, Pennsylvania
Only D/741 soldier to receive Silver Star Medal.

Leonard H. Trimpe, Service Company
Seymour, Indiana
Receives SSM from General Robertson, July 13, 1944
Battlefield commission April 17, 1945

Charles R. Young, Company C
Hackensack, New Jersey
Died of wounds February 3, 1945

Kurt Brödel, 3.Kompanie, SS-Panzer-Regiment 12
Annweiler, Germany
Killed December 18, 1944
Pictured as Army Panzerjäger lieutenant

Arnold Jürgensen,
I.Abteilung, SS-Panzer-Regiment 12
Pahlen/Tellingstedt, Germany
Died of wounds December 23, 1944

BELOW
Siegfried Müller,
SS-Panzer-Grenadier-Regiment 25
Krefeld, Germany

ABOVE
Helmut Zeiner,
1.Kompanie, SS-Panzerjäger-Abteilung 12
Linz, Austria; daughter Gerlinde born March 25, 1944

April 1942: Panoramic photo of Service Company at Fort George G. Meade, Maryland. (Author's Collection)

LEFT
October 3, 1942: Private Victor Valentin, a cook with Headquarters Company, pours water into a bag prior to departing for maneuvers at the Desert Training Center, Camp Young, California. The rations piled next to Valentin include fig bars, a staple in the desert diet of tank crews. Photo by Louis J. Murchio Jr., 161st Signal Photo Company. (NARA)

RIGHT
February 12, 1943: Sergeant Joseph M. Brost of Talbot, Indiana, commands an M4A1 from Company A during Third Army maneuvers at Camp Polk, Louisiana. Brost eventually transferred to Company D and had the sad distinction of being the last member of the 741st to die overseas, just three days before the war ended. Photo by Robert M. Wallace Jr., 165th Signal Photo Company. (NARA)

LEFT
February 11, 1943: First Lieutenant Jimmie Thornton grips the business end of a caliber .50 machine gun utilized in an antiaircraft role. Ten days afterward, Thornton assumed command of Company B, taking over from First Lieutenant George A. Yant who transferred to Fort Benning. Thornton remained in command until his death in September 1944. Photo by Robert M. Wallace Jr., 165th Signal Photo Company. (NARA)

BELOW
June 6, 1944: USS LCT(6) 625 carried two M4(75) tanks and nine men from Headquarters Company, 741st Tank Battalion. The vessel also transported eight jeeps from Headquarters Company, 16th Infantry Regiment, and one D-8 tractor from Company B, 348th Engineer Combat Battalion. In total, sixty-nine soldiers rode aboard the vessel, which sailed from Portland for an Easy Red landing at H+130. Photograph by William G. Willcox II, cameraman aboard USS PC 617. (NARA)

ABOVE
M4A1(75) DD from the 743rd Tank Battalion somewhere in England before D-Day. Companies B and C of this battalion operated DD tanks. No pictures are known of 741st DDs prior to the invasion. (DeMarco Collection)

RIGHT
June 7, 1944: Sergeant Robert D. Coaker's M4A1(75) named "Adeline II" photographed near the church at Colleville-sur-Mer, France. This tank from Able Company subsequently underwent repair work to replace a bogie wheel broken off by an antitank projectile. The Pressed Steel Car Company built the tank in Chicago. Service Company operated the T2 tank recovery vehicle. (NARA)

LEFT
September 12, 1944: Lieutenant Colonel Robert N. Skaggs and Staff Sergeant George A. Habib receive the Distinguished Service Cross from Major General Leonard T. Gerow at V Corps HQ in Bastogne, Belgium. Both men earned the DSC for their actions on D-Day. Photo made by Jack G. Kitzerow, 165th Signal Photo Company. (NARA)

LEFT
Service Company T2 on Omaha Beach. The man posing alongside was a member of the US Navy. (Christopher Nesbit, Archive of Modern Conflict, London, England)

RIGHT
Roger McDonough's M4A1(75), "Ace of Spades," registration number USA 3036894, produced by the Pressed Steel Car Company of Pittsburgh. The vehicle burned on Omaha beach after being struck by an antitank projectile. (DeMarco Collection and Archive of Modern Conflict)

Bronze propeller on an M4A1(75) DD tank recovered by Jacques Lemonchois in 1980 and displayed at the Musée des Épaves sous-marines du Débarquement, Port-en-Bessin, France. (Photo by author)

RIGHT AND BELOW
M4A1(75) DD tank recovered by Jacques Lemonchois in 1984 and displayed at the Musée des Épaves sous-marines du Débarquement. (Photos by author)

two routes and
one lead tank (A-5)
not shown
la Carosserie
(Ozouf farm)
A-4
A-15
A-14
A-10
A-9
Departure Line
1
2
3
4
D95
C/38 – A/38
German MLR
1 = Area where Hecox struck mine
2 = Hecox's M4 (A-9), USA 3039446
wreckage pushed off road
3 = Panzerspähwagen AB41 201(i)
1./Fallsch.Aufkl.Abt.12
4 = Fair's M4 (A-10), USA 3039496
5 = German Observation Tower
192
5
Dodge
Woods
Ford Woods
(Bois du Soulaire)
Planned Routes
Company A – July 11, 1944
meters
0
200
0
200
yards
Photo by U.S. 30th Photo Recon Sq, July 18, 1944

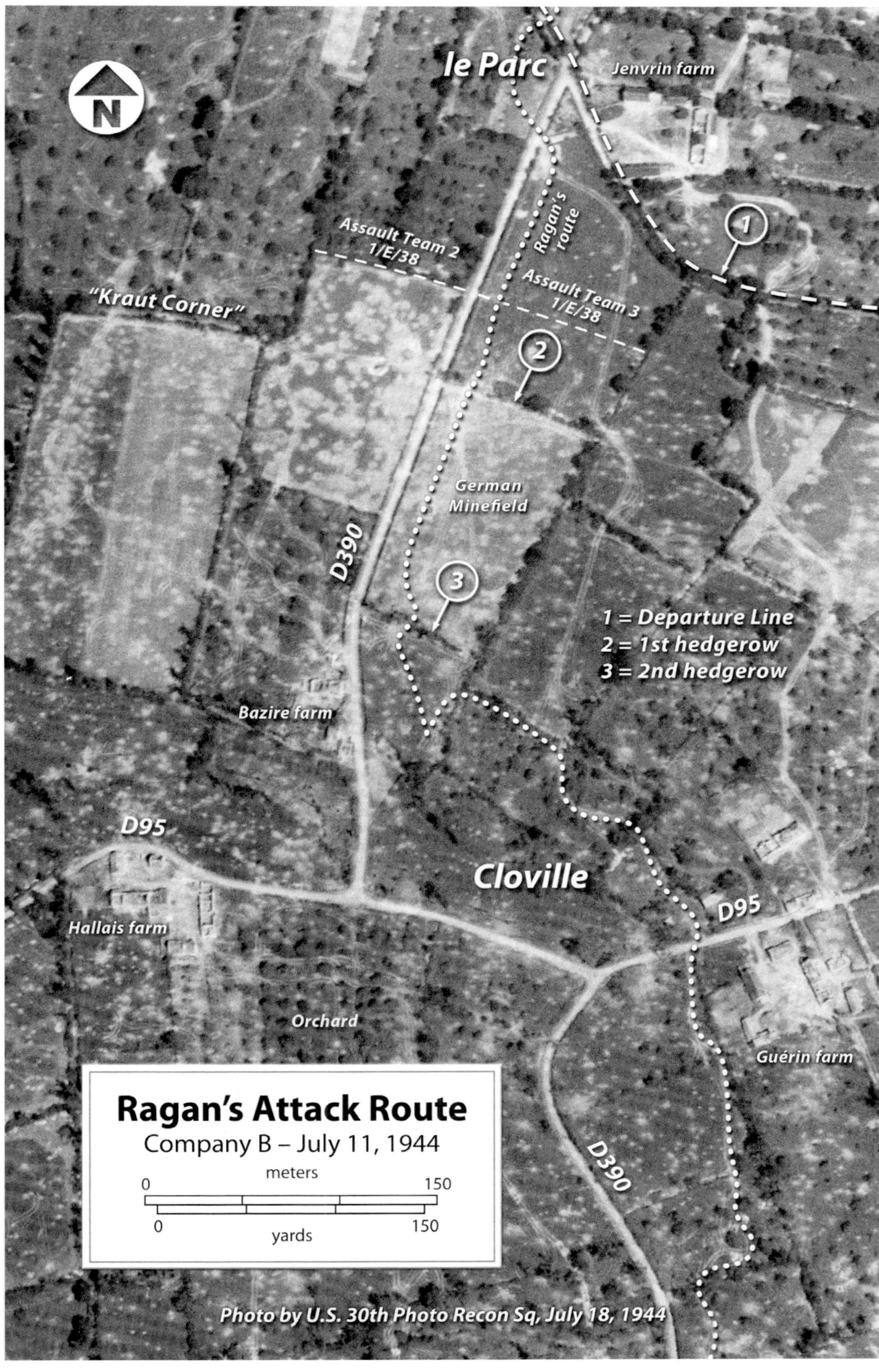

N
le Parc
Jenvrin farm
Ragan's route
1
Assault Team 2
1/E/38
Assault Team 3
1/E/38
"Kraut Corner"
2
German
Minefield
D390
3
1 = Departure Line
2 = 1st hedgerow
3 = 2nd hedgerow
Bazire farm
D95
Cloville
D95
Hallais farm
Orchard
Guérin farm
D390
Ragan's Attack Route
Company B – July 11, 1944
meters
0
150
0
150
yards
Photo by U.S. 30th Photo Recon Sq, July 18, 1944

LEFT
Chapelle Saint-Gerbold at Cerisy-la-Forêt served as battalion CP for the 741st. This Gothic structure included a dungeon. (Photo by author)

RIGHT
Hill 192: Thomas R. Fair's wrecked M4(75) and a German armored car (Panzerspähwagen AB41 201i) from Fallschirm-Aufklärungs-Abteilung 12. The unusual enemy vehicle was probably lost sometime prior to July 11, 1944. Hecox's M4(75) sits in the distant background, barely visible. Photo by Frank Scherschel. (LIFE Picture Collection/Shutterstock)

LEFT
Tommy Fair's destroyed M4(75), registration number USA W-3039496, built by the Pullman Standard Car Manufacturing Company. Photograph by Frank Scherschel. (LIFE Picture Collection/Shutterstock)

On July 12, 1944, an unidentified cameraman from the 165th SPC captured motion-picture footage of the armored car and the tanks of Fair and Hecox as well as dead Fallschirmjägers nearby.

RIGHT
Donald W. Hecox's destroyed M4(75), registration number USA W-3039446, built by the Pullman Standard Car Manufacturing Company. Hecox's driver was Russell M. Bradsher, whose remains have never been positively identified. (Fitts Collection)

Group burial for Fair, Nixon, and Pasqualini at Richmond National Cemetery, Virginia. (Photo by author)

July 1944: M4(105) assault gun assigned to the Headquarters Section of Able Company. Sergeant Adolf C. Rodriguez inspects the machine's newly installed "Rhino" device. (Fitts Collection)

LEFT
Captain Charles R. Young of Company C poses behind a Sturmgeschütz III, perhaps a machine from Fallschirm-Sturmgeschütz-Brigade 12. Young was athletic and stood 6' 1" in height. He received the Bronze Star Medal three times, the most BSM awards to any soldier of the 741st. (Fitts Collection)

RIGHT
July 27, 1944: M5A1 light tank of Company D. Named "Dare Devils" and carrying registration number USA 3046278-S, the tank was built by GM at its South Gate Assembly Plant in California. Photo by Josiah S. Carpenter, 165th Signal Photo Company. (NARA)

LEFT
August 13, 1944: Soldiers inspect a Sturmgeschütz III, whose crewmen flew the white flag after bazooka fire struck their vehicle, destroying a return roller and jamming a road wheel. This assault gun likely belonged to Sturmgeschütz-Brigade 394. Photo made at Saint-Germain-de-Tallevende, France, by Nelson E. Irving, 165th Signal Photo Company. (NARA)

LEFT
August 29, 1944: Medium tanks from the 741st Tank Battalion parade on the Avenue des Champs Élysées in Paris. Photo by Fred G. Poinsett, 165th Signal Photo Company. (NARA)

RIGHT
August 29, 1944: Light tanks from Company D motor along the Avenue de la Grande Armée toward the Arc de Triomphe. The boom assembly of a T2 from Service Company can be seen in the foreground. Photo by Leonard "Dick" Trimpe. (Trimpe Collection)

LEFT
February 9, 1945: First Lieutenant Joseph H. Dew receives the Distinguished Service Cross and Bronze Star Medal from General Robertson. Dew earned the DSC for his actions on September 14, 1944. Standing beside him, Lieutenant Colonel Skaggs waits to receive the Silver Star Medal. Photo made at the 2nd Infantry Division CP located at Wahlerscheid, Germany. The photographer was Leander P. Zwick Jr., 165th Signal Photo Company. (NARA)

RIGHT
Merwin Beck's identification tag recovered by a metal detectorist. The tag shows evidence of the fire that consumed Alfred Steineker's tank on September 20, 1944. Beck's remains have never been recovered or positively identified. (Manfred Klein)

RIGHT
Gap in the dragon's teeth created by Joe Dew and his crew on September 14, 1944. The Westwall historical marker next to the narrow passage says nothing about the gap and its significance. (Photo by author)

ABOVE
Vitamin Rear in Robertville is now a vacation rental property. (Photo by author)

RIGHT
Rear Command Post, 741st Tank Battalion (Vitamin Rear), located at Nr. 17 Robertville, a boarding house called the "Pension Bardenheuer." Marie Fagnoul-Bardenheuer purchased the house in 1922, nine years before her marriage to Franz Josef Bardenheuer. She operated the establishment until her death in 1967. Vitamin Rear operated on her premises from December 14, 1944, through February 1, 1945. (Rutyna Collection)

LEFT
Shermans commanded by Victor L. Miller Jr. and John W. Welch. Both machines met with destruction at the Ruppenvenn road junction on December 17, 1944. The tank on the right is an M4A1(75) with registration number USA W-3036771. The other tank, an M4(75), displays the tactical number C-5. Postwar photo by Roy N. Reece of the 393rd Infantry Regiment. (Freytag Collection)

LEFT
December 18, 1944: Panther 126 burns after being knocked out by an M10 TD commanded by Corporal (later Sergeant) Stanley V. Kepinski of Third Platoon, Company C, 644th Tank Destroyer Battalion. Masonry lies on the Panther's engine deck. This was debris from Nr. 41A Krinkelt, Frank Mildren's CP. Photo by James F. Clancy, 165th Signal Photo Company. (NARA)

RIGHT
February 1945: The shattered wreck of Panzer 617 rests near the church. Nr. 91 Krinkelt stands in the background. This was the home of Martin Küches, who operated a Gasthaus on the premises along with a small farm that included several cows. The pyramidal roof on his house was (and still is) unique in Krinkelt-Rocherath. (NARA)

ABOVE AND RIGHT
February 1945: Kurt Brödel's Panther 305 sits beside Nr. 93 Krinkelt, the ruined home of Joseph Schleck. American troops moved the machine there to clear it off the main street. Joe Dew and his crew destroyed 305 some sixty yards from this location. One armor-piercing shell created a "jagged hole about eighteen inches across," according to Dew. The penetration can be seen just below the sponson. In the far background stands Nr. 92 Krinkelt, the home of Joseph and Caroline Kalpers. (Rutyna Collection)

BELOW AND RIGHT
Lifeless SS-Sturmmann from Panther 318. On December 18, James Clancy of the 165th Signal Photo Company photographed this man sprawled on the tank's engine deck. The dead tanker lay there over six weeks until an explosion presumably swept his frozen corpse off the machine. The man's name is unconfirmed, but the best evidence points toward Kurt Eggers, a nineteen-year-old member of Brödel's company. (Green Collection)

BELOW
Kurt Eggers's grave at the Lommel German War Cemetery: Block 30, Grave 223. (Photo by author)

BELOW
Four soldiers from Company C beside Panther 318 in February 1945. They are, left to right: Peter J. Fardella, Ernest O. Padgett, Walter J. Fryer, and Frederick Neidrich. Padgett wears a German winter jacket. Johann Beutelhauser may have commanded this tank on December 18, his Panther 325 having been under repair. The turret damage occurred later when American troops recaptured the twin villages, the damage possibly caused by an artillery hit. (Guest Collection)

LEFT
Staff Sergeant Theodore Parker's M4(75) and Sergeant Walter J. Slodkowski's M4(75) lay derelict by the ruins of Saint John the Baptist Church in Krinkelt. In the center foreground rests the crumpled remains of a GMC CCKW 353 truck from Company A, 2nd Engineer Combat Battalion. On the right, sits the wreck of another GMC CCKW 353, its hoist superstructure prominently visible. (Geschichtsverein Rocherath-Krinkelt)

RIGHT
Helmut Zeiner and his Jagdpanzer crew destroyed this M4(75) commanded by Staff Sergeant Theodore Parker. Built by the Pressed Steel Car Company, the tank carried registration number USA-W-3036643 and tactical number A-6. Photo by James W. Love. (Love Collection)

LEFT
Postwar picture of Sergeant Walter J. Slodkowski's M4(75), registration number USA 3066230, built by the American Locomotive Company in Schenectady, New York. The apse end of the church stands behind the tank. (Geschichtsverein Rocherath-Krinkelt)

BELOW
Postwar photo of Sergeant Thomas Carlson's M4A1(75) tankdozer, its blade missing after being salvaged. The tank carried registration number USA-W-3036870, denoting a machine built in Chicago by the Pressed Steel Car Company. Another photo of this tank was the basis for the illustration below. (Geschichtsverein Rocherath-Krinkelt)

Illustration by Felipe Rodna

RIGHT
Postwar photo of Second Lieutenant George R. Coleman's M4(75), registration number USA-3066160, built by the American Locomotive Company. This was the Company B command tank, which Coleman commandeered on December 18, 1944. Its tactical number, B-1, appears on the upper rear hull plate. (Geschichtsverein Rocherath-Krinkelt)

LEFT
February 1945 view of Coleman's destroyed M4(75) with its cannon barrel severed. The sponson bears salvage markings made by the 463rd Ordnance Evacuation Company. (Warnock Collection)

RIGHT
Built by the Baldwin Locomotive Works, this M4(75) from Company A displayed the tactical number A-7 and registration number USA-W-3011019. This may have been Sergeant Edward L. Ayers's tank lost on December 18 in the vicinity of Lausdell. Or it could have been First Lieutenant Gaetano Barcellona's tank lost on December 19. (Geschichtsverein Rocherath-Krinkelt)

LEFT
February 1945: Nr. 41 Krinkelt, the ruined home of Martha Rauw-Kalpers, widow of Albert Kalpers (1893-1941). The building also served as a shoe store or Schuhwarenhandlung. On the right, a directional sign points toward the Forward CP of the 741st Tank Battalion, probably located at Nr. 81 Rocherath (as opposed to the earlier CP at Nr. 100 Rocherath). The sign features an unofficial 741st insignia, a horseshoe and cannonball. These represent the union of artillery and cavalry within the Armored Force. Photo by James W. Love. (Love Collection)

LEFT
January 11, 1945: Two members of Company B, Corporal Alfred Snike, Brooklyn, New York, and Private First Class Leonard F. Konert, Cedar Rapids, Nebraska, clean the barrel of their 76 mm gun at Berg, Belgium. Photo by Bernard J. Cook, 165th Signal Photo Company. (NARA)

RIGHT
January 11, 1945: Near Berg, Belgium, soldiers of Company B, 741st Tank Battalion take a break outside their living quarters. Left to right: Private First Class John C. Reedy, Bartlesville, Oklahoma; Technician Fourth Grade Tony Morgando, Mulkeytown, Illinois; and Technician Fourth Grade Brodie Ritchie, Vest, Kentucky. Ritchie died as a non-battle casualty in April 1945. Photo by Bernard J. Cook, 165th Signal Photo Company. (NARA)

LEFT
Able Company's kitchen crew poses at Elsenborn, Belgium. In the back row, left to right: Carroll M. Lenox, Sam H. Fuller, Edward Teffeteller, Glenn W. Kieffer. Front row, left to right: Edward O'Hara, Anthony D'Eliso, Willie E. George, Harry Bloom. George wears a captured P-38 pistol on a German belt with a Waffen-SS enlisted man's buckle. (Warnock Collection)

RIGHT
February 2, 1945: Men from Company E, 9th Infantry Regiment ride medium tanks to the front lines in the Monschau Forest. Photo by John Boretsky, 165th Signal Photo Company. Boretsky died in combat three weeks after making this picture. (NARA)

RIGHT
March 9, 1945: Mayne Youngblood's M4A1(75) named "Asp II" motors across a treadway bridge at Dümpelfeld, Germany. The five crewmen were, left to right: Arthur M. Najera, Mayne B. Youngblood, Steve J. Hoffer, Charles R. Freeman, and John Onuschak. Youngblood and Onuschak wear "ETO contract" tanker helmets made in Paris by Guéneau et Cie and supplied on a limited basis to First Army armored units. Photo by Louis Nemeth, 165th Signal Photo Company. (NARA) Photo insert below right, Mayne Youngblood.

LEFT
The US Army treadway bridge still stands today at Dümpelfeld, though only open to pedestrian traffic. (Photo by author)

RIGHT
Glenn Kieffer and Harry Bloom pose by "Asp II" in Pilsen, Czechoslovakia. This was the only Able Company tank that survived from D-Day until VE-Day. Built by the Pressed Steel Car Company, the tank carried registration number USA W-3036841. The number "12" appears on the rear of its turret. (Warnock Collection)

LEFT
"Victor Bridge," spanning the Rhine River between Niederbreisig to Hönningen. (Warnock Collection)

LEFT
March 1945: Across the Rhine, soldiers of the 9th Infantry Regiment mount tanks from Able Company. Photo by Glenn W. Kieffer. (Warnock Collection)

RIGHT
March 30, 1945: Men of the 9th Infantry Regiment mount tanks from Company A to form a combat team in Giessen, Germany. Anthony T. Martin of the 165th Signal Photo Company made this photo on Johannes-Strasse with the Stadt-Theater visible in the background. He walked to the other side of the theater and made the photo shown below. (NARA)

LEFT
March 30, 1945: First Lieutenant Elvin Ward Lowe, Lometa, Texas, left, aide to General Robertson, right, Commanding General, 2nd Infantry Division, holds map of the sector, as Robertson watches light tanks and infantry drive through Giessen, Germany. Photo by Anthony T. Martin, 165th Signal Photo Company. (Warnock Collection)

RIGHT
April 6, 1945: M4A3E8 from the 741st Tank Battalion passes through a demolished roadblock in the Altmünden area near the Weser River. Photo by Henry W. Gasiewicz, 165th Signal Photo Company. (Warnock Collection)

RIGHT
April 8, 1945: Outside Reinhausen, Company A discovered two King Tigers scuttled by their crews. The one in the background had its turret blown off. The Tiger in the foreground had one of its tracks broken, tow cables attached, and its cannon barrel bent by an internal explosion. These machines may have belonged to the 3.Kompanie of schwere Panzerabteilung 502. Glenn Kieffer made eight photos of the two tanks, and he gave the original negatives to the author. (Warnock Collection)

LEFT
The two Königstigers met their end on the road running between Diemarden and Reinhausen. The orchard on the right still exists today. Reinhausen can be seen in the background and has grown in size since April 1945. (Photo by author)

RIGHT
Panzer III knocked out by Staff Sergeant Lloyd Ball and his crew on April 6, 1945. This enemy tank had been a training vehicle at Sennelager before joining a company commanded by SS-Obersturmführer Max Tietz of SS-Panzerbrigade „Westfalen." Photo courtesy of Glenn W. Kieffer. (Warnock Collection)

LEFT
April 1945: 10.5 cm Flak gun located near Merseburg, Germany. This picture was on the same film roll as the King Tiger image shown above. Photo made by Glenn W. Kieffer. (Warnock Collection)

LEFT
April 14, 1945: Company B tanks knocked out by Flak shells fired from Hill 102. The shells destroyed three tanks: USA W-30103623 (Chrysler, M4 105 mm), USA W-30125832 (Pressed Steel Car Co., M4A1 76 mm), and USA W-3085258 (Pressed Steel Car Co., M4A1 76 mm). The latter two burned, one inadvertently set on fire by Harrell Gullatt, when he tried to start the motor after the battle. Photo by Richard W. Crampton, 165th Signal Photo Company. (Warnock Collection)

RIGHT
April 18, 1945: Sergeant George K. Cuthbert's M4(75), "Cleopatra," burns in Leipzig, Germany. He and the other four members of his crew died at about ten o'clock in the morning when a Panzerfaust struck the right sponson of their vehicle. Their unit, Company C, 741st, supported the First Battalion, 23rd Infantry Regiment. Photo by Richard W. Crampton, 165th Signal Photo Company. (NARA)

LEFT
April 18, 1945: Civilians watch George Cuthbert's tank burn on a street corner in Leipzig. The vehicle's registration number was a British War Department number with an X prefix. Photo by William C. Allen. (Northcliffe Collection/ANL/Shutterstock)

RIGHT
Summer 1945: Starý Klíčov, Czechoslovakia, left to right: Major Cecil D. Thomas, Captain James B. Eure, First Lieutenant Romolo P. Legnini, Captain Hurdle E. McDaniel Jr. (Legnini Collection)

RIGHT
May 1945: M4A1(76) commanded by Staff Sergeant Richard L. Maddock of Company B. Jack Boardman was his gunner. This tank, like many other vehicles, displayed a simplified 2nd Infantry Division insignia without an Indianhead. Unfortunately, the tank's registration number is illegible in this image. (Boardman Collection)

LEFT
Frank A. Klotz, Company A, resident of Johnstown, Pennsylvania. (Warnock Collection)

BELOW
Romolo P. Legnini displays the battalion colors. (Legnini Collection)

ABOVE
Edward Teffeteller posed beside the Mercedes sedan that he Frank Klotz commandeered after Soviet forces liberated them from Stalag IVC. The circumstances leading to their capture nearly resulted in court-martial proceedings. Both men served with Company A, but on D-Day, Klotz led the Dozer Platoon of Headquarters Company and earned a Silver Star Medal. (Warnock Collection)

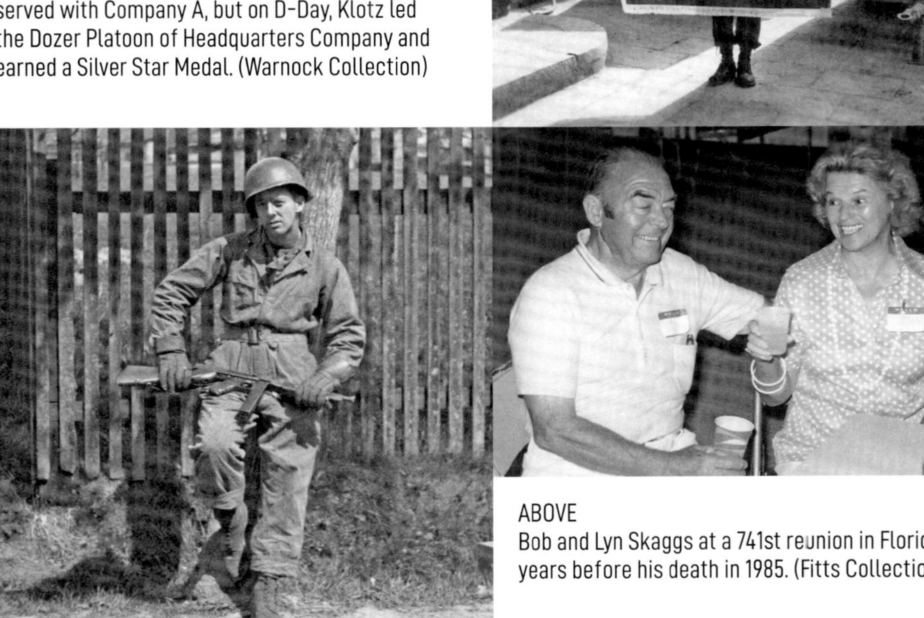

ABOVE
Bob and Lyn Skaggs at a 741st reunion in Florida two years before his death in 1985. (Fitts Collection)

Glenn Kieffer's portrait of John W. Call, Company A, native of Barlow, Kentucky. (Warnock Collection)

GERMAN ANTITANK WEAPONS

LEFT
September 15, 1944: Emplaced between two concrete bunkers of the Westwall, this camouflaged PAK 40 (75 mm antitank gun) fell into American hands in the Eifel region of Germany. Photograph made by Dale W. Franklin, 165th Signal Photo Company. (NARA)

RIGHT
Raketenwerfer 43 or Püppchen tested by a member of the 45th Ordnance Battalion. Soldiers of the 741st Tank Battalion encountered this weapon during the battle for Hill 192. Photo by Lawrence V. Emery, 196th Signal Photo Company. (NARA)

LEFT
Recoilless rocket launcher known as the Raketenpanzerbüchse 54, also called the Panzerschreck. Photo by Robert G. Edwards, 163rd Signal Photo Company. (NARA)

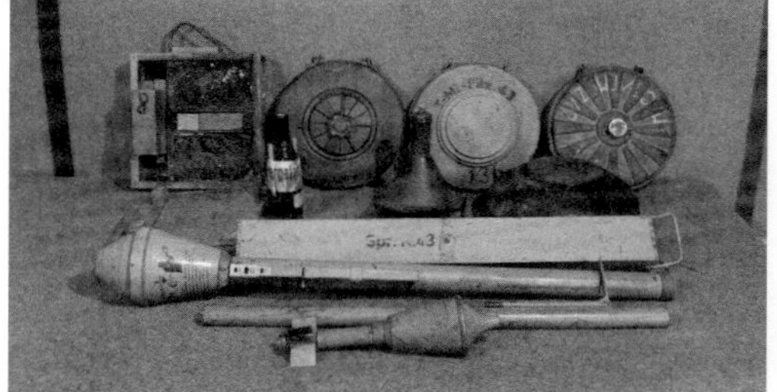

RIGHT
March 14, 1945: Wooden antitank mine; three Teller mines; Molotov cocktail; antitank grenade known as the Hafthohlladung or Panzerknacker; Topf ceramic mine; Riegel mine; Panzerfäuste, large (60 meters range) and small (30 meters range). Photo by J. Malan Heslop, 167th Signal Photo Company. (NARA)

LEFT
8.8 cm Flak 36 captured near Dülmen, Germany on March 29, 1945. The gun is mounted on a carriage comprising a platform carried on twin, two-wheeled limbers. (NARA)

Stielgranate 41 projectile for 3.7 cm antitank gun.

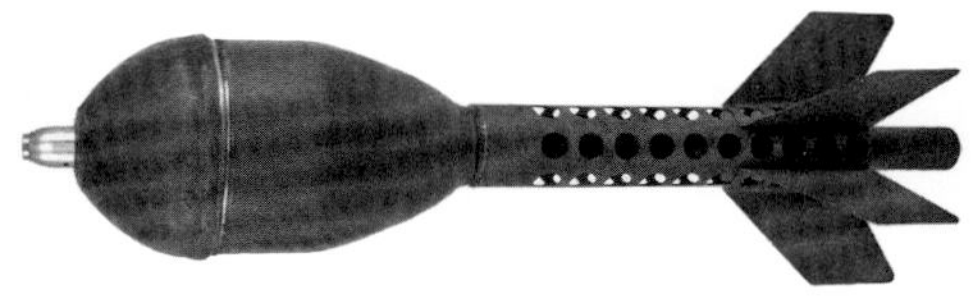

that point. Müller's men expected to romp through the American rear just like Kampfgruppe Peiper, but they would soon discover otherwise.

Evidence of an intelligence failure sat crouched in foxholes in the Krinkelter Forest. In all the minutiae of their preparations, the German planners had lost track of the 2nd Division after it relinquished the Schnee-Eifel to the 106th Division. On the situation map at their Supreme Headquarters (OB West), the 2nd appeared as an arrow pointed north from the Schnee-Eifel with a question mark penciled alongside. The 2nd Division had avoided detection thanks to rigid security measures, including an order to unstitch the Indianhead shoulder patch.

MacDonald's men unleashed a rude and bloody welcome to Kampfgruppe Müller. Enemy soldiers spun and toppled in a furious blast of bullets. The survivors fell back to gather for another try. Weapons held high, a second wave of SS men, more than before, stormed forward, but American fire bowled them over just the same. The Germans repeated the scene, seven suicidal assaults in all.

Müller had no base of fire to pin down the Third Battalion defenders and suppress their fire while his troops maneuvered to rout them. The SS men never attempted to strike from the flanks, only frontal assaults, infantrymen charging in lines. MacDonald's lads witnessed tactics meriting a flunk at Panzergrenadier school, but that soon changed.

The clatter of tank tracks heralded the punch that Müller's beleaguered grenadiers needed. The noise sent an electric current down the spines of even the bravest GIs, including Crisler and Padgett. Five Panzers rolled toward the American line. Broad and heavy, these were spanking-new Panthers from the 1.Kompanie of SS-Panzer-Regiment 12.[211]

MacDonald's troops had nothing to stop the armor-clad behemoths. He asked one of his platoon sergeants to fetch Crisler and Padgett, but the sergeant reported back with astonishing news. "They're gone, Cap'n. They pulled back to K Company."[212]

Cannon fire from the SS armor reverberated through the forest as MacDonald radioed Tuttle, beseeching him to order back the two tanks. MacDonald's position had begun to crumble. One after another, tank shells blew apart soldiers in their holes.

After again contacting Crisler and Padgett, Tuttle radioed back. The tankers had told him it would be suicide for them to face the Panthers. They would not budge unless he gave them a direct order, and he feared they would disobey it. Tuttle agreed they stood no chance.

The two tank commanders stuck to their plan. They would remain concealed among the spruce branches and fight on their terms.

The battle became an unequal contest between Panthers and foot soldiers armed with rifles, machine guns, and hand grenades. Within minutes, MacDonald witnessed the predictable outcome. His men fell back, flying for their lives through the trees.

The dense mass of conifers denied the Panthers freedom of movement, restricting their passage to the road. As they crept toward Ruppenvenn, the crewmen inside cranked their steel turrets right and left, blasting HE rounds between trees and down firebreaks. Whatever movement caught their eyes invited a shell. The bow machine gunners spewed bullets at any unfortunate who stumbled onto the road.

Outside the forest, Miller took heed of a report that "Tiger tanks" had overrun MacDonald and were ambling toward Ruppenvenn. The infantry and many tankers mistakenly referred to the enemy tanks as Tigers. In GI jargon, the word Tiger often described any piece of German armor larger than a Sherman.

Just before 3:00 P.M., Miller was "told to take up the best position possible and hold by all means."[213] Assuming the point position—as was his custom in combat—Miller led Vernon and Welch toward Ruppenvenn.

Private First Class Louis J. DiClementi Jr. served as the loader aboard Vernon's tank. Before reaching Ruppenvenn, he heard Miller order Welch and Vernon to halt. "He placed us on the left side of the road and then placed Welch on the left about fifty yards in front of us."

The lieutenant went ahead to Ruppenvenn alone. Through periscopes and telescopic sights, DiClementi and the other tankers eyeballed Miller's tank. "As soon as he pulled over, we saw a big, black flash and heard an explosion."[214] Faster than a flicker, an AP shell drilled the cast-steel turret. It killed Miller, his gunner, and his loader.

Two members of Miller's crew survived the destruction of their vehicle, though both men suffered injuries. The driver, Technician Fourth Grade

Michael W. Fox, had shrapnel lodged near his spine. Fragments also hit the bow gunner, Private First Class Richard Smith. His most severe wound resulted from metal that ripped into his right thigh and left him crippled. Fox helped Smith from the tank and to safety. The dauntless driver then grabbed a rifle and fought on with the infantry.[215]

Seconds after the enemy struck Miller's tank, DiClementi heard Welch exclaim over the radio, "The lieutenant's hit! I'm going over to see what I can do for him."[216] The sergeant followed in Miller's tracks until reaching Ruppenvenn. He then maneuvered alongside the motionless hulk and paused. The two tanks were no more than twenty feet apart when the Panther in the forest fired again, bashing Welch's right drive sprocket with an AP shell. The impact snapped the right track. As the driver jockeyed the crippled tank, the broken track peeled off its rollers, causing the machine to pivot right. A second shot stopped all movement. The crew needed no further persuasion, and all five men fled on foot.

After seeing two tanks destroyed in swift succession, Vernon never considered venturing forward. Longevity was foremost in his mind. After making a fast turnaround, the driver raced for cover behind an expanse of trees. The gunner swung the turret to the rear, facing Ruppenvenn. DiClementi then saw a "black flash go right over the turret." The shot flew high. Before the Panther crew could clear their cannon breech and ram another shell into the firing chamber, Vernon's tank escaped behind the trees where it halted.

Screened by conifers, DiClementi and his comrades hoped for a chance to shoot if the enemy showed himself. "We kept traversing and looking into the woods to see if maybe he would come out. Nothing showed up. We stayed there for fifteen minutes, then pulled back."[217]

Before Vernon's vehicle departed, Welch's tankless crew climbed aboard, as did a band of infantrymen.

Why had the Panther failed to finish the fight? Unknown to Vernon and his crew, Padgett's tank ensured the enemy machine never emerged from the forest. He and Crisler had retreated to Ruppenvenn, turned down the right fork, and then pulled off the road. Here they waited for the first enemy tank to reach the fork. When it did, its hull would be perpendicular to their muzzles. And so it happened.

The Panther that fired at Vernon's tank placed itself in Padgett's sights. The enemy machine succumbed to a flank shot at close range. Miller, Welch, and Vernon had been the unwitting bait.

Crisler and Padgett charged from the forest, weaving between the knocked-out Panther and its two American victims. The Germans dubbed the place *Sherman Ecke* [Sherman Corner].

Fleeing the battleground, Vernon's driver nearly lost control of the tank. Technician Fifth Grade Andrew W. Divers steered the machine toward Rocherath with a gaggle of foot soldiers and Welch's crew bunched atop the hull. "One of the infantrymen sat down on my periscope, and I went down in a ditch. I hollered to the tank commander, 'Get them guys the hell off my hatch. I can't see where I'm driving.'" Vernon shooed off the soldiers, but the tank nearly became stuck, according to Divers. "I was lucky to get out of that ditch."[218]

As platoon sergeant, Crisler inherited leadership of Third Platoon. Soon after he and Padgett departed Ruppenvenn, a GI machine gunner clad in a cut-down overcoat approached the two knocked-out tanks. The man jumped atop Miller's machine and pried open the turret hatch. His eyes found the slaughtered remains of three crewmen. He jerked away, then climbed down and hustled to nearby trees where he positioned his water-cooled machine gun, aiming it toward Ruppenvenn. When enemy grenadiers stepped gingerly from the forest and milled around the tanks, he cut into them with a sustained burst. The soldier was Private First Class Jose M. Lopez, and he was in the closing act of a drama that would earn him a Medal of Honor.

The fight for the Krinkelter Forest had reached its finale. Tuttle's thin line supported by Miller's five tanks had managed to slow the SS onslaught, but the Third Battalion lacked the strength to crush an armored attack. There had been no real chance to stop the Germans.

Tuttle's line was part of a larger stratagem, a defense in depth cobbled together by General Robertson. The unflappable 2nd Division commander planned to erect levees against the enemy flood. The Third Battalion had been the first levee behind a breakwater provided by the 393rd Infantry. The second levee would come from Hirschfelder's regiment still at Wahlerscheid.

CHAPTER 28 FLAMES IN THE CHURCHYARD

In muddy ditches, foxholes, and captured bunkers at Wahlerscheid, news of the withdrawal order fell heavy upon Hirschfelder's infantrymen. They stood open-mouthed in disbelief. Why? Who was to blame?

Many soldiers grasped for an easy explanation, the only one their limited perspective permitted. They had to abandon the fruit of their suffering because the 99th Division failed to contain a local attack. This notion increased during the coming hours when Hirschfelder's men encountered disheveled 99ers—joined by troopers from Tuttle's battalion—escaping westward, some without helmets or weapons.

Few who gazed upon this retreat realized the 99th had stymied an entire Panzer-Armee for more than twenty-four hours, a feat garnering one Medal of Honor, three Presidential Unit Citations, ten Distinguished Service Crosses, more than seventy-five Silver Star Medals, and many lesser decorations. The precious hours gained by the 99th were golden gifts to the 2nd Division, but, as so often in war, battlefield perceptions spun myths that eclipsed reality.

General Robertson personally oversaw the redeployment of all troops retreating from Wahlerscheid. He positioned Hirschfelder's First Battalion at Lausdell crossroads fifteen hundred yards northeast of Rocherath. The general also sent another of Hirschfelder's battalions into Krinkelt-Rocherath to hold the twin towns, along with most of the 38th Infantry.

Sergeant Edward L. Ayers and his tank crew from Able Company met Robertson face-to-face. Born in Texas, Ayers had rejoined the company and received his promotion to sergeant in August after recovering from his D-Day injuries. Robertson intercepted Ayers on the Wahlerscheid road as his crew motored toward the rear for repair work on a damaged track link. The general asked if the machine could still maneuver. Ayers said yes but added that sharp turns were impossible. Robertson instructed him to support the battalion at Lausdell until the troops finished digging in. The tank moved out.

Ayers and his men sat at Lausdell until the infantry entrenched themselves, but the infantry battalion commander ordered them to stay until further notice.

Inside Krinkelt-Rocherath, 741st soldiers from Service Company and Headquarters Company prepared to defend their battalion CP. They set up three strongpoints and eight observation posts. The battalion had fifteen operational tanks in and around the towns.

As darkness fell on December 17, everyone waited for the enemy's next move.

That enemy, the men and boys of Kampfgruppe Müller, had stopped in the Krinkelter Forest to mend their wounds after the bloodbath suffered in the woods. Their battle group commander hoped that all American resistance had collapsed. He planned to test that hope with Jagdpanzers from the TD battalion assigned to his group.

SS-Obersturmführer Helmut Zeiner had the task of leading the Jagdpanzers from the forest. The twenty-three-year-old officer from Linz, Austria (Hitler's hometown), had served three years in the Waffen-SS. His résumé included such nightmares as Kharkov and the Schlüsselburg Corridor near Leningrad. As a machine gunner, messenger, and squad leader, he had suffered three combat injuries. The last occurred at Kharkov when a bullet broke his lower jaw. After a hospital stay, he transferred to officer candidate school at Bad Tölz. He became a lieutenant and joined the Hitlerjugend Division in March 1944. Throughout the Normandy campaign, Zeiner served with SS-Panzerjäger-Abteilung 12. He took command of its 1.Kompanie in November.

His combat assignment in the Ardennes was different. In Normandy, his outfit had proved its worth as a tank destroyer unit, but attachment to Kampfgruppe Müller meant a different mission. Zeiner's Jagdpanzers no longer waited in ambush to kill tanks but assumed the function of assault guns, storming ahead with the infantry. Their high-explosive rounds would give the foot troops a base of fire and allow them to maneuver.

The TDs, turned assault guns, lacked the right design for close-quarter fighting in narrow forest lanes. Without a turret, it was impossible for a Jagdpanzer gunner to swing his cannon left or right to engage threats lurking on the flanks. Panthers had that ability and reigned supreme in the Krinkelter Forest, but with the woods clear, and open country ahead, the Jagdpanzers had a chance for success.

In the early evening, Zeiner led the way with his machine. Grenadiers strode behind each Jagdpanzer in the column. "At the forest edge a damaged Sherman could be seen," Zeiner recalled. "I don't know who destroyed it. In front of me, after marching through the forest, there were no German units on our route."[219]

There were actually two Shermans, the derelict tanks of Welch and Miller. Zeiner's column snaked past them and pushed through the darkness toward Lausdell.

The First Battalion, 9th Infantry occupied Lausdell. GIs digging foxholes heard the approaching armor. Many soldiers assumed the tanks were from the 741st but realized otherwise when the armor rolled through. The GIs stood dumbfounded as the enemy force passed, many of the grenadiers joking and laughing. Nobody fired a shot.

In the blackness of that December night, Zeiner and his party had no inkling they were smack in the middle of an American infantry battalion. After reaching a road fork at Lausdell, the Waffen-SS lieutenant and those in convoy behind him turned toward Rocherath.

"We drove to the left and, after approximately one kilometer, landed on the outskirts of Rocherath. Silence! We strained our ears, listening into the night, the engines turned off. Nothing. I sent a few foot troops ahead as a scouting party. They were to determine whether the enemy occupied the village or had already abandoned it."

Zeiner meanwhile attempted to make radio contact with battalion headquarters and the guns following behind him. The scouting party soon informed him that US troops held Rocherath. He also discovered that only three Jagdpanzers were with him. "The others had probably turned right at the fork," he imagined. "The important thing was to engage the unsuspecting enemy—as the scouting party had reported—and capture the village."[220]

The SS marauders rumbled into Rocherath. After rolling past a concrete water tower, they pushed onto the main street.

Across the twin communities, US infantrymen busied themselves fortifying selected homes. Tank crews from the 741st took up positions supporting the infantry. Sergeant Ray Dickson commanded a tank from the Second Platoon of Company B. His vehicle stood at a roadblock in Rocherath when Zeiner's Jagdpanzers lumbered out of the gloom.

"Suddenly I saw the silhouette of a tank approaching at about one hundred yards," Dickson recounted. "My gunner, Corporal Kroeger, traversed right, and Corporal Scurry had a shell ready to load when a doughboy yelled, 'Don't fire! That's our TDs and infantry withdrawing.' We held our fire.

"A few seconds later, the tank reached an intersection. When we saw for sure it was a Kraut tank, it was too late to fire. We didn't want to give our position away as we were outnumbered. The Kraut tank continued down the road into town on our right flank. It was supported by infantry and had good cover behind some buildings."[221]

The small German force had again bluffed its way forward. Zeiner sighted the apse end of the church after crossing the imperceptible boundary between Rocherath and Krinkelt. Suddenly he heard rifle shots. The 38th Infantry had detected the enemy intruders.

Zeiner recalled, "I pulled ahead to the church, the remaining guns right behind me. I directed the infantry-escort platoon to comb all the houses and to bring the soldiers they found, disarmed, to the street."

He instructed his driver to stop near the church and shut down the motor so he could assess the situation by listening.

"Now and then, I heard shots behind me, but, suddenly to the right and apparently behind the church, I heard the roar of a heavy engine. The only thing I could see was the corner of the church standing out from the snow-covered square. I suspected that an enemy tank had started to move behind the church."

Zeiner's driver restarted the motor and pivoted the tank destroyer ninety degrees to the right. The TD commander watched for the American vehicle.

"I spotted a Sherman pushing out from behind the church, in reverse gear, some eight to ten meters away. I ordered an AP shell and had my barrel lowered. I disappeared into my hatch when the giant was right in front of my gun. We knocked out the tank. It immediately burst into flames and lit up the battleground around the church square for a long time. Two men limped from the tank and fled into the church. We left them alone."[222]

Private Kelly B. Layman was one of those men. The Los Angeles resident had entered the army in April 1943 and joined the 2nd Division in Ireland as a rifleman with Company K, 38th but developed appendicitis in Normandy. After hospitalization, he landed in a replacement depot where a captain asked if anyone was qualified on the caliber .30 light machine gun. Layman piped up and said he qualified on it during basic training. The officer said the Armored Force needed such men, and the young rifleman volunteered.

Private Layman joined the 741st in October and became a bow gunner with Lieutenant Dudley's Second Platoon, Company B; yet on the night of December 17, the machine he abandoned under Zeiner's eyes was not his own.

Layman's assigned vehicle—Sergeant Walter J. Slodkowski's tank—had run out of gasoline several hours earlier. Its crewmen climbed out and endeavored to locate Baker Company headquarters several hundred yards from the church. Perhaps more fuel was there, but the search ended without finding gasoline or the headquarters. The luckless crew took refuge in the church.

Outside, the men heard two Company A machines: Staff Sergeant Theodore Parker's tank and Sergeant Thomas Carlson's tankdozer. Throughout Able Company, everyone knew Carlson as "Mother" on account of his advanced age (thirty-four), his broad shock of prematurely white hair, and his predilection for tidiness. That evening, he and Parker had orders to link up with Company K, 38th in Rocherath, but now lost in the dark, Parker dismounted to obtain directions.[223]

Inside the church, Layman heard the friendly tanks and decided to join them. He climbed aboard the dozer, but there was no room inside. He then tried Parker's vehicle and ended up in the bow gunner's seat. The staff sergeant had yet to return. As the newcomer settled in, the tankdozer motored across the churchyard. The men in Parker's tank had no idea there was a Jagdpanzer to their rear as they crept backward in reverse gear.

After the war, Layman offered his perspective on what happened next—an account that mirrors Zeiner's recollections.

"We had only backed up about ten feet when they let go with an HE and blew our right track off. We sat there for a couple of seconds trying to move, but it was no good. They hit us again with another HE on the turret, and we still sat there! But when they put an AP round through our motor compartment and set us on fire, it was time to bail out, which we did.

"An A Company man and I ran into the church, and a medic put shaving cream on our burns. It was the only thing he had; he said he was sorry he couldn't do more, but it sure felt soothing."[224]

The flames had scorched Layman's face, leaving him without eyebrows, but he counted himself fortunate to be inside the church and its heavy stone walls. The Company A man with Layman was Corporal Virgil Michael from Harlan, Kentucky. He too suffered facial burns.

Layman later explained that at least one crewman perished after escaping the tank. "Somebody else got out, but instead of running into the church with me and the other guy, he went down the left-hand side of the church and got killed, machine gunned."[225]

This was Private Charles R. Ireson from Bluefield, West Virginia. He had already earned one Purple Heart for wounds sustained on Hill 192. His second decoration would be posthumous.[226]

Zeiner and his crew spotted Carlson's tankdozer in the churchyard and threaded a shot between Parker's wrecked machine and the church, putting the dozer in flames.[227]

"We caught another Sherman at the far end of the churchyard. A third Sherman was knocked out by a tank destroyer following behind us and which had swung to the right. Now absolute silence."[228]

The "third Sherman" claimed by Zeiner's force may have been Lieutenant Dudley's tank, which sustained a hit that night. His vehicle and the others in his platoon (minus Slodkowski's tank) covered a road junction in Krinkelt-Rocherath when Zeiner's group first arrived.

Dudley's gunner, Sergeant Ray M. Wilson, swung the turret toward the nearest enemy machine and opened up with AP rounds. George Beeley, the bow gunner, watched them fly as he too cut loose, firing his machine gun toward the tank and its infantry escort. Three AP shells bounced off the Jagdpanzer like ping-pong balls.

Wilson persevered, hollering for Beeley to pass more AP rounds to the loader. He quit firing his machine gun and turned to open the thirty-round stowage bin beneath the turret basket. "My hands were so cold I couldn't open the compartment right away," he recalled. "Some shells had come loose, which made the door hard to open."[229]

After finally opening the door, Beeley had to grab the right type of ammunition, but it was too dark to see the markings on the shells. When he switched on his flashlight and found the right ones, he made a shocking discovery. Except for the driver, Ivan Schmidt, the rest of the crew had abandoned the tank. Why? A half century passed before Beeley learned the reason.

In a handwritten note to his wartime buddy, Wilson explained that an enemy shell had glanced off the turret. "That's when Dudley shouted, 'Bail out! We're hit!'"[230]

After reading the note, Beeley concluded, "I guess the reason my driver and I never heard the order was that we were temporarily deaf from our 75 firing at the enemy tank."[231] Whatever the reason, the two remaining tankers decided to bail out.

"Schmidty got out first," Beeley remembered. "I almost crapped when I couldn't lift my hatch. They had left the gun over it."

Beeley dove over the transmission housing and pulled himself out through Schmidt's hatch. As the two men ran from their tank, enemy machine-gun bullets chased them. The pair reached the other tanks in their platoon, hidden from enemy view behind a wall.

"I found Dudley and asked him what he wanted Schmidty and me to do. He told us to try to get back to our lines."[232]

Beeley and Schmidt took off in what they hoped was the right direction. It soon became evident that no frontline existed in the twin villages, no demarcation between foe and friend. Americans held this house and Germans that one. Homes and barns became defensive outposts in a wilderness of chaos.

Illuminated by the two blazing Shermans near the church, Zeiner dismounted his Jagdpanzer to huddle with the leader of his infantry escort and fellow TD commanders. The infantry meanwhile brought a collection of prisoners they had prodded from buildings along the main street. The SS lieutenant held the captured Americans under guard in a nearby house. He called the meeting with his subordinates to learn how much ammunition and gasoline remained. Zeiner instructed them to report back with the information, and he headed back toward his Jagdpanzer.

As he returned to his vehicle, he came under fire from a butcher shop on the main street in Krinkelt. The building was the CP of First Battalion, 38th Infantry. Its commander, Lieutenant Colonel Frank T. Mildren, and his headquarters crew defended their CP in Wild West fashion. They assailed Zeiner's group with rifles, carbines, an air-cooled machine gun, one caliber .50 machine gun, and whatever other killing tools they could muster. Joining in the shootout, an M8 armored car rained fire on the Germans in an almost suicidal attack.[233]

Some of Mildren's men grabbed weapons and ventured into the street, wading into enemy bullets. Others opened up from windowsills. Jagdpanzer shells roared through the air, booming against building facades and destroying the M8 seconds after its crew bailed out. Orange flames from the burning Shermans lit the scene with a bright bonfire glow.

Inside the butcher shop, Mildren shouted into his radio for armored backup from the 741st. He reported German tanks right outside his door.

But the threat receded. The SS force backed off, unable to sustain a pitched fight. Zeiner's subordinates had reported their petrol and ammunition too low for a battle. The group moved a short distance to a quiet spot and established an "all-around defense."[234]

CHAPTER 29 **HELLFIRE AT LAUSDELL**

The GIs at Lausdell resolved to bar more Germans from passing through, unlike Zeiner's men who slid through unchallenged. The Americans slapped antitank mines on the road and readied bazookas for action.

Sergeant Ayers and his tank crew had also watched the enemy armor and infantry pass through unchallenged. The incident astounded everyone. Ayers and his gunner dismounted their machine. One man awakened soldiers sleeping in nearby foxholes, and the other disappeared into the night to alert 741st personnel in Rocherath. Three men remained in the tank: Technician Fifth Grade Joe B. Hall, the driver; Private First Class Virgil H. Lohman, the loader; and Private Niles N. Nutter, the bow gunner. Nutter took the gunner's seat to keep the cannon ready for action.

The next enemy force to approach Lausdell was a Panther column from the 1.Kompanie of SS-Panzer-Regiment 12. While turning left toward Rocherath, the leading machine detonated a mine, and its left track snapped in the explosive flash.

The maimed Panther began blasting its cannon and fanning machine-gun bullets from its bow gun. Cannon shells set alight a barn nearby. Three American sergeants heaved an open can of gasoline onto the disabled Panther and lit the fuel with a phosphorus grenade. Bullets cut down all but one of the German crewmen as they abandoned their burning battlewagon. A second Panther negotiated the left turn, but a bazooka rocket stopped it dead.[235]

Ayers's tank maneuvered on the Panther column, and Nutter caught one enemy tank in his gunsight. He scored two hits before his crew decided to back off. The driver headed toward a defilade position. He used the burning barn and an adjacent farmhouse for cover, but an AP shell hit the tank and damaged a track.

The crew sprang from their disabled machine as a second shell hit and bore into the crew compartment and engine compartment. Steel fragments flew into the driver's legs and buttocks, and one piece lodged beneath Nutter's left kneecap. He and the loader carried the driver a half-mile to an aid station as hell boiled over at Lausdell.

The artillery-liaison officer supporting the US defenders vectored a torrent of howitzer shells against the approaching enemy. As the rounds burst, stunning explosions of light danced across the fields and roadways. Each blast unleashed a pressure wave that rocked the ground and hammered eardrums as steel fragments flew in every direction. In that maelstrom, all German attempts to break through disintegrated.

When the fight waned sometime before midnight, the two Panther carcasses blocked the crossroads. Nothing German had advanced past Lausdell, except Zeiner's vanguard force, which remained alone and cut off in Krinkelt. The Germans suspended all offensive operations until dawn, but the American artillerymen offered them no respite. Throughout the night, they pummeled the enemy-held forest to the east.

Lost in the chaos, Sergeant Ayers and his gunner never returned to their tank. Ayers landed in German captivity and ended up at Stalag IID.[236] The gunner's identity remains uncertain, but it may have been Corporal Charles F. Urbanas.

To his dying day in 2005, the loader, Virgil Lohman, claimed that an American TD knocked out his tank.[237]

★★★

Inside the twin villages, George Beeley and Ivan Schmidt searched for a house to make their haven. They approached a building and, uncertain who dwelled inside, threw their hands in the air, shouting, "Don't shoot! We're Americans!"[238]

Fortune smiled on the two tankers. Soldiers from the 38th Infantry garrisoned the stone farmhouse. These men belonged to Antitank Company and were under the direction of First Lieutenant Charles E. Gamble Jr. Once allowed to enter the makeshift fortress, Beeley and Schmidt underwent an impromptu interrogation.

"They questioned us about the States to make sure we were not disguised Germans who had been reported infiltrating our lines. When they were satisfied we were Americans, the commander, a Lieutenant Gamble, asked me if he could have my Tommy gun, which I was glad to give him. ... They were as glad to see us as we were to see them. They gave me a rifle and sent me to the second floor as a lookout. Schmidty had a similar job downstairs."[239]

Back where their abandoned tank sat, a venture was underway to retrieve the still-operable machine. Lieutenant Dudley went after it with a tank driver from his platoon, Technician Fourth Grade George J. Clement. The pair slithered inside the hull and found Beeley's flashlight still burning. Clement positioned himself in the driver's seat, cranked up the motor, and the two drove off under German fire. Impressed by the recovery, their company commander later recommended both men for decorations. Clement received a Bronze Star Medal and Dudley a Silver Star Medal.

December 18, about 2:00 A.M.: Helmut Zeiner's radio operator searched the airwaves for a familiar voice.

"We tried, time and again, to establish radio contact with our own tank destroyer battalion. Suddenly, we had a short period of contact with the radio station of a Waffen-SS armored regiment. The conversation was inconclusive since there was constant interference."

Zeiner pondered his situation as daybreak approached.

"I still had no radio contact with battalion, did not have a motorcycle messenger available, did not know where my battalion was situated, probably two or three kilometers behind us in the woods we had broken through the previous day. I only had about forty infantrymen and my tank destroyers only about ten HE shells each. Also, our fuel supply was low."[240]

He concluded that his position would be hopeless once the Americans pinpointed it and realized his actual strength. Zeiner decided to abandon Krinkelt and set up another hedgehog position three hundred meters east.

As Zeiner's force moved out, along with its prisoners, an American captain from the 644th TD Battalion witnessed the exodus. "At about 5:00 A.M., the Germans noisily prepared and ate their breakfast seemingly confident that no American GIs were anywhere near. They then climbed into their tanks and moved back east."[241]

The battle resumed once Zeiner's group reached its new position.

"We had barely arrived there when a thick hail of enemy shells fell. But, another sensation was also in the making. Watching the edge of the forest, we saw tank after tank leaving it in a wide formation. They were German. We waved pieces of cloth so as not to be fired upon by our own people, forgetting to watch out for the white-phosphorus shells exploding near us. Our prisoners crawled under the tank destroyers for cover. Our infantry did the same."[242]

The tanks that Zeiner witnessed were Panthers heading toward Lausdell. The SS onslaught resumed at 6:45 A.M.

American artillery lambasted the attackers as they rolled forward, but, after repeated assaults, their numbers proved too great, and the Germans penetrated the US line. Hand-to-hand fighting and grenade duels abounded in the morning mist. Enemy armor rolled within touching distance of the defenders and pumped shells into one foxhole after another.

Lieutenant Colonel William D. McKinley commanded the defending force. His radio carried painful reports of a hopeless cause. One company commander called for artillery fire on his own position. The forthcoming barrage lasted thirty minutes. The artillerymen dropped one shell on a Panther. The projectile punched a fist-sized hole through the turret roof and detonated, sparking a volcanic eruption that shattered the vehicle. Colossal hunks of armor plate soared through the air, but this lucky hit did little to stem the enemy flood.

It was now 11:00 A.M., and McKinley had permission to withdraw. The Panthers and grenadiers had already overpowered two of his rifle companies. The enemy would blast his remaining men in their backsides if they tried to disengage. They needed help. McKinley's antitank platoon leader, First Lieutenant Eugene V. Hiniski, spotted a rescue force.

He saw four US tanks on the Rocherath-Wahlerscheid road. The antitank officer dashed to intercept them. He hollered at First Lieutenant Gaetano Barcellona, leader of Second Platoon, Company A.

"Do you want to fight?" Hiniski said.

"Hell yes! That's what I'm here for," Barcellona said.[243]

Excited hands pointed Barcellona toward McKinley's dugout. The colonel and his operations officer briefed Barcellona. He then braved enemy fire to reconnoiter the terrain. He scratched out a plan to spring the remaining Lausdell defenders. Barcellona split his platoon, two tanks north of McKinley's planned withdrawal route and two tanks south of it. The northern pair moved first, decoys to attract German eyes.

The ploy worked. Enemy heads turned north while Barcellona's other two tanks moved in from the south. Howitzer crews laid down covering fire as the tanks edged close enough to make their AP shells effective. Barcellona claimed two Panthers in quick succession. Another pair bolted toward Rocherath, and the lieutenant wanted to chase them, but a sunken road blocked the way. His tanks pulled back and circumvented the obstacle to pursue the escaping Panthers. McKinley later described the engagement as a "savage attack" by Barcellona's platoon.[244]

With the Panzer menace temporarily dispersed, McKinley's men—those not already overrun—pulled their noses from the muck and fell back. The fight for Lausdell ended.

The first head count afterward found 217 officers and enlisted men, a sharp reduction from the six hundred soldiers who defended Lausdell the previous evening. The SS had captured 146 men, including the company commander who called for artillery on his own position. Some two hundred other men straggled home during the coming days. The least fortunate remained at Lausdell until Graves Registration teams eventually relieved them from their posts.

McKinley later credited Barcellona with making the withdrawal possible.

"Due to Lieutenant Barcellona's superior knowledge of tank tactics, because of his superb courage, decisive action and bold leadership, under heavy enemy fire, the First Battalion, 9th Infantry was able to withdraw a large majority of its troops."[245]

McKinley and Colonel Skaggs recommended Barcellona for a Silver Star Medal, but division headquarters downgraded it to a Bronze Star Medal. No reason was given.

CHAPTER 30 PANTHER SLAYERS

Despite a rain of steel from American artillery, Panthers from SS-Panzer-Regiment 12 muscled through Lausdell. The first Panthers to enter the twin villages that morning were battlewagons from the 1.Kompanie. Its commander, Hauptmann Walter Hils, was an army officer, who had transferred to the company from a Panther training unit.[246]

Waiting to meet Hils and his armor, Lieutenant Dew's platoon had orders to help protect Rocherath, especially the 741st command post. Three tanks guarded the CP, their gun barrels aiming toward the main street, positioned to ambush the enemy.

Though still mobile, two of the tanks were "deadlined" for repair work. Sergeant Neidrich commanded one of these machines, and the other was Dew's vehicle but with Corporal Curtis Hall in charge. The vehicle had a faulty electrical generator, so Dew had taken over the Charley Company command tank with its crew: Corporal Uleyse G. White, gunner; Corporal Richard J. Dorsett, loader; Technician Fourth Grade Hascal Combs, driver; and Private First Class Gordon R. Conner, bow gunner. Hidden at Nr. 90 Rocherath, their tank sat inside a stone-built barn with a clear view of the main street, its muddy roadbed bordered with leafless hedgerows. Company C's 105 mm assault gun also sat near the 741st CP. Inside the command post, a rifle squad from the

38th Infantry stood ready to defend the premises along with men from battalion headquarters.

Everyone heard the raspy rumble of Maybach motors as enemy tanks advanced into the heart of the twin villages. Dew and his tankers strained to see their opponents through shifting fog. The mist concealed two Panthers that bulled past the waiting Shermans and approached Frank Mildren's CP in the butcher shop as its occupants jumped to defend their ground.

When the lead tank, Panther 125, neared a four-way intersection just past the shop, a bazooka rocket flew into its left side. The detonation severed a track link, and the entire track unspooled from the running gear, causing the tank to pivot right. Those inside lashed back with their cannon and machine guns. The enemy tankers kept shooting despite another bazooka hit.

Mildren's motor officer, First Lieutenant Howard A. Emmerich, encountered the wounded brute while making his way toward the butcher shop. Thinking fast, he hustled back to a 741st tank that he had just seen. After guiding the Sherman into position, he watched as the tankmen loosed an AP round that bored into the turret of 125. Scratch one Panther!

Mortar and bazooka fire had meanwhile subdued the second enemy tank, but another one—Panther 126—proved far more stubborn than its predecessors. It too had slipped past Dew's tanks in the fog. The men inside 126 saw danger ahead. Their periscopes reflected fire and doom. Rather than meet the same end, the crafty Panzer Kommandant ordered his driver to turn onto a side street just past the church. The Panther circled behind the butcher shop before heading back toward the main street. The Kommandant thought he had executed a sly maneuver around the heart of American resistance, but the wreckage of 125 blocked the way.

Seeing this, Mildren's intelligence officer and a sergeant rousted six cows from a nearby barn and drove them toward 126 from behind. The officer shot one of the animals, causing the others to panic. Seeing this unusual roadblock, the enemy crew dodged the cattle by crashing through gardens and yards behind the butcher shop. Caliber .50 bullets meanwhile rained down on 126 from the upper floor of the shop. An empty lot next to the building offered an avenue to the main street, and the Panther plunged through the gap.[247]

Once on the street, 126 rolled back and forth over three jeeps parked on the curb. Their flimsy frames pancaked under the Panther's weight. The Kommandant ordered his driver to charge the butcher shop and Mildren's men inside it. The gun tube of 126 rammed through a first-floor window, but no HE round followed. Why? The GIs inside had no answer.

The enemy tank backed off, damaging a corner of the building as stone and mortar crashed onto the machine's engine deck. Anxious to escape, the enemy tank bounced through Krinkelt like a pinball, penetrating deeper inside the American enclave. The tank veered onto the Büllingen-Krinkelt road, and the crew of 126 claimed a fourth jeep, squashing the vehicle as its passengers scrambled for safety. American onlookers watched in giant-eyed amazement as the jeep struggled to recover from the crushing blow. The crumpled chassis rose from the mud as if trying to unfold itself, like something from a Looney Tunes cartoon.

Nearby, the crew of a 57 mm antitank gun took a crack at 126. The American peashooter heaved one of its pathetically undersized rounds at the enemy machine and scored a hit, but the shell failed to penetrate, though it miraculously welded the turret to the hull. Frantic to break the bond, the driver smacked the cannon barrel against utility poles until the turret broke free. To observers, 126 crashed about as if its driver were drunk.

The assistant operations officer for Third Battalion, 38th Infantry witnessed the mad drama, sprinted to a Company B tank, and guided it into firing position. Sergeant John J. Angeletti commanded the vehicle. His gunner responded with an AP shell, but the projectile missed. The enemy vehicle disappeared from view before the gunner could fire again.

Panther 126 motored ahead, rumbling downhill toward the last house in town—Nr.1 Krinkelt. The crew had run the gauntlet from one end of the twin villages to the other, but now the German tankers blundered in front of an M10 tank destroyer from the 644th TD Battalion.

The TD commander, Corporal (later Sergeant) Stanley V. Kepinski, marked the time as 11:00 A.M. when his crew engaged 126. Their first two AP rounds ricocheted off its armor. The gunner aimed and fired again, but the shot sailed high. The loader rammed another AP shell in the breech, and the gunner tried once more. The round penetrated the rear of 126, straight into its engine compartment, dicing its iron guts. Another shell compounded the hurt.

The dying Panther drifted off the road and rolled to a halt. Kepinski's gunner polished off his work with a round of his hottest stuff—HVAP (High Velocity Armor Piercing). This coup de grâce set the beast alight with flames whipping back and forth on its engine deck.[248]

At the center of town, Kelly Layman also witnessed the helter-skelter fight with Panthers from the 1.Kompanie. The young tanker from Company B began his day inside the church where he had sought refuge during the night. The holy place had become an improvised aid station, its aisles and wooden pews awash with bleeding GIs rescued from the mayhem outside. Medics and doctors from the 99th Division tended to the hurt and dying.

While in the church, Layman chanced upon Sergeant Kenneth V. Williams, a member of his tank crew (Slodkowski's crew). As morning dawned, the situation outside seemed stable, and the two men set out to locate their company CP. About fifty yards from the church, they stopped to ask directions from 2nd Division soldiers, who shrugged their shoulders. Just then someone shouted, "Here comes some Panther tanks!"[249]

Layman and Williams hustled to the nearest house, hoping to find a cellar for protection, but they failed to find one. As the Panthers rolled forward, engines gunning and roaring, infantrymen inside the house prepared their antitank weaponry. The scene harked back to some prehistoric drama with cave dwellers stalking woolly mammoths. Across the millennia, human hands had exchanged spears for bazookas, stone knives for thermite grenades and antitank mines. Yet no twentieth century innovations had replaced animal cunning and good fortune.

The modern-day hunters stood ready with three loaded bazookas. As the first Panther lumbered through the fog, one of the infantrymen let go with a rocket that bounced off the tank's right side. Frustrated, he fired at the second Panther, and that projectile also ricocheted, wobbling off into the air. Determined to score a penetrating hit, the soldier grabbed his last bazooka and charged outdoors to stalk the third Panther at close range. Again, success eluded him.

Huddled together, Layman and Williams sweated out the minutes of terror. Cannon shots reverberated through the house when a fourth Panther stopped and fired on an infantry CP across the street. The two men suddenly heard the tank reverse gear and back up toward their door. Guts twisting with dread, the pair imagined the machine blasting them to blood-red hamburger, but the enemy crew changed direction and drove down the street, giving Layman and Williams an opportunity to flee the house.[250]

"We finally found our company and reported in," Layman recalled. "It was sure good to see them again. I didn't stay long because of my burns. I was evacuated that same afternoon."[251]

Layman's first stop outside the 2nd Division was the 77th Evacuation Hospital, and from there he traveled to a Paris hospital. He spent eighteen days recuperating and undergoing penicillin therapy to prevent infection.

Behind the 1.Kompanie, Panther crews from the 3.Kompanie of SS-Panzer-Regiment 12 embarked on their first day of action in the Ardennes.[252]

The company moved into the twin villages under the tested leadership of SS-Hauptsturmführer Kurt Brödel, a former German Army captain. He had commanded a self-propelled TD company on the Eastern Front until joining the Hitlerjugend Division in March 1944, the same month he became an SS officer. Brödel had over eight years in uniform, half spent in combat. The Soviets, British, and Canadians were familiar foes to the thirty-year-old commander, but today was his first clash with American armor and infantry.[253]

SS-Unterscharführer Wilhelm "Willi" Fischer commanded a tank in Brödel's second platoon. He recollected the nightmare he witnessed on the main street in Krinkelt-Rocherath.

"When we reached a point near the church, we encountered a cruel sight: Beutelhauser was knocked out in front of me, just after both our tanks had crossed the second intersection. When Beutelhauser was hit, I spotted the approximate location of twin enemy antitank guns. Beutelhauser managed to bail out and escape to safety. Rifle bullets killed his loader as he tried the same. I moved my tank into position behind a house, which provided cover and concealment, without knowing, right then, what to do next."

The bewildered Fischer glanced around and spotted Panther 305, his company commander's tank.

"Beside me, I noticed Brödel's vehicle burning gently. Brödel sat in the turret—lifeless. Ahead of me on the street, all our Panzers had been knocked out, some burning."[254]

The gunner aboard 305 survived although injured. While being evacuated, he reported the commander's death to the 3.Kompanie rear echelon. SS-Untersturmführer Willi Engel heard the news while waiting for maintenance personnel to fix his tank. Anxious to fight, he grabbed Panther 325, a newly repaired vehicle that had struck a mine.

At the outskirts of Rocherath, Engel defied a hand signal from the regimental aide-de-camp, who urged him to follow the main street into town. Experience told Engel to avoid that course. He veered left and ran parallel to the street, plunging through barnyards and pastures before swinging right toward the street. Across that muddy thoroughfare, he spotted his battalion commander, SS-Sturmbannführer Arnold Jürgensen, in conference with other officers. Engel hustled over to report. The dire expression on Jürgensen's face underscored the situation.

"At that moment, a single Panzer approached the battalion command post. Suddenly, about one hundred meters away, it turned into a flaming torch."

This was Panther 327 commanded by SS-Unterscharführer Heinz Freiberg. He soon appeared at the CP, his head wrapped in a bandage. Engel added a final thought. "No doubt, I would have suffered the same fate had I followed the directions given at the village entrance."[255]

Staff Sergeant Parker from Company A participated in the battle that morning, having relocated to the battalion CP after losing his tank during the night.

Rifle in hand, the sergeant took position at a window. He barricaded himself behind cardboard boxes filled with 10-in-1 rations and sniped at German infantrymen riding atop Panzers on the main street 175 yards away.[256]

Outside the CP, the two deadlined Shermans commanded by Hall and Neidrich sat about twenty-five yards apart. Hall had a makeshift crew with Private Frank J. Fiorito as the gunner, Technician Fifth Grade Shirley R. Schuh as the loader, and Technician Fourth Grade Sam E. Walker as the driver. Normally assigned to Company C headquarters, Schuh was a cook and Walker a maintenance man, but they had volunteered that morning to replace two sick tankers. In the other deadlined tank, Neidrich had an experienced crew with Corporal Herman A. Johnson as the gunner, Private Lee C. Rains as the loader, Private Howard C. Rowe as the driver, and Private George R. Cambron, the bow gunner. Firing shell after shell, the two crews reportedly knocked out five enemy tanks with flanking shots. One of their victims was the Panther commanded by Hauptmann Hils, though he escaped on foot. All nine men received Bronze Star Medals for their achievement.

Dew and each man in the Charley Company command tank also earned Bronze Stars. They delivered the deathblow to Kurt Brödel's Panther 305 with an AP shell that struck at an angle below the right sponson and cracked the hull armor, leaving a large, jagged hole.

Another Bronze Star went to Sergeant Padgett, who shared credit for destroying Heinz Freiberg's Panther 327. The kill was a joint effort between the sergeant and Captain James W. Love, commander of Antitank Company, 38th Infantry. It was Love who spotted 327 entering Rocherath and brought Padgett's tank to bear.

Love explained what happened. "I led the tank into position and stood not over thirty yards away while it fired to its left front."[257] Two AP rounds brought down the Panther. Its cannon barrel went limp, angling down to the muddy street. Bullets flew toward the crewmen when they emerged, but Freiberg survived to report his narrow getaway.

Helmut Zeiner and his crew returned to the twin villages at midday on December 18. "The devil was loose as we arrived in Krinkelt. To my right, there were small houses and a sizable orchard that ascended to a row of fruit trees. Enemy-tank silhouettes suddenly appeared there, but these particular tanks did not have their sights set on us. I swung to the right to engage the first tank when a shell struck the engine deck of my vehicle. Fragments hit me in the head and right arm while standing in the open hatch. The hatch cover protected my chest and back, thus my wounds weren't life-threatening. I was bleeding and unconscious, so my comrades lugged me into a nearby cellar where one bandaged my head."

The shell splinters that hit Zeiner amounted to his second injury that day. "It happened that I had already been wounded by a shell fragment during the early morning hours. This first injury initially looked harmless, but it turned serious because phosphorus or another toxic substance came into play. I suffered from blood poisoning for three months in a hospital."[258]

Lieutenant Coleman from Company B received a call for help when SS armor first rolled into the twin villages that morning. The call came by radio from 741st headquarters.

The lieutenant, First Platoon leader, and his four tanks stood guard atop a hill south of Krinkelt watching German traffic pass through Büllingen. They had arrived there the previous day after the Company B tank-recovery vehicle cleared aside destroyed armor from Kampfgruppe Peiper following that unit's abortive probe from Büllingen on December 17.

After the call for help, Coleman and his men pulled out and headed to rendezvous with their company commander, Captain Cinquina, who had positioned himself at a road junction next to the Krinkelt cemetery.

Coleman led the way followed by Sergeants Edward Mazzio, Richard Maddock, and Wilbert Mortzfield. As the tank column entered Krinkelt, an excited voice shouted over the company radio channel, "Look out for enemy tanks around the church!"[259]

Corporal Wardell Hopper, Coleman's gunner, reacted to the warning and spun his turret toward the church. At the cemetery, the tank's cannon barrel struck a thick, stone wall surrounding the burial ground. Hopper never saw it coming. The impact hung the traversing mechanism, freezing the turret.

Clad in his familiar olive-drab sweater, Coleman sprang from his damaged tank and charged over to Captain Cinquina, who was holding a conference by his jeep. The captain's command tank sat nearby, and he relinquished it to Coleman along with its crew.[260]

Corporal Frank Manganaro was the gunner; Private First Class Robert H. Aldrich, the loader; Technician Fourth Grade Ernest S. Reed, the driver; Private Roy A. Garber, the bow gunner and assistant driver. The four men usually found their place in the rear with the captain, but today would be different.

Cinquina told Coleman to establish a blocking position in Rocherath with First Platoon and engage enemy armor reported there. The lieutenant dashed to each of his tanks, shouted instructions to the commanders, then jumped into his new vehicle and drove toward Rocherath.

About nine hundred yards up the road, Coleman's tank crossed a four-way intersection and traveled another one hundred yards along a dirt track. The tank shuddered to a stop as it crested a gentle rise. In Maddock's tank, Jack Boardman heard his platoon leader on the company radio channel. There was a dreadful shriek, then silence. An enemy AP round had struck home.

Smoke from a burning German halftrack obscured the scene. "The smoke was blowing across the road," Boardman remembered. "I think that's what saved us because at least two cannon tracers came right between our tank and the one just in front. I could see them through my gunsight."[261]

Maddock's driver threw the tank's transmission into reverse gear, churned backward, and turned off the road into a garden. Mazzio and Mortzfield—fore and aft of Maddock—also scampered for cover like mice scurrying from a cat. Mazzio crawled between two houses. Mortzfield backed into a barn, closing the door with only his cannon muzzle peeking out.

Coleman's tank sat motionless. Boardman listened as his radio operator and others tried to raise the lieutenant. There was no response, only a sea of crackling static. "We kept calling him over the radio. Just nothing. It was sad."[262]

The platoon leader and his crew had perished, though witnesses said the lieutenant bailed out of his vehicle only to be cut down as he fled.

Six weeks later, Boardman returned to Rocherath and inspected his platoon leader's tank. Its cannon barrel was missing, somehow blown off. The inquisitive corporal also discovered a single hole in the tank's frontal armor just above the transmission housing. There was no exit hole, and Boardman observed little damage inside except four dead crewmen. The driver and assistant driver sat slumped forward in their seats with their heads down. The loader lay sprawled on the floor. Manganaro rested in the gunner's seat, his left hand on the elevating wheel, and all the fingers on that hand gone. The lieutenant had somehow escaped the tank. Boardman failed to locate his body, but Colonel Skaggs and Major Browder found him, his boots protruding from the snow. Graves Registration men later determined he died from a gunshot wound to the head.

Coleman's parents in Elgin, Illinois, received the news three weeks later. Western Union delivered a telegram from the Adjutant General: THE SECRETARY OF WAR ASKS THAT I ASSURE YOU OF HIS DEEP SYMPATHY.[263]

The Adjutant General also dispatched telegrams to the other four families, including Arthur and Irene Garber in Boston. The Garbers passed the sad news of Roy's death to his brother Ralph, who occupied a hospital bed in the United States, still recovering from his September 1944 injuries. Ralph later learned that Ernest Reed perished with Roy. Ernest had driven Ralph's tank in September when the Germans destroyed it. The twenty-four-year-old driver escaped that calamity unharmed, but his luck ended in Rocherath.

Coleman's entire crew lay in graves at the Henri-Chapelle temporary cemetery. Workers there inventoried personnel effects removed from the dead soldiers. Coleman's possessions included a knife, collar insignia, an identification bracelet, and a gold signet ring, the one his father wore in combat during the First World War. The US Army returned the mementos to the fallen officer's heartbroken parents.

Lieutenant Coleman was one of 19,246 Americans whose lives tragically ended during the six weeks of the Ardennes-Alsace campaign.

CHAPTER 31 KRINKELT-ROCHERATH ABANDONED

Arnold Jürgensen hurled his two Panzer IV companies into the fight. These tanks reached Krinkelt-Rocherath on December 18 behind the two Panther companies. Together they constituted the whole of Jürgensen's battalion, which became the armored battering ram for Kampfgruppe Müller.

Jürgensen was a devoted career soldier contracted to serve until 1955, and he wore the trappings of success, including a Knight's Cross at his neck, a German Cross in Gold on his breast, and a Death's Head ring on his left hand. Although a respected commander, his stony façade and air of patrician authority led to a nickname—*Der Herzog* (The Duke). In Normandy, Jürgensen's battalion claimed 321 British and Canadian tanks, but all that seemed like a half-forgotten dream compared to the present bloodbath.

He established his CP in the local police station on the main street in Rocherath, only four hundred yards from Nr. 100 Rocherath, where his counterpart, Colonel Skaggs, had his CP.[264] With the SS roosting so near, Skaggs felt growing pressure to shift his headquarters to a safer locale, and he reluctantly gave the order. At 5:00 P.M., in the late afternoon darkness, his operations officer began orchestrating the move to a new CP in the southwest corner of Krinkelt, the least embattled part of town.

Outside Nr. 100 Rocherath, the 105 mm assault gun from Company C stood guard in the dark. Corporal Ralph Woodward was its gunner. He entered the house and witnessed his battalion commander in despair. "Skaggs broke down," Woodward recalled, "I remember him sitting in the basement, on that dirt floor, crying like a baby."[265]

The battle pushed many men toward combat exhaustion. Their efforts had exacted a fearsome toll from the enemy, but they had failed to halt the attack. Despite all, only four 741st soldiers developed neuro-psychiatric symptoms severe enough to warrant evacuation to the 622nd Clearing Company. Most stayed on duty, including Skaggs, who pulled himself together.[266]

The old CP remained in use for a short time as a supply and evacuation point. First Lieutenant Wilbur T. Latshaw, the transportation officer, had earlier delivered three truckloads of gasoline, two loads of AP ammunition, one load of HE shells, and another truck with rations and potable water. After dark, the company commanders received word to begin sending their tanks in small groups to the old CP for fuel and ammunition.

Turner Sheppard, the newly commissioned lieutenant from Company B, headed security for the operation. The Intelligence and Reconnaissance Platoon from the 38th Infantry assisted him. "Spare men" from Headquarters Company remained behind to help as did several crewmen from disabled tanks. Flames from nearby house fires cast a flickering orange glow on the scene, making it too bright for safety. Turner shifted refueling and rearming to the cemetery.

Back at Nr. 100, Staff Sergeant Crisler's tank sat near the 105 mm assault gun. Four crewmen snoozed inside while the gunner, Corporal Charles W. Eubanks, kept watch from the commander's seat. He poked his head through the turret hatch, exposing only his eyes and helmet. As he peered into the darkness, his ears registered something untoward, a commotion nearby, but he could see nothing.

The driver's voice suddenly rang out below, "Somebody's pouring gas on us!"[267] Combustible liquid seeped into the crew compartment, soaking the driver and bow gunner while filling the interior with noxious fumes. An enemy flamethrower operator had tried to burn the tank, but human error or mechanical malfunction caused the flamethrower to spew nothing but its fuel, a tar-and-gasoline mixture called Flammöl.

Eubanks snatched a hand grenade and prepared to toss it toward the enemy. He yanked the pin just as Crisler bolted up through the hatch. The two men became wedged in the circular opening. Eubanks hollered, "I've got a live grenade in my hand!"[268] Crisler dropped back inside, and Eubanks pitched his grenade. He had no idea if the explosion caused damage, but the flamethrower operator disappeared into the darkness.

Days later, Crisler described the incident to a *Chicago Tribune* reporter. "Boy, oh boy, did that stuff stink. It smelled so bad we could hardly stand it. We thought it might have been a new V-weapon."[269] Humor was only possible in hindsight.

That same night, the Company B tankdozer crew also confronted enemy foot soldiers. The crew had thus far spent the battle parked outside the twin villages, never firing a shot. The dozer commander, Sergeant George Cora, suddenly excused himself from the vehicle, saying he would return. The four men Cora left behind wondered why their sergeant departed, perhaps to obtain instructions. The crew waited, but Cora never reappeared. Fearing the worst, the men drove off to rejoin their company in Krinkelt-Rocherath, which they thought lay entirely in American hands.

The error of that assumption struck hard when an enemy mine or projectile halted the dozer. The four crewmen emerged from their hatches, tossed their submachine guns over the side, and jumped to the ground. They never had a chance to grab their weapons. German infantrymen captured Corporal Joseph T. Walsh, the gunner; Private James P. Coffin, the loader; Technician Fifth Grade Clayton A. Howard, the driver; and Private Neal F. Hockenberry, the bow gunner.

The Germans hustled the unlucky quartet into a cellar and separated the prisoners from one another. Each captive underwent a cursory interrogation with questions regarding his rank, name, and unit. Walsh volunteered that he was from Company B, 741st. He reasoned there was little point in withholding that information since the Germans possessed the dozer and could see the white stenciling on its rear hull that identified the company and battalion.

Besides undergoing interrogation, the prisoners forcibly relinquished their valuables. Walsh lost his wristwatch and class ring from Saint Mary's High School in Berkeley, California. For the next couple of days, his captors, whom he identified as regular army (perhaps soldiers from the 277.Volksgrenadier-Division), kept him confined in the cellar along with his fellow crewmen, whose voices he occasionally heard. The Germans eventually brought him upstairs, and he began a long foot-and-railway journey to Stalag IID at Stargard in Pomerania. His comrades endured much the same. Coffin traveled to Stalag IVB at Mühlberg, while Howard and Hockenberry went to XIB at Fallingbostel.

Walsh remained at Stargard until the Red Army drew near in March 1945. The Germans compelled the internees to vacate IID and trudge west across the Elbe River. In April, following a night spent in a barn, he and the other prisoners awoke to find their German guards had vanished. Soon afterward, the 29th Infantry Division arrived, and so ended their captivity.

As for Sergeant Cora, his company listed him as missing in action but later discovered that medics had evacuated him to the 47th Field Hospital.

December 19, 4:10 A.M.: Staff Sergeant Crisler heard the distant clatter of tank tracks, and he notified battalion headquarters that he heard German armor assembling in the fog and darkness northeast of Rocherath. Jürgensen's battalion and Kampfgruppe Müller were repositioning for a final thrust to dislodge the GIs still clinging to the streets, houses, and farmyards of Krinkelt-Rocherath.

At 8:00 A.M., Sergeant Maddock from Company B reported "five enemy tanks were approaching from the northeast."[270] The whine and thunder of incoming artillery followed his report. The shells reignited hostilities in the villages.

Maddock's tank and the other two from Coleman's platoon joined Lieutenant Dudley's platoon, which now numbered eight tanks, all hidden in Rocherath. While waiting for instructions, Maddock monitored the battalion radio net and heard that the enemy had knocked out Lieutenant Barcellona's tank, though mercifully the lieutenant and his crew survived.

Lieutenant Dudley meanwhile conferred with the infantry, and, at 9:29 A.M., he radioed that tanks from his platoon would shift position to help blunt the attack. He sent out two of his sergeants on foot with an infantry officer to assess the situation. Sergeants Case and Dickson accompanied the officer into a house and up to a second-floor window. The officer pointed toward a lone Panzer. Case and Dickson brought their tanks around to destroy it but discovered four enemy machines where there had been one. An infantry captain rushed over and ordered the tankmen to blast the enemy armor.

"To attract the Krauts' attention," Dickson recounted, "Case fired on the left flank from a covered position, while we moved on the right, between two buildings about forty-five yards from the enemy."

Dickson's gunner, Corporal Kroeger, fired four AP rounds, each bouncing off the target. The enemy returned fire and narrowly missed Dickson's tank, instead hitting a nearby building, though a shell fragment pierced Dickson's steel helmet but left him unharmed. He later recalled what transpired next. "We immediately backed up and took cover behind a building, and the Krauts started an encircling movement."

The jig appeared up for Dickson and his crew. "We called for artillery and additional support but thought we'd never get it in time. A few minutes earlier everyone was excited, now we thought this was the end, and nobody said a word."

Dickson watched smoke roll from a burning house until the pall grew thick enough to use for concealment. "We took off like a bat from hell," he remembered.[271]

Minutes later, a Jagdpanzer crashed through a stone wall, its crew hoping to complete the encirclement, but Dickson and his crew were gone.

Hidden alongside a building, Sergeant Mazzio's tank had its cannon barrel pointed in that direction. The gunner, Corporal Snike, followed a cow through his gunsight and suddenly felt hungry, his mind fixated on prime rib. Just then, the animal wandered past the Jagdpanzer, its squatty hull partly concealed from view. Snike fired at the machine and set it on fire. Cat-and-mouse contests between US and German armor continued into the afternoon. The jumbled battlefield was unlike anything the Americans had experienced. No contiguous frontline existed, just scattered strongpoints established by friend and foe.

Lieutenant Dew from Company C reported at 1:40 P.M. that an enemy tank was near the old battalion CP, and he intended to maneuver for a side shot. He also expressed uncertainty whether the Panzer was still operational because it was stationary and perhaps already dead in its tracks. Ten minutes later, he transmitted news that his crew had scored six hits, and the enemy vehicle stood blazing.

At the highest point on the battlefield, a German machine gunner waged war from the church bell tower. He sat snug in his stone fortress, firing from the eight louvered windows surrounding the belfry. The infantry failed to stop him with bullets, and the situation required greater firepower. Company C had the answer in its 105 mm assault gun, which was in the neighborhood. Sergeant Leonard B. White commanded the vehicle, and he accepted the challenge along with his gunner, Ralph Woodward.

Just above the belfry hung the central clockwork, which controlled four clock faces. Woodward reasoned that if this hefty mechanism could be dislodged, it would crash down, tear out the belfry's wooden floor, and eliminate the machine gunner. All Woodward had to do was hit one clock face. He chose the one on the front of the bell tower and fired several HE rounds before hitting the target dead center. The ripping blast ended the enemy shooter's reign over the battleground.[272]

At the new battalion CP, Nelson Hall followed the fight for the villages in baseball fashion with a radio receiver providing the play-by-play. "We listened continually to tanker talk and cheered silently as we heard a B Company tank commander tell another commander to back up between two houses where he could get a side shot at a German tank that was moving up the road. A moment later we knew the move had succeeded, when someone said, 'You got the sonofabitch!'"

During the radio drama, "company commanders came and went, getting advice and orders from Colonel Skaggs. But through it all there came a great feeling of certainty that even though we had held off the attack for three days, time was not with us."[273]

At 2:40 P.M., Skaggs learned from 2nd Division headquarters that all US forces were withdrawing from Krinkelt-Rocherath after dark. His tank battalion and TDs from the 644th received instructions to form a rearguard.

Inside the battalion CP, everyone knew that bringing up the rear meant possible envelopment by the enemy. Skaggs ordered Nelson Hall to burn all his documents, including a unit diary he had kept since June. The sergeant emptied his field desk and incinerated all paperwork in the kitchen stove.[274]

Hall listened as Skaggs instructed his staff to be prepared for capture. "All of us in this room should be ready to tear off our insignia if we see we're going to be taken prisoner. It might make it a little easier for us if nobody knows anything about us."[275]

Skaggs also briefed his company commanders. Captain Sledge from Able Company bristled when he heard comments about capture and surrender. The captain leaned into Skaggs. "I told him, 'In a pig's arse you're gonna surrender! You can surrender B and C Companies, but you're not gettin' A Company.'"[276]

Sledge stormed away and returned to his men.

Captain Cecil Thomas, the operations officer, a quiet, courageous soldier with a master's degree in agricultural economics, directed preparations for the retreat and rearguard action.

At 5:30 P.M., the first 2nd Division elements began withdrawing along the Krinkelt-Wirtzfeld road, accompanied by armor from the 741st. Two machines from Sledge's company bogged down while trying to circumnavigate a roadblock. Stuck in mud and with no time for help from a recovery vehicle, the two crews abandoned their tanks but scuttled them before departing.

The procession of haggard faces and mud-caked vehicles lasted two and a half hours before all had cleared Krinkelt. The only American combatants still in town were Lieutenant Dudley's eight tanks, a few riflemen, and a combat engineer platoon. The engineers had remained behind to lace the road with antitank mines, thwarting any effort by the Germans to rush the retreating column from behind.

After the engineers finished sowing the mines, Dudley's tankers drove toward Wirtzfeld from where they would move to Berg, a lakeside village where the battalion was assembling (less Company D at Bütgenbach). As they rolled west, the tankers encountered cooks and clerks from their company shuffling in the opposite direction. Outfitted with carbines and sullen faces, they were substitute infantrymen sent to help cover the withdrawal. The tankers gave the unhappy foot soldiers a hearty cheer, but their facetious gesture drew no response.

Dudley's tank was the last to leave Krinkelt. On its engine deck were leftover mines, which he deposited along the rutted road while driving west. Every few hundred yards his driver stopped, and Dudley planted four or five mines.

In Wirtzfeld, Dudley's machine passed six tanks from Company A positioned to protect the US withdrawal from the town. Three were under First Lieutenant McDonough and three under First Lieutenant Thomas A. Brooks. This screening force—later joined by Second Lieutenant Hugh M. Helm's Third Platoon from Company B—remained in place until early on December 20. They were the last GIs in Wirtzfeld. Brooks's men, who brought up the rear, set out antitank mines before departing, but no SS armor and infantry pursued.

The survivors of Jürgensen's battalion and Kampfgruppe Müller proceeded to the Büllingen sector by way of Mürringen. They were to hurl their remaining energy and firepower at a new adversary, the 1st Infantry Division. Foot soldiers and assault guns from the 3.Panzergrenadier-Division took over the tortured streets and homes of the twin villages.

The battle for Krinkelt-Rocherath cost the 741st one T2 recovery vehicle and eleven medium tanks, three of the latter destroyed by their crews. On the enemy side, no loss records survived the war, but photographic evidence confirms the destruction of sixteen machines: eleven Panthers, three Panzer IVs, and two Jagdpanzers. Others were knocked out but made operational again.[277]

Siegfried Müller earned a Knight's Cross for his leadership, but the German victory at Krinkelt-Rocherath brought no exultant hurrah from men of the Hitlerjugend Division and no triumphant communiqué from Panzer-Armee headquarters.

Forty-eight hours of horrendous street fighting reduced the twin farming towns to a sprawling bloodstain. The two foes clashed with unmitigated ferocity, but, as George Beeley noted fifty years later, chroniclers of the battle often took a slanted view of the determination shown by the men in olive drab and their counterparts in Feldgrau.

Beeley wrote, "History books tell us the Germans fought fanatically and we fought bravely. In reality, both sides were the same."[278]

CHAPTER 32 **WINTER BASTION**

Panzer-Armee headquarters ordered Generalmajor Walter Denkert's 3.Panzergrenadier-Division to attack Elsenborn Ridge. The 2nd and 99th Divisions had made this their new bastion. Here, as everywhere in the Ardennes, Hitler's minions still chased his fantasy of triumph in the West, a dream fast turning to frozen corpses and burned-out Panzer hulls.

Denkert's troops mounted a headlong thrust to inaugurate their contribution to the hopeless offensive. They crept from the morning fog on December 20, striking the ridge and entrenched soldiers from the 99th Division. The attack began around 10:00 A.M. with an exchange of small-arms fire. The main effort started ninety minutes later. With their steel tracks clanking, enemy assault guns rolled forward, but then came an attempt at deception—German soldiers carrying white flags (a gambit familiar to the 741st from Normandy). Whatever advantage the enemy hoped to gain by this hoax quickly petered out. The 99ers opened fire.

American artillery pounded Denkert's men and armor and kicked them backward in disarray. The artillery reaped a cruel harvest of dead and maimed Germans. Six assault guns lay derelict, bogged down in a marshy bottomland facing the ridge.

12:55 P.M.: 741st headquarters received notification that thirty enemy tanks had reportedly massed opposite the ridge. The 2nd Division operations officer asked that all 741st tanks be prepared to launch a counterattack.

Less than thirty minutes later, General Robertson rang Skaggs to request that he dispatch ten tanks from Berg to Elsenborn, where they could better respond to any German penetration in the 99th sector. The mission fell to Lieutenants McDonough and Covington from Companies A and C, respectively. McDonough had four tanks, and Covington's platoon was at full strength with five tanks. It had supported the 395th Infantry (99th Division) during the slugging match for Krinkelt-Rocherath and had stayed well north of the hottest combat.

The nine tanks under the two lieutenants arrived at Elsenborn. McDonough reported to Brigadier General Hugh T. Mayberry, assistant 99th Division commander. Both platoons joined the 99th and had riflemen assigned to protect against the handheld antitank weaponry ever present among German infantry formations. For added firepower, the Assault Gun Platoon from Headquarters Company moved to Elsenborn.

After sundown, the two tank platoons expected a call after the enemy launched a two-pronged attack. Grenadiers and two assault guns from Denkert's division pierced the 99th line, but, like the previous attack, this attempt failed. American artillery bled the attackers to death. The 99th needed no 741st tanks to curtail the breakthrough.

All the while, Company B stayed on alert, but none of its tanks departed Berg. The entire company sheltered in a barn. Throughout the night, its tin roof rattled with the metallic clink of falling shrapnel. The clatter caused at least one tanker, Jack Boardman, to surrender his spot in the comfy hayloft, grab his sleeping bag, and dart to the icy but protected space beneath his tank.

Less than a mile away in Bütgenbach, Company D also stayed on alert, its light tanks supporting the 1st Infantry Division. One of its regiments now faced the Hitlerjugend Division, hurling back two vicious attacks by the division on December 20.

After the second attack, Private Mitchell F. Adamek, a short, muscular bow gunner with Company D, spotted SS troops on a nearby hill. They had pinned down three GI linemen sent to repair telephone wires torn up by artillery fire. Adamek grabbed a submachine gun and sprang out of his light tank. He bolted across fire-swept pastureland and sprayed the SS men with bullets, an act that allowed the linemen to escape.

Concussion from a shell burst flattened the courageous tanker, leaving him unconscious. After awakening, he carried on the fight, acting as an observer for an M10 tank destroyer. Adamek and the gun crew claimed two enemy tanks destroyed.

His fighting spirit earned him a Bronze Star Medal, sergeant's chevrons, and command of a light tank.

December 21. All day long, enemy batteries lashed rear-area targets and administered sporadic punishment along the frontline. Shells smashed into Berg where the 741st had its base camp and forward CP. The rear CP was still in Robertville along with Service Company.

Enemy artillerymen also pounded Elsenborn. One explosion killed four crewmen from the Assault Gun Platoon, who were in the open replenishing their ammunition stock. When the shell detonated, Sergeant John Unterseher and three others stood outside their vehicle passing howitzer shells to a man inside who escaped harm. Private Herbert F. Rhode died on the spot. The others later perished from their wounds at the 2nd Medical Battalion's Collecting Station set up in a tavern just north of Sourbrodt. Shell splinters also wounded two men. Sergeant Joseph T. Rutyna's right arm had a compound fracture and nerve damage. Private Clinton H. Kelley had steel fragments in his legs. His injuries required skin grafts and left him permanently disfigured. Doctors eventually discharged both men from the army.

Amid the shell fire, Lieutenant Covington's platoon returned to Berg. McDonough's platoon remained in Elsenborn, soon joined by the rest of Company A. Together, they formed a mobile reserve for the 99th should Denkert launch another assault.

Little German movement occurred on December 21 until the afternoon, when an enemy soldier waved a white flag while advancing toward the 99th line. Two armed men followed him. The onlooking GIs considered this another *ruse d'guerre* and called for howitzer shells. As the artillery thundered, enemy assault guns tried to spring an attack.

American cannoneers broke apart all their efforts. Like the day before, there was no call for 741st armor.

Everyone anticipated another attack by the 3.Panzergrenadier-Division, but none came on December 22. The division was done for, its armor and infantry units depleted. Only its artillerymen remained active, many of their rounds falling on Berg. Shell fragments killed Technical Sergeant Elmer W. Becker, crew chief for the Company C maintenance section.

Called the "Battle of the Bulge" by journalists, the epic struggle in the Ardennes kept the 741st personnel officer laboring to replace men. He and the battalion commander temporarily demobilized one platoon from Company D to help mitigate the personnel shortage. Its men joined the medium tank companies to operate vehicles with missing crewmen or none altogether.

The weather had thus far remained one or two degrees above freezing during the day. That now ended. Thermometers fell, and barometers rose as a high-pressure system eased over the Ardennes. December 23 was bitterly cold but sunny and cloudless.

The clear weather brought out warbirds from both sides. Allied and German aircraft scribbled the sky with vapor trails as they battled in the blue. For most ground troops, the air war was beyond the range of their weapons, but that changed for an instant on Christmas Day.

About noon, a lone enemy fighter buzzed the 99th sector at treetop level and invited bullets from nearly every automatic weapon on the ridge. Lieutenant Frank Klotz, the Company A maintenance officer, attacked the German flier with a caliber .30 machine gun and claimed partial credit for downing the aircraft. The pilot unholstered a pistol as he floated down in his parachute and began firing. GIs everywhere returned fire and riddled the German.

The 741st, too, had a death on Christmas Day. Sergeant Clarence W. Blankenship from Company A died at the 2nd Evacuation Hospital in Eupen. Shell fragments had ripped out brain tissue six days earlier. Surgeons performed a craniotomy, but the prognosis was hopeless.

Christmas brought cheerful moments, too. Cooks throughout the 741st prepared turkey dinners, the first hot food many tankers had enjoyed since the battle started. There were also Christmas parcels from home. Yet the holiday was anything but carefree. The men continued to battle the knife-edge cold as well as mortar and artillery shells.

Corporal Valentine Fister, like many in Able Company (and the battalion), had nothing but an overcoat and a blanket for warmth. No matter where he hid from the weather, inside his tank or in a foxhole, the frigid temperatures reached him. When he awoke on December 31, both his legs felt like icicles. Desperate to find warmth, the corporal set out for the company kitchen in Elsenborn. After a painful hike, he tramped into the two-story dwelling, which housed the kitchen. While eating breakfast, Fister complained in a weary, disgusted voice that he "couldn't sleep in that damn tank anymore."[279] As he espoused his intention to go on sick call, one of his compadres interrupted. Technician Fifth Grade William J. Myers, the company commander's jeep driver, offered Fister a space upstairs near a wood-burning stove. Fister considered the offer, but wisdom whispered a warning: avoid the top floor of any building within range of German artillery. Rather than report for sick call, he found a spot on the ground floor of a nearby house.

That night, enemy shells swept down on Elsenborn, and one pierced the roof above Myers. The blast killed him and left others bleeding. A clock in the house died too, its hands frozen at the midnight hour.

CHAPTER 33 REPLACEMENT TANKS

New Year's Day. First Army Ordnance had replacement tanks for the 741st, an up-gunned variant of the Sherman—the M4A1(76)W with a long-barreled 76 mm cannon.

The new machines were among 254 US medium tanks supplied to the First Army from British sources after an urgent request for replacement armor to help offset losses suffered in the Ardennes.[280] Only five M4A1s with 76 mm cannons were available for the 741st, enough to outfit one platoon. Colonel Skaggs selected Lieutenant Dudley's Second Platoon, Company B to receive the tanks.

On January 1, Dudley and several tank drivers set out for the 72nd Ordnance Group to fetch the tanks and ferry them to the company bivouac at Berg. The next evening, the lieutenant and his drivers returned with the machines but only reached the rear CP in Robertville. All roads leading forward were closed. Something was brewing.

The next day, 2nd and 99th Division troops (and other V Corps troops) launched patrols supported by artillery fire. These coordinated stabs at the enemy were a diversion to persuade the Germans that a general attack had erupted along the entire First Army front. The real attack zone was less broad, with five divisions plunging into enemy territory to capture a bomb-blasted city named Houffalize. First

Army planned to link up there with Third Army troops driving north from Bastogne, a north-south squeeze to erase the enemy intrusion into Belgium.

As part of the diversion, the 38th Infantry sent three patrols through knee-deep snow to collect intelligence information and torment the Germans holding Wirtzfeld. Positioned to support the patrols, the tanks of Company C joined an artillery bombardment, which lasted for thirty minutes beginning at H-hour (8:00 A.M.).

The only difficulty was the gripping cold. Several tank crews discovered that their cannons had frozen tight after ice formed in their firing mechanisms. Following the bombardment, the day's work ended for the tankers, and they returned to Berg having fired 458 high-explosive rounds and 305 white-phosphorus projectiles.

German artillerymen reacted to the diversion with shell fire of their own, but it petered out by noontime. Normal road traffic resumed, and Dudley's new tanks moved out to join Company B in Berg. Soldiers from the 741st billeted in the town emerged from barns and houses to see the new tanks built in Chicago by the Pressed Steel Car Company.

The machines had "wet stowage," a new arrangement for housing cannon shells in racks surrounded by a hollow jacket filled with "Ammudamp," a liquid quenching solution concocted from water and ethyl sodium potassium phosphate. Theoretically, if steel fragments punctured the jacket, its liquid contents would quench the white-hot pieces of metal and saturate any propellant spilled from torn shell casings. Should flames erupt elsewhere in the crew compartment, the liquid might briefly protect the shells from blazing temperatures. Such was the thinking behind Ammudamp, but its actual effectiveness remained unproven.

The racks themselves lay on the hull floor below the turret basket except for a ready box with six shells just beneath the cannon. Earlier Sherman versions had ammunition storage under the turret basket, but shells also rested in sponson racks on either side of the hull. Steel slabs welded to the exterior hull protected these areas from direct hits, but catastrophic fires and explosions often started in the sponson racks. The enemy had little difficulty penetrating the steel slabs with AP ammunition. The Ordnance Department conducted tests at Aberdeen Proving

Ground and confirmed this fact, something already known to veteran tankers. Ammunition storage in the sponsons ended. The liquid-protected racks occupied a less vulnerable region under the turret basket, and therein lay the chief advantage of wet stowage. As part of the redesign, draftsmen more than doubled the storage capacity below the basket and encased the racks in quarter-inch armor, including the ready box. All this represented an improvement, though tank crews often foiled the system by stuffing every nook with extra rounds.

Along with wet stowage, the new tanks mounted a high-velocity, 76 mm gun that afforded greater accuracy than its 75 mm predecessor. The larger weapon could also penetrate an additional inch of armor plate, but against the best Panzers, this was a difference without a distinction. In Normandy, outfits such as the 3rd Armored Division first received the M4A1(76)W and learned that the heaviest German armor remained impenetrable. The 76 mm shell could gouge a deeper furrow in a Panther's front armor but could never penetrate at distances greater than two hundred yards. In more vulnerable spots, the sides and rear, penetration performance was almost unchanged.

The larger cannon necessitated a redesigned turret. This included a hatch for the loader, who had no escape hatch on most previous variants (DD tanks were an exception). The lack of a second hatch could be lethal in a fire. With no quick way out, the loader had to duck under the cannon breech and scramble out through the commander's hatch, usually having to wait for the gunner and commander to exit first. That delay cost lives.

Alongside the 76 mm gun, there was an improved telescopic sight: the M71D, a 5-power instrument with a 13° field of view. This was nearly equivalent to the sights found in the Tiger and Panther. The M71D was superior to the M55 found in most 75 mm Shermans, which was a 3-power device with a 12° field of view.[281]

The army also introduced a simple device to improve drivability. On New Year's Day, the 741st received its first shipment of extended-end connectors for use on tank tracks. These metal castings, locally manufactured, replaced the standard connectors that joined one track link to another. The new connectors had a distinctive shape, which brought about their nickname—duckbills. They added about four inches in width.

The added width improved flotation on marshy ground. Ironically, the Tiger and Panther, which far outweighed the Sherman, had greater flotation due to their massive tracks. American tanks bellied in the mud while their enemies rolled across sodden fields. Duckbills reduced this disparity but never eliminated it. These devices unfortunately suffered from excessive breakage. They nevertheless worked better than nothing on medium tanks, although it was impossible to install them on tankdozers.[282]

Duckbills, along with new cannons, better ammunition storage, and a loader's hatch, increased an American tank crew's ability to survive and succeed on the battlefield.

Prior to these improvements, the Sherman had advantages. It was easy to maintain, especially its powertrain. Motor and transmission replacement were simple tasks compared to the Tiger or Panther. It burned less oil and gasoline than the heavier German machines, giving the Sherman greater range for swift-moving offensive operations. The American medium tank also employed a power traverse that rotated its turret faster than any system in enemy tanks, and the traverse functioned with or without the motor running. Unlike German armor, the Sherman included a gyrostabilizer integral to its cannon. This innovation helped a well-trained gunner keep the reticle in his gunsight on target while on the move, allowing him to engage the enemy quicker after stopping to fire. The ability to shoot first often meant the difference between life and death.

These advantages earned praise from 741st tankers but evoked no trepidation among their enemies. In many minds, little about the Sherman made it a worthy opponent for Germany's best. Yet it proved effective at the twin villages amid the largest tank battle during the "Bulge."

Intangible qualities made a difference, too. These included crew training, experience, and resourcefulness. Soldiers in the 741st also followed tactics prescribed in *Armored Force Field Manual 17-33*, the handbook for tank battalions. Originally published in 1942, the manual expounded on "tank versus tank action." The preferred tactic was to withdraw and funnel enemy armor into one's own antitank defenses. When none existed, the manual advised maneuvering to within effective range while screened by smoke or the terrain. The manual also advocated flank attacks, the closer, the better.[283]

These tactics helped 741st crews destroy German vehicles with thicker armor and greater firepower.

CHAPTER 34 **THE ONDENVAL GAP**

Sustained attacks by the First and Third Armies squeezed the German salient in the Ardennes to less than half of its original size. As the "Bulge" imploded on its makers, the First Army commander wanted to hasten its end.

He issued orders directing the 1st Infantry Division to lunge south and seize the Ondenval-Faymonville-Schoppen area. This operation required the 2nd Division to lend its 23rd Infantry to the attack force. Once the 1st Division crushed the enemy's defenses, the 7th Armored Division had orders to drive toward Saint Vith, a city it had failed to hold in December.

Colonel Skaggs reviewed the plan with General Robertson after dark on January 11. Two hours before midnight, Skaggs asked Captain Young to send a liaison officer to the 23rd Infantry in the morning. Young's company would assist the 23rd.

While planning continued, tankers from Company C used house-painting brushes and brooms to smear white coloring on their tanks for camouflage. Calcimine served that purpose as did a homemade recipe: two quarts salt, four gallons water, and six gallons lime—enough to coat one tank. The men had to work quickly before the soupy mixture froze.

Each platoon leader also had an additional radio installed in his tank. The new radios were the SCR 300 type, commonly called a "walkie-talkie."

These devices allowed the tank officers to communicate directly with their infantry counterparts.

On January 13, tanks from Charley Company departed Berg by ones and twos and drove to Robertville, where they joined the 23rd. The crewmen drew ammunition from 1st Division stocks, while Captain Young reported to Colonel Lovless, the regimental commander. The colonel said his men would attack on January 15 at 7:00 A.M. The axis of advance was a two-lane road called N76 that ran south from Waimes to Saint Vith. To open it, the 23rd had to seize the strategic Ondenval Gap from the 3.Fallschirmjäger-Division. Since its Normandy defeat, the enemy division had doctored its losses with replacement troops having little or no ground combat experience—the same story throughout Hitler's armies. The revamped division seemed unfit to mount a protracted defense, or so the Americans hoped.

The attack plan split Young's company. Two platoons would stand ready in Waimes, while the other platoon would sit in defilade on a hill called Kreuzberg. Acting like self-propelled artillery, the latter platoon would provide indirect fire against three hamlets occupied by the enemy—Steinbach, Ondenval, and Remonval—as well as high ground beyond Remonval, where the Fallschirmjägers had carved out a strongpoint. The 23rd had to capture all four places to reach the Ondenval Gap.

Before dawn on January 15, Lieutenant Covington and his four tanks moved from Robertville and linked with the 23rd in Waimes. They climbed the Kreuzberg and waited to begin shelling Steinbach and Remonval.

After sunup, the tankers received their first request for fire, but they were unable to respond. Fog cloaked much of the area, so the infantry switched to a target at higher elevation, one partly visible—the strongpoint beyond Remonval. Covington's men fired, but the destructive effect was hard to judge through the fog.

At half past 8:00 A.M., a new mission beckoned. Company G, 23rd radioed for tanks to help seize two houses at a road junction on the southern slope of the Kreuzberg. Battle maps identified the place as "Point 68."[284]

Covington and Sergeant Fitts departed with their crews to render assistance. Captain John M. Stephens, the Company G commander, already had one rifle platoon pinned to the snow by enemy bullets.

Stephens sent word for Covington and Fitts to fire on the houses. This bombardment was the only protection for his men. The barren, snow-blasted terrain offered no hiding places or covered avenues of approach. Behind a curtain of 75 mm shells, the captain's men crunched across the white openness without casualties. The tanks fired white-phosphorus rounds through basement windows, forcing out thirty-seven troopers from Fallschirmjäger-Regiment 9. The enemy soldiers stumbled into the open, their hands raised and choking from the noxious smoke.

Stephens recommended Fitts and Covington for Silver Star Medals, though, seven weeks later, only the lieutenant received the decoration. Skaggs explained that the action merited only one Silver Star, so the two men would have to "wear it together." Like many soldiers, Fitts perceived little value in the glitter of medals. They offered no insurance against death. He accepted his commander's words with an understanding smile.[285]

The contest for Point 68 ended before noon. With that task complete, Covington waited for orders from regimental headquarters. Lovless wanted to continue the attack, but German paratroopers in Remonval had stalled the 23rd. The colonel applied more pressure, and, soon after 1:00 P.M., the two tanks and Company G swung east and moved downhill toward Remonval.

Minefields initially slowed the advance. Machine-gun bullets then forced Stephens's men to their bellies. The tanks held their fire until Covington received a call saying that no friendly troops occupied the area ahead. The tank crews silenced the hostile guns.

Togged in white bedsheets for camouflage, the infantrymen trudged forward, their captain leading the way. They reached a protected area southwest of Remonval, and Stephens crept through the snow to reconnoiter the next objective, a rise overlooking the town. He called his platoons forward, but two machine guns spat in their direction from houses in Remonval. Once more the infantrymen pressed themselves into the snow.

Stephens crawled along an exposed ridge, squeezing his way under the bullets. He waved the tanks ahead. One tank provided cover fire while the infantry officer prodded two of his rifle platoons forward. They advanced into the village with the other tank at their side.

Covington's tank moved into Remonval with the two platoons. "We started assaulting houses containing German troops. Many of the enemy were observed leaving via a draw to the east. We placed machine-gun fire from our tanks on the area with good results."[286]

At 6:00 P.M., on Remonval's darkened streets, the other two tanks from Covington's platoon arrived and relieved his tank and Fitts's tank. The pair returned to Waimes for more shells, gasoline, and caliber .30 ammunition. The tanks in Remonval remained with Company G all night. Anxious infantry officers asked the tank crews to play the role of TD men and take up antitank positions. The infantry feared a counterthrust by German armor.

Besides Covington's platoon, Lieutenant Dew's platoon joined the January 15 attack. His four tanks left Waimes during the morning to help dislodge enemy defenders on the outskirts of Steinbach. While plowing toward the village, the platoon shoved aside an M18 tank destroyer from the 612th TD Battalion that had detonated an antitank mine. The platoon moved down the road to bolster footsloggers from the First Battalion, 23rd, who had encountered stout resistance from Fallschirmjägers holding fast inside a stone farmhouse.

While moving forward to blast the structure, Staff Sergeant Suydam's tank rolled over a mine concealed in the snow. The men inside had no idea what caused the thunderous crash that reverberated through their bones. Suydam and three other men scrambled out within seconds, Suydam with a broken nose. One man failed to emerge—the bow gunner, Private Ralph G. Kelley. His comrades shouted for him, and they heard muffled words emanating from beneath the tank. Kelley had popped the escape hatch and was tunneling through the snow. He surfaced from under the tank without his shoes, having removed them earlier to wrap his feet in burlap for warmth. When the mine exploded, he had spent precious seconds fumbling for his Superman comic book instead of the shoes.

Due to the mine, Suydam's tank sustained a broken track and required a tow back to Waimes for repairs. Dew's platoon continued and quelled all gunfire coming from the house. Steinbach soon fell, and one rifle company secured the town, while two others pressed ahead after dark and easily captured the Fallschirmjäger strongpoint behind Remonval. Its defenders had silently withdrawn to defend forested high ground that flanked the Ondenval Gap.

Throughout the morning of January 15, Company C kept one platoon in Waimes with Norman Crisler in charge. He had relinquished his sergeant's stripes on January 7, accepted a battlefield commission, and succeeded Victor Miller as Third Platoon leader. The new lieutenant's men had no contact with the enemy and remained in town until the afternoon, when they were dispatched to the Steinbach area on a TD mission. Like Covington's two tanks in Remonval, Crisler's men held antitank positions throughout the night. Dew's men did the same. The infantry felt more secure, but the tankers had misgivings. They easily imagined what might be creeping in the darkness—enemy tank hunters, each with a Panzerfaust tucked under his arm.

On January 16, the Second Battalion, 23rd pressed south from Remonval toward Ondenval. Covington's platoon started moving at 6:00 A.M., about thirty minutes after the infantry pushed off. The foot soldiers crossed open snow and attracted no enemy attention. After a quiet crossing, the Americans broke into Ondenval without a fight. The main German force had pulled out that night. The enemy numbered no more than a few snipers left behind to slow the advance.

Covington's tanks kept to high ground outside of town. Around noon, Second Battalion soldiers hiked south across windswept snow to secure a ninety-degree bend in N76 at the center of the Ondenval Gap. The gap constituted the only passageway through the forested high ground.

Artillery blasted away in support, while the tanks moved ahead with the infantry, yet the road bend remained out of reach. Machine-gun fire thwarted the infantry, and a torn-up railroad crossing blocked the tanks. Covington ordered his driver to circumvent the obstacle, but the driver threw a track when he tried. American artillery dropped right and left of the disabled vehicle, killing and wounding Second Battalion soldiers. Shrapnel flailed the tank, while Covington and his men sat inside, unable to venture outside and repair their machine. Immobile, they continued firing until the artillery abated after dark. The men remounted the track with help from another tank crew. The platoon traded places with relieving tanks and returned to Waimes for gasoline and ammunition.

After Ondenval fell during the morning, Dew's platoon and Company C's tankdozer entered the town. The dozer spent all day plowing mines from snowbound roads. Dew's tanks helped guard the town, though the lieutenant's vehicle was absent due to mechanical trouble. After dark, Dew and his crew rejoined the platoon with their repaired tank.

Crisler's platoon attacked at 7:30 A.M. that day with the First Battalion, 23rd. The infantrymen seized a crossroads east of Ondenval during the morning. In the early afternoon, two rifle companies bore down on the Rohrbusch, a forested area flanking the Ondenval Gap. German bullets sent the foot soldiers diving headlong into the snow. One company managed to penetrate the forest by sundown. Crisler's tanks remained in defilade and supported that company with fire before two other tanks replaced them after nightfall.

The 23rd Infantry planned to push south again on January 17 in an effort to reach Iveldingen and Eibertingen, two villages on the opposite side of the Ondenval Gap.

After spending the night in Waimes, Covington and his four tanks moved into Ondenval at daybreak to assume antitank positions and to stand ready for fire missions against the Rohrbusch. Dense fog promised to make observing the fire a guessing game. When the call came, the four tanks slammed away, adjusting their fire on tracer shells from the platoon leader's tank. The Germans responded with flat-trajectory fire, but their rounds missed. The regimental commander eventually ordered the tanks to stop firing and remain just behind the battle area until needed again. Just as Captain Young relayed this message to Covington, American artillery hurtled overhead and fell throughout the Rohrbusch. The rain of shells pounded steadily for ten minutes and then shifted deeper into German territory. It was the prelude to an attack by the 23rd.

Shortly after 3:00 P.M., infantrymen from the Third Battalion passed through the First Battalion line. The Third Battalion infiltrated the forest through dense fog and flushed the enemy from his defenses. The attackers wiped out several outposts and reached all their initial objectives. The regimental commander ordered the men onward and sent the First Battalion into the Rohrbusch as a backup force.

As the First Battalion moved forward, enemy fire erupted from a strongpoint built around three isolated farmhouses. The resistance halted a rifle company commanded by First Lieutenant Marvin H. Prinds. Two tanks from Crisler's platoon arrived to help.

Prinds leaped atop one tank and rode it like a chariot. Three enemy machine gunners shifted their attention toward him, but he ignored their murderous fire as it glanced off the steel hull. He waved his men forward and directed the tank's fire against a machine gun fifty yards from the middle house. He succeeded in destroying the weapon, but an enemy shooter honed in on the selfless officer. Slugs tore into his intestines, and he tumbled from the tank, leaving his men and the tanks to finish off the strongpoint. The next day, Prinds would die in a hospital.

Inside the Rohrbusch, Third Battalion riflemen tried to finish clearing the forest but suddenly ran into a storm of bullets. Some GIs hunkered down amid the pandemonium, while others fell back through the trees to the First Battalion. Cold rain soaked everyone and later turned to sleet and snow as temperatures dropped that night.

News circulated that the Third Battalion had won the Rohrbusch, and minesweepers moved forward to clear N76 now that it was supposedly safe. The Charley Company tankdozer joined the minesweepers, who were from Antitank Company, 23rd and the 2nd Engineer Combat Battalion. The group passed Third Battalion soldiers and pressed ahead until meeting more soldiers. The Antitank Company captain who led the expedition felt no cause for alarm. "We saw some men by the road and thought surely they were from I Company. We pulled up to talk to them, and they opened up with a bazooka. It grazed the tank and caused some light casualties."[287] Before the minesweepers retreated, they grabbed three prisoners after shooting an enemy officer along with seven of his subordinates.

At first light on January 18, the Germans punched back in force. The 23rd Infantry faced several hundred paratroopers and assault guns from Panzerjäger-Abteilung 348.

Major Johann Schweigert commanded the enemy armored unit, which fielded one company equipped with the Sturmgeschütz IV. This dual-purpose weapon served as a tank destroyer and infantry-support gun. Oberleutnant Wilhelm Schmid led the company.

The attackers stormed the Rohrbusch from the front and flanks. Under pressure, the Third Battalion troops, who had held their ground during the night, fell back to the First Battalion line. Sergeants and platoon leaders from First Battalion exhorted their men to blast the enemy, while retreating Third Battalion soldiers fell in with them, plugging gaps in the line.

Crisler and Covington assembled their platoons in Ondenval as the German counterattack erupted. The tank crews rolled forward at around 9:00 A.M. to help dislodge the enemy from the Rohrbusch, but the narrow spaces between the tree trunks offered no room for armor. This constricted world belonged to the foot soldier and his rifle.

The forest battle seesawed until midday, when the First Battalion plunged riflemen and machine gunners deep into the trees amid blizzard conditions. Artillery shells meanwhile repelled Schmid's assault guns, which approached through the Ondenval Gap.

The tide turned in the Rohrbusch, and the Americans prevailed.

Slaughtered Fallschirmjägers cluttered the snowy forest floor. Ninety-nine survivors called it quits and surrendered to the 23rd Infantry. After interrogation, they boarded trucks bound for a distant PW cage, a place without terror, a place warmer than any foxhole in the Ardennes. Their GIs captors lamented the irony. The victors had no reprieve from the terror and stinging cold.

As darkness fell, Covington's platoon relieved Dew's tanks at the railroad crossing. Covington and his men set up antitank defenses should the assault guns reappear, but the weather erased that possibility.

The black sky blew its bitter worst as the tankers positioned themselves for the night. Low-hanging clouds dumped snow for hours. Each tank crew became isolated from friend and foe. The crewmen huddled together for warmth inside their vehicles, the freezing air as deadly as any bullet or artillery shell.

Fresh crews departed in the morning to relieve Covington's men. Surrounded by the cold colors of winter, the replacements plodded across the arctic wilderness on foot and struggled to locate the tanks beneath the snowfall. It took hours to find them. The new crews discovered turrets frozen, breechblocks frosted solid, and crewmen stiff from the frigid temperatures. The replacements pried Covington's men from the tanks.

The 23rd Infantry resumed the battle on January 19 at 7:30 A.M. with the regiment, having mastered the Rohrbusch, focused on Iveldingen and Eibertingen. Two battalions moved on the villages. Dew's platoon accompanied the Third Battalion toward Iveldingen, and Crisler's platoon escorted the Second Battalion toward Eibertingen.

The weather hindered the Iveldingen assault more than the Germans, who offered only token resistance. What shooting erupted quickly died under return fire. Dew's tanks blasted a couple of buildings, and the infantry secured the place before noon.

Eibertingen was another matter. The enemy possessed several assault guns and had transformed houses into forts to block the Second Battalion. Combat spilled into the afternoon as Crisler's tanks pounded houses and sparred with the assault guns. One AP round whooshed over Sergeant Padgett's tank and smashed against a building, which sent masonry showering down into his open turret hatch. Tankmen nearby chuckled as Padgett tossed out bricks.

Crisler's platoon lost only one tank when it sank to its sponsons in the Schinderbach, a small stream deceptively covered with snow. All attempts to rescue the M4A1 ended in failure, and in the process, the deep snow claimed a Service Company tank-recovery vehicle. Five days later, Service Company mechanics and a recovery vehicle from the 463rd Ordnance Evacuation Company extracted both machines. GI scavengers had looted the tank's radio equipment, auxiliary generator, and personal items abandoned by the crew. Its front caked with ice and frozen mud, the 463rd towed the machine to Waimes for repair and cleaning.

The hard-boiled defenders in Eibertingen forced Crisler to call for assistance. In response, Colonel Lovless redirected Dew's tanks there after the Iveldingen fight ended. Together the two tank platoons cracked one strongpoint after another with high explosives and white phosphorus. The crews fired HE through windows to make kindling inside before they shot WP to ignite the broken mess. Aside from busting enemy forts, Staff Sergeant Neidrich's tank stopped an assault gun with AP shells, while two other German armored vehicles fled the village.[288] American artillery observers called down HE on German foot troops trudging away from town, and later another group received identical punishment. Eibertingen now belonged to the 23rd Infantry.

The 23rd had reached all its objectives and opened the Ondenval Gap. The troops now had the dicey task of removing mines and booby-traps along N76. The effort cleared the way for the 7th Armored Division, which rolled through the gap on January 20 as the drive for Saint Vith gathered steam. The Germans ceded their Ardennes salient one town after another.

Soldiers from the XVIII Airborne Corps relieved the 23rd Infantry, and the regiment marched to bivouac areas in the Camp Elsenborn area. Captain Young's tank company retired along with the 23rd but briefly laid over in Robertville—home of the 741st rear echelon and Dog Company—before moving to Berg for a rest.

The battle to open the Ondenval Gap was the first significant action involving the 741st that cost no lives among the tankers. Young understood that such good fortune was only happenstance. Mistakes had occurred. He set to work typing a critique for Colonel Lovless.

CHAPTER 35 **RETURN TO KRINKELT-ROCHERATH**

Armor from Companies B and D returned to frontline duty on January 22, 1945. They joined the 9th Infantry as it relieved the 26th Infantry Regiment (1st Division) at Domäne Bütgenbach. Company B contributed one platoon, and Company D lent two.

The tanks assumed defensive positions while the departed 26th Infantry gathered at an assembly area near Bütgenbach. The 26th prepared for an attack that would take it south through its former sector now maintained by the 9th Infantry.

The newcomers at Domäne Bütgenbach were still making the place home when the 26th returned the next day on the attack. Its riflemen and machine gunners passed through at 4:00 A.M. on January 24, and they hacked away at German defenses. On January 29, the 26th swung east through snowdrifts and machine-gun bullets to grab Büllingen. The attackers cleared the ragtag garrison defending the town and opened a path for the first offensive move by the 2nd Division since the heart-ripping misery at Wahlerscheid in December.

Corps headquarters conveyed typewritten field orders down the ladder of command. The document directed the 2nd Division and 741st to seize Schleiden, a town inside Germany. After its fall, the advance would continue toward Zülpich. The operation was part of a First Army campaign aimed at pushing across the Rhineland to the Rhine River.

The overall scheme involved the 2nd, 9th, and 99th Infantry Divisions, with the 7th Armored Division in reserve. The latter was now in bivouac after capturing Saint Vith on January 23. The three infantry divisions planned to attack during the wee hours of January 30. The 9th would go for Gemünd, the 2nd for Schleiden, and the 99th would launch a supporting attack between the two divisions.

Less than ten hours before the attack started, the Company B tankdozer, commanded by Sergeant Donald Lucey, struck an antitank mine while plowing roads. Maintenance men towed it away for repairs, and the crew waited for a replacement. They leaped to work when it arrived. The gunner, Corporal John C. Reedy, explained what happened.

"Presented with a new dozer six hours before the push off, we worked hurriedly cleaning ammunition and guns and putting on grousers. After getting the dozer in shape for service, we demonstrated it to the engineers so they could see what it could and could not do, and then determine how it should be used. This finished, we waited for orders to move out. At dark, we moved up to join the infantry and engineers we were to work with."[289]

The operation started after midnight. Reedy and his comrades in the dozer trailed rifle squads from the 9th Infantry and engineers from the 2nd Division. The infantrymen struggled through the snowdrifts and blackness toward Büllingen. The engineers swept the roads for mines, an arduous task due to the snow depth.

Lucey's dozer pushed its way along N32, the main thoroughfare leading into Büllingen. The surface and shoulders bore evidence of the failed attempt by the Hitlerjugend Division to seize Domäne Bütgenbach in December. The dozer blade churned up frozen bodies and broken weapons. Work progressed well according to Reedy, but the day's labor ended in a mishap at the village limits.

"Everything went nicely until we almost reached our initial objective, which was the town of Büllingen. Here we ran into a roadblock of one American and one German truck, both knocked out, but pulled across the road to hamper our advance. No mines had been encountered up to this point. But the Jerries had been very clever to place mines under these trucks, knowing that it would be impossible to sweep them clear. We had only pushed the trucks a short distance when the blade hit two Riegel mines and damaged it beyond further service. So, being of no more use, we returned to the company in the evening and the next morning went to Service Company for a new blade."[290]

The dozer was the only Company B loss that morning. Ten other tanks broke into Büllingen and secured a hill northeast of town. The high ground now occupied by Company B and the 9th Infantry was the same place where an armored probe from Kampfgruppe Peiper caught hell on the morning of December 17.

Like the 9th, the 38th Infantry attacked that morning but with only one tank from the 741st. The Company A tankdozer cleared roads behind the attackers, who claimed Wirtzfeld from its German defenders. While the dozer crew worked, the rest of Company A—and all of Company C—remained in bivouac, waiting for a call to battle.

In the Wirtzfeld area, the Company A dozer met the same hidden hazards encountered by Lucey's dozer. The enemy had concealed mines in and under everything imaginable. The Company A dozer set off an antitank mine and split a track. After repairs, it detonated another mine with the same result, but this time the damage ended its day.

The 38th Infantry encountered light resistance in Wirtzfeld. Mines and hip-deep snow presented the chief obstacles, but that changed when the attack reached Krinkelt. House-to-house fighting ensued, and the infantry needed tank support. The regimental commander called for two platoons from Company A and two from Company C.

Along the road between Wirtzfeld and Krinkelt, the tanks passed a T2 recovery vehicle lost by Able Company in December. Someone had used explosives to rupture its two cannon barrels (37 mm and 75 mm), splitting them open like peeled bananas. Whoever was responsible had no idea that both guns were dummies, nothing more than steel tubes welded in place. The sight drew chuckles from the passing tankers.

Inside the decimated village, enemy soldiers used buildings as strongholds. HE shells and machine-gun bullets did little damage to farmhouses built from heavy fieldstones. The tankers used a proven method to solve the problem. They fired HE combined with white phosphorus to drive out the occupants. The Americans thus prosecuted the fight in Krinkelt, cellar by cellar, house by house.

The tanks and infantry controlled the village by 9:00 P.M., but Rocherath still lay ahead. Its reduction had to wait until daylight. The two platoons from Company A buttoned up for the night, and the two from Company C returned to Berg. The men bedded down there in a barn and farmhouse rather than outdoors as was usually their lot.

On January 31, Company A tanks worked with the 38th to eliminate the last German defenders in Rocherath. With business settled in the twin villages, an infantry battalion pushed seven hundred yards north to secure a line of departure for the 23rd Infantry.

One by one, the Company C platoons linked up with the 23rd and journeyed through Wirtzfeld to the twin villages. In the afternoon, Lieutenant Crisler's platoon and infantrymen from the Second Battalion, 23rd rolled north from the line established by the 38th.

Progress halted at a crossroads called Rocherather Baracken, where the Germans decided to fight for three buildings. The bruising struggle cost one tank when an enemy soldier hit Lieutenant Crisler's vehicle with a Panzerfaust. Corporal Curtis Hall, the gunner, fled the crippled machine, but machine-gun bullets killed him. Crisler and the other crewmen fared better, though the lieutenant and the bow gunner suffered minor injuries. Both required evacuation for medical care. Crisler returned five days later.

Later in the evening, Lieutenant Covington's platoon helped finish the chore, weeding out the enemy from Rocherather Baracken. Even though daylight had turned to darkness, the tanks and infantry pressed ahead.

They reached a road fork eighteen hundred yards past Rocherather Baracken and found three machine guns defending the area. The tanks softened up the enemy emplacements, and a rifle company destroyed them. The tanks and infantry pushed ahead into the Monschau Forest on a dirt lane called the Hasselpat trail. Inside the woods, this lane ran parallel to the Wahlerscheid road. The 23rd advanced on the trail, while the 9th Infantry drove straight up the road. Once again, the 2nd Division aimed its infantry at Wahlerscheid. It was now February 1.

Covington's tank entered the forest followed by Sergeant Fitts and his tank. Fitts saw a flash beneath the lieutenant's vehicle, and everyone presumed the worst—an enemy antitank gun. Fitts instructed his driver to pull closer so he could locate a muzzle flash if the enemy fired again. Covington and his crew escaped injury, though their vehicle sat dead in its tracks.

Unaware that a mine was the culprit, Fitts called to his driver and told him to halt. They were close enough. Fitts searched the darkness, vigilant for the fiery wink from a cannon barrel. Nothing happened. Then *boom*!

While settling into the snow, his vehicle had detonated a mine. The blast left Fitts with a sore neck and trauma to his right arm caused by the ulna bone, which punctured the skin at his elbow. Fitts remained with his crew despite the pain. The tank itself sustained internal and external damage, including the loss of several end connectors on the right track. In the crew compartment, the blast sheared off bolts on the transmission housing, and hot oil flowed onto the floor. The driver steered the still-operable machine back to the twin villages, where the Company C maintenance section took over. The other tanks in Covington's platoon also returned to the villages after Dew's platoon relieved them.

Captain Young moved his headquarters to Rocherath during the predawn hours and established a CP in the northernmost cluster of homes.[291] That afternoon, he traveled to the forest, where Dew's platoon was clawing its way forward with the Third Battalion, 23rd.

The tanks, roughly abreast, weaved between tall conifers and steered away from the Hasselpat trail for fear of mines. The foot troops were several hundred yards ahead and presumably able to overpower any Germans intent on a Panzerfaust ambush. Progress in the woods abruptly ended at a broad, snowy clearing. The tanks and infantry halted.

Who would cross first?

The infantry officer in charge, First Lieutenant Samuel D. Boone from Company K, motioned for the tanks to roll into the open ahead of his men. Young intervened. Not his tanks. According to one onlooker, Private First Class Phil Zodda, the two officers stood in plain view by the trail and argued for several minutes.

Zodda kept watch, his fingertips tingling as a sense of peril washed over him. "It was so strong I wanted to open my turret and yell at Captain Young to get the hell out of there. I just held on, though I kept looking at them through my periscope. I could see them clear as day."[292]

An enemy observer also saw them, and a mortar shell crashed down. Steel splinters lacerated Young's forearm and sliced open his abdomen, spilling out his entrails.

The effort to save Captain Young began on the forest floor when jeep-borne medics hustled to the scene. Two days later, the drama ended at the 2nd Evacuation Hospital, where surgery and penicillin failed to keep him alive. The hospital occupied a former sanatorium on a hill overlooking Eupen, Belgium. Orderlies wrapped the dead captain in a woolen blanket and slipped it inside a mattress cover. On February 5, a burial detail at Henri-Chapelle lowered his body into a temporary grave. No tears or ceremony marked the event. The Graves Registration men tacked one of his dog tags to a wooden cross, having left the other tag with his body.

Minutes after the fateful mortar shell exploded, the Company C radio channel crackled with calls for Lieutenant Covington. He reported to the CP posthaste and took temporary command of the company as its ranking officer. The day was also his thirtieth birthday.

Drafted from his home in Santa Ana, California, Herb Covington entered the army as a buck private in 1942. He later received a commission through OCS at Fort Knox and joined the 741st as a second lieutenant. His wife, Eva, lived in Santa Ana with their daughter, a child Herb had never seen.

As February 1 ended, Covington reported one assault gun and nine tanks functional, though two were stuck in a ditch. Six other tanks were inoperative for mechanical reasons. Dew's platoon returned to Rocherath for the night, having left the 23rd Infantry after dark.

The platoon had crossed the grassy clearing *behind* the infantry.

★★★

The main effort on February 2 fell to the 9th Infantry, which had a new commander, Colonel Philip Dewitt Ginder. (The old warhorse, Colonel

Hirschfelder, was gone. His pleuritic condition forced an end to his combat service when orders from above compelled him to relinquish command on January 10 and return Stateside.)

Ginder's men had fought two miles north the day before and met GIs from the 9th Infantry Division. The 9th Division had swooped down and snatched Wahlerscheid from behind.

An order from 2nd Division headquarters augmented Ginder's force with tanks uniquely equipped to eradicate mines. These belonged to a platoon from the 738th Tank Battalion now attached to the 741st. Each special Sherman featured a T1E3 mine exploder. This bizarre appliance detonated mines with massive steel rollers. "Aunt Jemima" was the nickname given to the rollers due to their prodigious size and weight, nearly thirty tons.

At 7:00 A.M., Company B, along with the 9th Infantry, pushed through the Monschau Forest headed toward Wahlerscheid. Schöneseiffen was their first target beyond the forest.

As they chugged north to Wahlerscheid, the tanks entered an apocalyptic landscape. Artillery shells had stripped the bark and branches from most trees, and their bare trunks stood like giant exclamation marks punctuating the stark ruin. Antitank mines abounded, and the Company B armor advanced at a timid pace. Mine exploders from the 738th had yet to reach the area, so engineers scanned the road with mine detectors and used their fingers and bayonets to extract the deadly plantings. Enemy holdouts also abounded. Soldiers from the 9th Infantry fanned out and prodded Germans from bunkers among the tattered trees. The captives trudged to a makeshift PW cage at the Wahlerscheid customs house. The Germans had originally constructed the enclosure to hold American prisoners.

As the roundup swept from bunker to bunker, Lieutenant Helm from Company B ventured to the forest edge with several tanks and the company assault gun. The lieutenant's group unleashed fire on Schöneseiffen to reduce the enemy's ability to buck the coming onslaught.

Once the bombardment let up, the Company B tanks moved on Schöneseiffen with infantrymen riding aboard the machines. The attackers embarked on a journey across a swath of open snow separating them from their objective. They used the only road between Wahlerscheid and Schöneseiffen, Reichsstrasse 258.

The Americans had nearly reached the village when the Germans reacted. Mortar shells plunged down among the attackers. The tanks answered, their shells clobbering barns, houses, and anything else that looked like it might conceal an enemy gun or observation post. The infantrymen leaped from their armored transport and double-timed the final yards on foot. Darkness fell as they finished wringing the last enemy troops from Schöneseiffen. The tanks rolled into town along the mud-plastered main street.

The hour was 5:00 P.M., and one more objective lay ahead, Harperscheid. The infantry advanced all the way on foot, while the tanks crept along behind, still using 258. Nightfall made the crews wary. They struggled to see ahead, a situation ripe for a Panzerfaust ambush. Suddenly, an enemy cannon barked at them, firing high-velocity shells that barely missed. Tank gunners from Baker Company returned fire and detected at least three enemy armored vehicles. One of these was a self-propelled gun, which an infantry private disabled with a rifle grenade. When the crew of a Panzer attempted to retrieve the gun using tow cables, another infantryman rushed forward and banged on the tank with his rifle butt. He demanded that the crew surrender, and three Germans complied, emerging with their hands raised. The fourth crewman came out with his pistol drawn, and the GIs shot him dead before dropping a hand grenade into the vehicle, killing a fifth man inside. Meanwhile, other American soldiers dislodged German troops from attics, cellars, and sundry hiding places. The fight for Harperscheid ended by 9:00 P.M. with the tanks and infantry forming an all-around defense.

Soldiers from the 23rd Infantry marched to Harperscheid in the morning. One battalion planned to strike east to capture Bronsfeld. Ten tanks from Company C and the company maintenance section left the twin villages and joined the 23rd at Harperscheid. Company A and the 9th Infantry aimed to capture Berescheid, Ettelscheid, and Schuren.

Enemy resistance withered after the 9th and 23rd pushed out from Harperscheid. The Company C tanks reached Bronsfeld by 1:00 P.M. and waited for relief before retiring to Rocherath, where Company B had already returned for a rest. Only Company A stayed with the infantry and helped claim Berescheid and Ettelscheid.

Combat that day levied few casualties on the infantry and none on the tankmen.

The crews who returned to the twin villages traipsed about the ruins, assessing the damage. Recent rainfall had melted the snowy shroud and unmasked a congress of horrors. Unburied dead lingered in the alleys, streets, and barnyards. Corpses also lay inside houses. Corporal Jack Boardman entered one such place after returning to Krinkelt from Harperscheid.

"We parked our tanks in the churchyard, and I started hunting for a place to stay, one that was in reasonable shape. I found a building that had its ground floor covered with dead bodies, German and American. Apparently, it had been an aid station."[293]

Boardman climbed upstairs and found no bodies. He made himself at home, shrugging from his mind the charnel scene below. Months of combat had made death all too commonplace. Its power to shock had long since vanished.

The only things that aroused anyone's interest were the defunct war machines scattered about the twin towns. Tankers and 2nd Division soldiers wandered about, surveying the wrecked Panzers and American tanks. Several men clicked away with cameras. The most popular subjects were lifeless Panthers, the cream of SS-Panzer-Regiment 12. These dead and burned monsters offered bracing testimony to the dark defeat wrought upon Hitler's armies in the West.[294]

It had been a winter of ruin for the Germans.

CHAPTER 36 "EASY EIGHTS"

Field Marshal Bernard Law Montgomery derailed the American push toward the Rhine River, at least temporarily. The British and Americans lacked sufficient gasoline, ammunition, and replacements to sustain an offensive along the entire Allied front.

In early November, Eisenhower had promised Montgomery that if American troops failed to reach the Rhine by January 1, the field marshal could launch a drive for the Rhine with his armies in the north, which now included the Ninth US Army. The German Ardennes offensive had given Eisenhower an excuse to defer the deadline, but more than a month had passed, and Monty held Ike to his word.

First Canadian Army developed plans to drive down into the Rhineland from Holland, and Ninth Army would cross the Rhineland to link up with First Canadian. Second British Army would then advance and vault the Rhine.

As part of Monty's offensive, the Ninth Army GIs had to leap the Roer River. The four dams that had eluded capture in December again became critical. As with the first attempt to seize them, the second try fell to First Army.

The 9th Division aborted its effort to capture Gemünd and focused on the Urfttalsperre Dam, while the 78th Division concentrated on the Schwammenauel Dam. The 2nd Division had only an ancillary role, providing flank protection for the 9th Division. This meant continued efforts to eject the Germans from Schleiden and capturing nearby Hellenthal.

★★★

Early on February 4, Lieutenant Covington summoned Sergeant Fitts to the Charley Company CP located in a Rocherath cellar. The lieutenant had officially become the commander.

Fitts found it difficult to salute when he reported, his injured elbow growing stiff. Covington surprised the sergeant by offering him two choices: accept a battlefield commission and take over the lieutenant's former platoon, or travel Stateside on ninety days leave as part of the home-furlough program. Fitts made a quick decision. "I had not seen my wife in thirty months, and I had seen and done so much battlefield action that I did not feel like a shirker, so I took the furlough."[295] Moments later, Colonel Skaggs rushed down the cellar steps and said, "Covington, I want to see this Sergeant Fitts who you're sending home."[296] The lieutenant had already submitted Fitts's name for the leave. Leonard Henkelman took the leaderless platoon, having received a commission in late January.

Skaggs congratulated Fitts and arranged for travel orders to Fort Meade, where he could catch a bus to his wife's home in Washington, DC. The lucky soldier shied away from medical attention for his elbow, lest he end up hospitalized.[297] Before shoving off from Krinkelt-Rocherath, Fitts said goodbye to his crew and ordered them to stay alive until he returned.

After his departure, Company C also left the twin villages. The company moved to Schöneseiffen, where its tanks could help the 38th Infantry smash the enemy nest in Hellenthal.

Two battalions from the 38th marched toward Hellenthal on February 5, advancing downhill into the town without any meaningful resistance. German shells and bullets greeted the leading elements as they neared Olef Creek, which sliced through the town center. The only bridge was in ruins, but the infantrymen managed to wade across by midmorning. Enemy fire soon drove them back across. With his troops stuck among houses on the near bank, Colonel Francis H. Boos, the 38th commander, called for tank support.

Ten tanks from Company C left Schöneseiffen to join the fight. While descending the single road leading into Hellenthal—the same route used by the infantry—one tank blew a track on an antitank mine. Nobody died, but the immobilized machine blocked the road, and the ground on either side was too soft for armor. Boos demanded the tanks push ahead. No go. They remained behind the disabled tank and fired into town at enemy strongpoints. Division artillery blasted away to keep the Germans from reinforcing the town.

Boos requested more armor, as if upping the tank count would help. Company B sent two platoons from the twin villages. Nothing changed. By evening, the colonel conceded to reality and authorized Company C to withdraw into Schöneseiffen. He kept the Company B armor in place at Hellenthal.

After dark, Company C maintenance men repaired the disabled tank. The infantry meanwhile pressed its attack and again crossed the creek under fire. This time the Americans held firm to their gain. Behind them, 2nd Division engineers cleared mines and labored to bridge the creek. Company B had orders to throw one platoon across at the first opportunity.

Dawn found a seventy-foot Bailey bridge spanning the creek as well as one built from timber and steel treadway. Neither structure was strong enough to support armor.

Without tank support beyond the creek, troops from the 38th finished clearing Hellenthal themselves on February 6. It was light work compared to the previous day. Most enemy defenders had withdrawn during the night after failing to upend the American effort to bridge the creek. Only a delaying force stayed behind to harass the Americans.

Elsewhere, 9th Infantry soldiers stole into Scheuren during the night. Company A tanks tried to follow, but darkness and challenging topography foiled all but those on foot. After first light, the tanks moved in, the infantry having secured the town.

★★★

Late on February 6, division headquarters placed the 741st in reserve except for Companies A and B. These two companies kept several tanks in Bronsfeld, Scheuren, and Hellenthal. The crewmen rotated with other crews parked behind the front.

Charley Company departed Schöneseiffen for Krinkelt-Rocherath on February 7, and Headquarters Company did the same.

Once at the twin villages, Company C began building for future combat operations. The unit folded in new men and machines. One of its platoon leaders, Lieutenant Dew, traveled to the Ivanhoe CP. The lieutenant received a Distinguished Service Cross for his actions on September 14, when he breached the dragon's teeth near Kesfeld. General Robertson presented the decoration during a ceremony held in a muddy yard behind the forest warden's house at Wahlerscheid. Robertson also presented Silver Star Medals to Colonel Skaggs and John Barner.

Two days later, on February 11, the company received four tanks mounting 76 mm cannons. Two were M4A1s identical to those obtained by Company B the previous month. The other pair were M4A3E8s, something new to the battalion. The E8 model, nicknamed the "Easy Eight," had a suspension system known as Horizontal Volute Spring Suspension, or HVSS. With broader tracks, the E8 offered a better ride and had superior flotation compared to previous Sherman variants. The day after their arrival, the Easy Eights toured all three 2nd Division regiments to familiarize the troops with the new tank's appearance and capabilities. Recent rainfall and warmer temperatures created ample mud and soggy terrain on which to demonstrate the improved flotation.

★★★

The day before Company C received its new tanks, the two largest Roer dams fell to American soldiers. The retreating Germans managed to destroy control machinery and discharge valves. Snowmelt and water escaping from the dams pushed the river above flood stage, and Montgomery's Ninth Army postponed all river crossings until the flood subsided.

Meanwhile, preparations continued for offensive operations once the Roer receded within its banks. The 2nd Division constructed a task force from several infantry battalions and supporting units. This was

Task Force Stokes. The name referred to its leader, John H. Stokes Jr., now a brigadier general and assistant division commander. His ad hoc unit would help protect the right flank of Ninth Army.

On February 18, Skaggs received temporary bivouac assignments for his battalion: Dreiborn for Able Company, Wollseifen for Charley Company, Einruhr for Headquarters Company, and Höfen for Dog and Baker Companies as well as Service Company.

Each company dispatched an advance party on February 19 to secure billets. Only Company C moved that day. The tankmen arrived after dark in Wollseifen and hid their vehicles among the barns and houses. The crews occupied homes and nearby Westwall bunkers.

In the morning, Headquarters Company left Krinkelt for Einruhr. Service Company reached Höfen during the afternoon, but nobody else from the 741st moved. Task Force Stokes had priority on the rutted, muddy roads. Company D finally reached Höfen on February 23. The next day, Companies A and B rolled into Dreiborn and Höfen, respectively.

Captain Sledge wrote to his mother and described the goings-on in Dreiborn:

"We have a couple of different houses for the company now. All the officers are in one room, and it's really a madhouse. Lt. Klotz, my maintenance officer, is playing a captured accordion. He calls it playing, but I don't. McDonough, Call, and I are having a big argument while writing letters. Klotz, Mac, and I have been together in the same company since Camp Polk, so we're all old buddies. The three of us have all been wounded a couple times, so we are the company Purple Heart Club. It is the three of us who gave 'A' Co the reputation of being the loudest in the battalion and also the biggest bunch of gripers."[298]

In Höfen, the tank crews from Company B found a town battered by sledgehammer blows from a Volksgrenadier division that tried to capture the place during the "Bulge." A single 99th Division battalion held the village throughout the battle. This was the only frontline sector in the Ardennes where the enemy failed to break through in December. As in the twin villages, the February thaw revealed ghastly sights. German bodies lent legitimacy to the accolades given the 99ers who defended the town.

The 741st men occupying Höfen slept in wrecked houses used as forts during the battle. Throughout the day, tankers from Company B used the now-quiet streets and fields to give their latest replacements driving instruction and gunnery practice. The company had orders to join Task Force Stokes.

Everyone hoped the coming campaign would amount to little more than rolling over enemy territory as fast as tanks could conquer.

CHAPTER 37 ROLLING UP THE RHINELAND

When the Ardennes offensive lost all momentum in late December, sound military judgment dictated a retreat to the Westwall. The Germans might have stalled there long enough to rally and defend the Rhine River, the most formidable defensive barrier in Northwest Europe.

Hitler instead opted to prolong the Ardennes misadventure, frivoling away lives and material. "Where the German soldier sets foot," he once trumpeted, "there he remains!"[299] That mindset destroyed any chance for successfully defending the Rhine.

By February, the Nazi leader's ability to hold the Rhineland had greatly diminished. His generals committed their last reserves on the Western Front to battle the First Canadian Army attacking from Holland. With every spare German unit focused there, the Ninth Army attack promised success, but the Roer River flooding caused a postponement for safety's sake. The need for action soon took priority, and the Americans had to cross regardless of the water level.

Ninth Army's assault crossings on February 23 caught the Germans unprepared, as they had expected an Allied offensive after the flooding ended. That same day, Task Force Stokes put troops across the Roer at Heimbach. One after another, its units crossed the rushing river on a

Bailey span that engineers had wrestled into place. Once across, the task force assembled at Mariawald.

Tanks from Company B reached the other side on March 3 and joined the task force at Mariawald. After the entire task force had gathered at the assembly area, the advance continued. Around 2:00 P.M., Lieutenant Dudley's platoon pushed off toward Wolfgarten with the Second Battalion, 38th.

Booby-trapped roadblocks stalled the advancing armor until engineers cleared the obstructions. With the way open, the tanks rumbled into Wolfgarten, where German gunners opened fire from two houses. In response, Dudley's 76 mm cannons pounded both buildings.

Less than a mile beyond the village, the road formed a T-intersection with Reichsstrasse 265, the main avenue leading to the next objective—Gemünd. The enemy halted the Americans by seeding the road with antitank mines. Explosions disabled two tanks, blowing apart their tracks. Dudley's platoon took up defensive positions for the night and waited for maintenance men to repair the disabled vehicles. Cloaked in darkness, engineers and infantrymen swept the road and finished their work by morning. The maintenance men never arrived.

At 9:00 A.M., Lieutenant Sheppard's platoon prepared to move from Mariawald to Dudley's position in case the infantry needed help. Within the hour, his tanks moved out, but a call from the infantry never came. The footsloggers advanced alone, conquering pillboxes at the west end of Gemünd. In the afternoon, Dudley's operable tanks crept a half-mile down 265 to another intersection, where his men again assumed defensive positions. Sheppard's men remained put. Baker Company maintenance men meanwhile arrived to repair Dudley's damaged tanks. The mechanics repaired one vehicle by nightfall.

With Task Force Stokes gnawing at Gemünd, the 9th Infantry (less its Third Battalion with Stokes) broke into the town from another direction. By this time, enemy strength had shriveled to nothing, but a labyrinth of mines and explosive traps remained in place.

After conquering Gemünd, Task Force Stokes continued to oust Wehrmacht soldiers from pillboxes in the surrounding hills. The task force completed the job on March 5 and disbanded, its purpose served.

★★★

The task force had finished up just as Supreme Headquarters reopened offensive operations on a broad front beyond Montgomery's armies. The 2nd Division and other units received orders to chase the fleeing enemy and deny him any pause to gather strength. The 9th and 23rd pursued with full energy. It was the Normandy breakout all over again (minus the gleeful civilians). American tanks and infantry spilled into the Rhineland.

After having crossed the Roer, Company A assembled at Wolfgarten and took up the pursuit on March 5 along with infantrymen from the 9th. Lieutenant Brooks led his tank platoon into Glehn with foot soldiers from the First Battalion, while Sergeant Youngblood led three tanks into Bergbuir with the Second Battalion. The rest of Company A converged on Hergarten.

Company C left its bivouac at Wollseifen and joined the 23rd. Its tanks crossed the Roer at Heimbach and reached Vlatten by late afternoon. The tankers picked out billets among the houses, while the company kitchen rolled into town and doled out dinner to hungry mouths. The men had barely gulped down their chow when orders came to mount up and head to Floisdorf. As the tanks swept into this tidy hamlet, its streets and architecture displayed few scars from shells and bullets. The backpedaling Germans had no time to fight or to sow mines and booby traps. The absence of even rudimentary delaying tactics signaled an anemic foe.

After spending the night at Floisdorf, Company C departed on March 6, breezing through Eicks before entering Kommern. The tank crews joined the Second Battalion, 23rd. Foot soldiers climbed aboard the tanks and roared off toward Mechernich. Outside the city limits, the infantrymen dismounted and cautiously edged into town without finding opposition. The tankers formed a shuttle service in Mechernich and transported the battalion to the next community, Breitenbenden. They repeated this process and delivered the foot troops to Holzheim.

Reconnaissance reports indicated that hostile troops occupied the next village. Two tank platoons, each paired with a rifle company, attacked in battle formation and captured Nöthen by sundown. The place teemed with German captives, but somewhere nearby less passive Germans sporadically lobbed mortar shells into Nöthen throughout the night.

The day passed much the same for Company A operating with the 9th Infantry. One platoon blitzed through Scheven, another through Wallenthal, and the third through Strempt. Once these three towns fell, the platoon in Strempt drove north through Mechernich, where it turned south to assist the infantry in capturing Bergheim and Weyer.

In preparation for the next day (March 7), division headquarters attached Company D to the 23rd Infantry and ordered the light tanks to assemble near Mechernich.

Remagen became the name forever linked with March 7, 1945. Spearhead units from Combat Command B, 9th Armored Division reached that city on the Rhine River. All along the waterway, German engineers had blown bridge after bridge, but armored infantry from CCB interrupted their demolition work at Remagen, where a large railway bridge stood.

Erected during the First World War, the structure bore the name of Erich Ludendorff, Quartermaster General for the Imperial German Army. The CCB commander had no orders to cross the river—his unit's mission was to gain Ahr River bridges—but he recognized the momentous windfall and sent his men charging across to establish a bridgehead.

Days earlier, First Army turned southeast to link with Third Army. When this occurred, the 9th Armored Division sliced south and pinched off the 2nd Division and other units, blocking them from the Rhine. The 2nd turned away from the river and began mopping up German forces north of the Ahr. No clear battle lines existed. Enemy soldiers surrendered by the thousands and relinquished war provisions in vast quantities. The human haul included troops from a pitiful adjunct to the German military—the Volkssturm, a home guard composed of aged men and juveniles. Many wore civilian clothes with a thin armband bearing the words *Deutscher Volkssturm*. To inspire these conscripts, Heinrich Himmler offered blood-stirring words: "Our cursed enemies must see and realize that a break-in to Germany, even if successful anywhere, will cost the attacker sacrifices equal to national suicide."[300]

Despite Himmler's rhetoric, the prisoners revealed that saving one's hide eclipsed all concern for protecting the Reich. The surrendering masses represented a docile majority, but a minority remained steadfast. Some of these bitter-enders appeared on March 7, when Company A helped the 9th Infantry clear Eicherscheid and Schönau. This resulted in the hottest action for the 741st in the Rhineland.

One tank detonated a mine near Eicherscheid. The crew escaped their disabled vehicle, but the blast injured the tank commander, Second Lieutenant John W. Call, a former noncom who had recently received a battlefield commission.

The other tank crews brushed off the loss and pressed ahead with the infantry. The attackers secured Eicherscheid and made for Schönau with Sergeant Mayne Youngblood commanding the lead tank. He had previously served with the Second Platoon, but an event earlier that day brought a change. Fragments from an enemy machine-gun bullet wounded Captain Sledge in the right hand, and Lieutenant McDonough from First Platoon took over the captain's post. Youngblood transferred to First Platoon and filled McDonough's vacant position, no lieutenant being available.

Youngblood's new platoon motored toward Schönau on a dirt road that meandered along the bottom of a hollow toward a road fork where he had instructions to bear right. In this constricted terrain, the sergeant smelled danger and cautioned his crew over the interphone. They forged on and gritted their teeth. When his tank nosed around the last curve before the fork, the Germans struck. "That's when they lowered the boom on us," Youngblood recalled.[301]

An enemy antitank gun stung his vehicle with two AP shells. One struck the gun shield on the turret, and the other bored through the frontal armor near the driver, killing Technician Fifth Grade Ernest M. Meads. Corporal Frank W. Peake Jr. climbed out injured, and Youngblood bled from cuts on his face. The two helped extricate Private First Class Rocco J. Strazza, who suffered a badly fractured leg. Luckily, they had no bow gunner aboard. The tank smoldered for awhile but never blazed up. It was nevertheless a total loss.

The tanks behind Youngblood withdrew to Eicherscheid, leaving Schönau in enemy hands for the night. Peake and Strazza departed for medical attention, which meant a leg amputation for Strazza. Remaining on duty, Youngblood received a replacement tank and crew some twelve hours later. The new tank was one he had previously commanded. Named "Asp II," the vehicle had just returned to Company A after receiving a new motor. The crew consisted of the gunner, Steve J. Hoffer, the loader, Arthur M. Najera, the bow gunner, Charles R. Freeman, and the driver, John Onuschak.

March 7 passed without casualties for Companies C and D. They operated with the 23rd Infantry. Before sunrise that day, the light tanks moved from Mechernich to join Company C and the Second Battalion, 23rd at Nöthen. The attack force pushed toward Münstereifel with two Stuarts on point. The usual routine ensued with the infantry climbing off the tanks and advancing ahead into town. Münstereifel fell without a fight. Then in rapid succession, the attackers also seized Lethert, Effelsberg, and Kreuzberg. The retreating enemy answered their opponents with bullets and artillery shells at Kreuzberg, but the GIs captured an intact bridge over the Ahr River. Once across, the attackers grabbed Brück, Pützfeld, and Hönningen. Altogether for the day, the 2nd Division swallowed fifteen towns.

On March 8, many tankers and infantrymen received a break to satiate stomachs and gasoline tanks, but Company A and the 9th Infantry enjoyed no such respite and kept attacking. Three tanks and foot troops captured Schönau and threaded their way through a half-dozen other villages before approaching Insul on the Ahr. Sniper fire rang out here and there, but nothing substantial greeted the tanks until a lone Panzer cut loose. It scored no hits before 2nd Division howitzers pounded it to a smoking ruin. Meanwhile, one tank platoon moved onto Sierscheid and another to Pitscheid.

On March 9, during an afternoon drizzle, Company A backtracked and assembled at Rupperath. The company pulled back farther to Ohlerath, where it married with the First Battalion, 9th. Robed in ponchos and raincoats, the infantry mounted the tanks, and they rode forward toward the Ahr River and a tributary called Adenauer Creek. The company split into six groups and fanned out to cross both waterways at points where engineers intended to slap down temporary bridges to replace ones blown by the Germans.

Dümpelfeld presented the hottest crossing spot. Mobile Flak guns and machine guns forced the leading patrols to scamper backward. Artillerymen from the 2nd Division tore up the town and settled the contest, thus enabling engineers to move in and bridge Adenauer Creek, which cut through the village.

Sergeant Youngblood and his new crew reached Dümpelfeld during the morning. He dismounted Asp II and joined two roving infantrymen searching for eggs. He brought along a flashlight but forgot to bring a pistol or submachine gun for protection. The egg hunt led Youngblood into a home with a cellar. Perhaps here was the jackpot.

"I shoved the door," he reminisced. "But it didn't open. I reared back and laid my big number twelve against that door and kicked it open."

Youngblood beamed his flashlight into the blackness. He froze in cold fright.

"There was a German sitting on a table with a burp gun lying across his lap. I shone my light right in his face. He was waiting, thank God, to give up. He laid down his gun and put up his hands."

The prisoner emptied his pockets and offered his captor ersatz cigarettes. Youngblood shook his head and held out American smokes. The enemy soldier heartily accepted the gift. Youngblood then inquired—drawing on his scant German—where he could find eggs. The prisoner gestured toward a room upstairs, Youngblood gave an approving nod, and his captive darted away to fetch them. With eggs in hand, the pair joined the two infantrymen and searched more buildings, while poking along toward an infantry CP that also served as a collection point for prisoners. It became prudent to drop the folksy air as the group neared the CP. The German understood his role.

"He took one last draw on his cigarette," Youngblood remembered, "then put it out. He stepped out in front about three steps, put his hands behind his head, and marched around the corner to the CP as our prisoner."

His American acquaintances last saw him in the backseat of a jeep, destined for a PW cage. As the vehicle roared off, the German bid a discreet farewell. "With his hand down low, he waved like a little kid."[302]

Youngblood returned to his tank and found Signal Corps photographers Corporal Louis Nemeth and Staff Sergeant Robert R. Rowe. The engineers had just laid a treadway bridge atop a ruined railroad bridge that spanned Adenauer Creek. The two cameramen wanted to photograph "Asp II" rolling over it and requested help. Youngblood consented, but first one of his men wiped away grime that covered the name painted on their tank. Nemeth captured the scene with his hefty Speed Graphic, while Rowe filmed the crossing with his Eyemo motion-picture camera. In years to come, Nemeth's still photograph became the most published 741st image.[303]

The same day, with the mop-up complete north of the Ahr, Companies C and D rejoined the Second Battalion, 23rd for a short push to the Rhine. M18 tank destroyers from the 612th TD Battalion joined the group. Their goal was Sinzig on the Rhine.

The tanks and infantry sliced ahead through the rain. Towns fell like dominoes as the attackers knocked over Kesseling, Staffel, Nieder-Heckenbach, Ober-Heckenbach, Beilstein, Blasweiler, Ramersbach, Obervinxt, Mittelvinxt, Untervinxt, and Schalkenbach. Opposition remained soft until the advancing column reached Königsfeld, where German troops had a roadblock covered by machine guns and a concealed antitank gun. Cannon fire from Company C blew apart the obstacle, but the antitank gun struck the company tankdozer.

Peter Fardella, the Charley Company radio repairman, recounted what transpired:

"At Königsfeld, as our column was tearing through, an AT gun opened up from the right flank and put two holes into the side of the tankdozer about four inches from the bottom, just behind the bow gunner's seat. The shots went along the floor without hurting anyone. Everyone in the column took care of the Jerry. The dozer crew—Norwood, commander; Wetzel, driver; Reser, gunner; Dwinchick, bow gunner; and the [loader]—had to abandon the dozer. This being the fifth that Dwinchick had shot out from under him, he was pretty shaky. The opinion in the company was that he had risked his life enough. From then on, Raleigh Kraft [the First Sergeant] kept him helping out in the kitchen."[304]

The shooting at Königsfeld halted progress for an hour until German resistance faded, and the advance resumed. The column reached Sinzig at 6:00 P.M. Fardella was well behind and arrived in town later with company headquarters.

"It was dark when we entered Sinzig, and it seemed as if hell were loose. Shells kept whistling in and hitting nearby, and we fell asleep with their sounds in our ears. Our billets were in several houses near the edge of the city, and some of the boys had beds to sleep in. The next day the civilians were chased out as we had orders not to stay in the same buildings with them. One platoon was sent to clean up the town to the south, Niederbreisig, with the First Battalion, 23rd Infantry. Colonel Hightower, the battalion commander, was wounded, and that ended our association with him, which dated back to Normandy."[305]

Lieutenant Crisler's platoon entered the action on March 10 and helped seize Niederbreisig. Fortified with wine, cognac, schnapps, and other goodies plundered in Sinzig, the tankers pushed off in midafternoon with the infantry and cleared Niederbreisig.

Company B headed from Königsfeld toward the Rhine after a four-day absence from the front. Its destination was Rheineck, the next community upstream from Niederbreisig. Lieutenant Dudley's platoon with its 76 mm cannons supported the Third Battalion, 9th Infantry.

Sergeant Joseph A. Brennan commanded a tank in the platoon. "Before we started up," he recalled, "our tanks tested their machine guns by firing into a wooded hillside. When we ceased firing, out walked a Jerry, surrendering."

The armor and foot soldiers pushed through Waldorf and Gönnersdorf while heading toward Rheineck. The tanks encountered Panzerfaust projectiles at almost every road bend. The weapons caused no casualties. The Germans fired indirectly and missed with every shot. This also happened outside Rheineck, and Brennan recalled a close call. "Doughboys pulled a Jerry out of a hole about six yards from us. He had a rifle and the usual Panzerfaust with the pin pulled and ready to fire."[306]

The Panzerfaust men prevented the tanks from entering Rheineck that day. One projectile exploded against a tree near Dudley's vehicle, and a fragment conked him in the head but left him unharmed. Three tanks withdrew to Gönnersdorf for the night, and the rest stayed at a roadblock outside Rheineck.

At 7:00 A.M. the next day, the three tanks that had pulled back to Gönnersdorf joined riflemen from Company K, 9th. After an uphill trek, they arrived at Niederbreisig, from where they descended to Rheineck. Meanwhile, the tanks at the roadblock joined Company I, 9th and also entered Rheineck. The two groups snaked through the town until they met. All five vehicles remained in the area until the afternoon, when they returned to Königsfeld, the Company B bivouac area.

The fight west of the Rhine closed for the 741st at Niederbreisig and Rheineck.

★★★

Traffic bottlenecked at Remagen. Since its capture, the Ludendorff Bridge had been center stage in Northwest Europe. Vehicles crossed the bridge in a single column around the clock. To increase the flow, engineers set up ferries and stretched a pontoon bridge across the river on March 10, but they had to construct more bridges to convey all the traffic. The 741st and 2nd Division waited.

Battalion headquarters settled in Königsfeld; Companies A and B in Niederzissen; Companies C and D in Sinzig; and Service Company in Ramersbach. The tankers made use of the combat recess to undertake training, cleaning, and vehicle maintenance.

Peter Fardella washed up and found time for amusement. "Our battalion shower, built in an old German trailer, helped us clean up. Our vehicles also got a good scrubbing. Outside of the usual maintenance, we enjoyed the warm sunshine and amused ourselves chasing away the German brats since we didn't care to pay the $65 fine for fraternizing with kids."[307]

Serious matters also gained attention. Company A conducted training to educate the 9th Infantry on the M4 series. Tank crews and their vehicles drove to one infantry platoon after another, reciting pertinent facts and demonstrating operational capabilities. Question-and-answer sessions followed.

Company C and the 23rd Infantry practiced loading infantrymen onto tanks. They determined how many soldiers with full equipment could ride on one vehicle for long hauls. The optimal number was fifteen, although they tried thirty-three soldiers in one trial. All held on, but the large crowd hampered rapid dismounting. Maintenance men from the 741st welded grab rails to the tanks to help the troops climb on and off.

Training and other outdoor activities turned perilous on occasion due to falling shells. Antiaircraft gunners in the Remagen area peppered the heavens with projectiles each time the Luftwaffe launched a bombing sortie against the Rhine bridges. Spent bullets and shell fragments tumbled to earth in profusion. The aerial fallout caused one casualty on March 11 in Sinzig, when a chunk of steel struck First Lieutenant Clifford Shirk in the neck. The Company D platoon leader died within thirty minutes.[308]

On March 14, division headquarters directed the tankers to search for ordnance tossed aside by the enemy. The entire 741st (except Company A) along with the 612th TD Battalion and 2nd Reconnaissance Troop engaged in the task.

Carrying picks, shovels, and sidearms, soldiers from the three units beat the hills and forests from March 15 to March 18. They weeded out ordnance and destroyed it or salvaged it for future use. During the sweep, men from the 612th found a radio station with a high-powered transmitter, albeit wrecked by the Germans. Tankers from Lieutenant Dew's platoon discovered a smashed P-47 including its pilot, his burned remains a grotesque sight.[309]

The division operations officer telephoned Major Browder on March 20 with news that Company B and the 38th Infantry would jump the Rhine within twenty-four hours. They had orders to relieve a 99th Division regiment and a cavalry squadron at Hönningen. The rest of the 741st and the 2nd Division would cross after the 38th.

Navy personnel planned to ferry the 38th across the river in Higgins boats. First Lieutenant Howard F. Sharpless, the Baker Company motor officer, reconnoitered a route to a pontoon bridge at Remagen. It was a floating Bailey bridge erected by the 148th Engineer Combat Battalion. This was the sixth bridge over the Rhine, five of them built by US Army engineers.

The Ludendorff was the only German-built span, and it no longer stood. On March 17, the celebrated bridge collapsed into the river after engineers had closed it for repairs.

★★★

Far removed from the bridge and the battlefront, Captain Sledge informed his parents about his hand injury. "Luckily it wasn't serious," he told them. "So, don't worry about me as all is well. I now have two oak leaf clusters on the Purple Heart."

He expressed pleasure in his present circumstances:

"I am in a general hospital near Paris, and I really do like this rear-echelon life. I'd be content to stay here for all time. Today we heard the Glenn Miller Band, and it was really enjoyable. It was almost like being in the States except for the surroundings."[310]

CHAPTER 38 REMAGEN BRIDGEHEAD

March 21, 3:15 A.M.: Under the leadership of Lieutenant Sharpless, a truck convoy from Company B journeyed through the darkness ahead of the company's tanks. Technical Sergeant William A. Hunt and his kitchen truck headed up the column.

After dawn broke, Company B led the 741st over the Rhine on the bridge built by the 148th Engineers. The company made its bivouac in a field outside Arienheller. Headquarters Company and Service Company crossed the river and settled in Hönningen during the first hour of March 22. Company A followed and passed through Hönningen before arriving at the riverside town of Rheinbrohl just after 3:00 A.M.

Even before sunrise cast its rosy glow that day, Baker Company broke camp and pushed into the Rheinbrohler Forest, where wooden watchtowers once marked the Roman Empire's outermost limit. The forest now stood at the tattered fringe of another empire, one dying in a firestorm of its own making.

Teamed up with the 38th Infantry, Company B headed toward the Wied River, but deep defiles, wooded hills, and winding dirt roads forbade open deployment. One forest road teemed with Topf mines. Their wood-pulp casings made them an impossible challenge for mine detectors, but the tanks found them.

Sergeant Angeletti's machine blew a track after striking a mine. No injuries resulted, but the explosion stopped the column. An infantry jeep also detonated a mine.

The column lumbered ahead again until another booming report resounded through the leafless trees. Sergeant Mortzfield's tank shuddered to a standstill. The crew remained unharmed, but that was it for the road. The other tanks turned into the forest, squeezing their way down narrow firebreaks and logging trails.

Lieutenant Helm's platoon stole into Rodenbach and gained a bridge over the Wied, while Sheppard's platoon reached the river at Niederbieber but found no bridges able to bear armor. Remaining in the forest until March 23, Dudley's platoon, minus one tank deadlined for mechanical problems, crossed the river at Au.

The tanks supported the Second Battalion, 38th that day. The GIs soon heard an all too familiar noise in the distance. Enemy mortar shells flew into the sky and arched back toward earth. Despite the shells and more mines, the attackers advanced until evening, when fading light and steep terrain slowed progress, as did German soldiers inside a crumbling castle.

To dislodge these holdouts, an infantry company besieged the hilltop fortress known as Burg Braunsberg. The castle defenders numbered thirty to forty men, including a headstrong lieutenant who "made them swear they would fight to the last drop of blood."[311]

One of their rifle slugs injured Sergeant Brennan as he stood bolt upright in the commander's hatch of his tank. The bullet pierced his helmet, lacerated his neck, and he fell back into the tank, screaming in panic. His gunner, Corporal Harrell E. Gullatt, grabbed him—"Joe, you're just grazed!"[312] Embarrassed but relieved, Brennan collected himself as Gullatt applied a cotton dressing. The tank driver backed up to a safe spot, and Brennan waved goodbye as he walked away toward the nearest aid station. His war was over, at least for awhile.

Gullatt wanted to blast the castle, but the tank crews had orders to spare the medieval structure. The United States had signed an agreement with twenty-one other nations in 1935, which protected cultural treasures from destruction during war. This agreement, known as the

Roerich Pact, granted neutrality to historic buildings, monuments, and institutions, but there was an exception. The neutrality privilege ended if one side used a protected site for military purposes. This provision afforded Gullatt and his fellow tankers every right to fire, but someone high in the chain of command had to make the decision. The situation baffled those who saw the castle as little more than a bastion for enemy gunmen.

The deadlock dragged on for eighteen hours before frustration gave way to high explosives. Shell fire from the 612th TD Battalion ended the episode when a round killed the enemy lieutenant. His men surrendered.

As events at the castle unfolded, Company C crossed the Rhine on a steel treadway bridge recently stretched across the water from Niederbreisig to Hönningen. Constructed by the 254th Engineer Combat Battalion, the bridge had a large wooden sign at either end. VICTOR BRIDGE, each placard announced, LONGEST TACTICAL BRIDGE IN THE WORLD. Company C rode across the span during the afternoon, rumbled several miles upstream, and climbed a high hill to a farmstead called Gebranntehof. Here, the company spent the night.

Daylight on March 24 found Company C and the 741st out of action but for one platoon from Company B. Four tanks under Lieutenant Helm rolled forward with the First Battalion, 38th. Sergeants Lucey, Warren, and Kammeyer commanded the other tanks in the lieutenant's platoon. Gladbach was the next town ahead.

The tanks first assaulted a hill, firing 76 mm HE rounds to flush out enemy troops from a house near the top. The defenders hid in a ditch, and machine-gun fire from the tanks kept them pinned there. Three times the enemy hoisted a white flag, and the tankers stopped shooting, but they had no infantry escort to handle prisoners. The Germans resumed firing, their bullets nothing but a nuisance to the tank crewmen, like buzzing flies.

The tankers had no idea that danger lurked behind the trapped riflemen. The hoarse cough of an enemy antitank gun sounded from Gladbach. One round struck Kammeyer's tank, and his driver immediately backed up to a safe location. Corporal Harmon W. Tusten and Private Leslie E. Brandenburg suffered compound fractures.

The other tanks reversed direction, backing away from the exposed position. Another orange flash winked from Gladbach. Lieutenant Helm picked up the enemy gun and directed everyone's attention to the spot, but their cannon shells had no effect.

The enemy also hit Warren's tank, which left Private First Class Jess V. Cuny dead and three crewmen burned: Technician Fourth Grade Marshall Martin, Corporal Corlin Legler, and Private First Class Walter R. Jackson. The latter two also agonized with broken bones.

Overwhelmed by his hairbreadth escape—his second since D-Day—Warren lost hold of himself. "I think I went into shock. I didn't know heads or tails."[313] Medics later diagnosed the tank commander with a concussion.

The enemy continued to hammer AP shells at the tanks. Nobody among the crews knew how many and what type of antitank weaponry the Germans had in Gladbach.

Helm's tank fell victim to an enemy shell, wounding the lieutenant and his gunner, Homer F. Cupit. The driver, Private First Class Cleo B. Keller, perished.

After an AP round struck Lucey's tank, everyone inside bailed out and hid behind a turnip bank. Private Charles Miller clutched an eye, partially blinded by steel fragments. The tank remained in gear and slowly rolled backward, receiving two more hits before it reached a protected point where the crew remounted their still-moving vehicle. Although the main gun no longer functioned, the crew kept their machine in operation and helped evacuate wounded.

Still at Gebranntehof, Company C learned about the deadly scrape near Gladbach. One platoon moved forward to a reserve position at Niederbieber, while the other two platoons joined the Second Battalion, 38th.

The enemy's backbone suddenly snapped. Four self-propelled guns emerged from Gladbach and retreated into the Heimbacher Forest as artillery shells chased them.

Charley Company tanks and infantrymen from the Second Battalion captured Gladbach. At 2:00 P.M., the armor withdrew to Niederbieber for the night.

Baker Company mended its losses. The tanks commanded by Helm and Warren amounted to complete write-offs, burned beyond worth. The other pair traveled back under their own power. Kammeyer's machine had taken a shot to its turret, damaging the sight for the main gun. Lucey's machine had sustained three hits, but only one projectile penetrated.

Beyond Gladbach lay the twin towns of Weis and Heimbach. Before sunrise on March 25, the Second Battalion, 23rd Infantry took over from the Second Battalion, 38th, and prepared to seize the towns assisted by two platoons from Company C.

Minutes before beginning the attack, Sergeant Herman Johnson's tank detonated a mine as it chugged downhill to the departure line.

The captain leading Company G, 23rd witnessed the blast:

"A deafening explosion broke the monotony of the tank motors. A streak of flame sprang from beneath the left track of the lead tank. For a moment I thought an artillery barrage was beginning to fall, and I started instinctively to fall to the ground. Then I realized what had happened. The lead tank had hit a mine.

"Men milled around the tank as I reached it. Two were climbing from the turret.

'Anybody hurt?' I asked.

'Nobody hurt,' one of the tankmen said. 'Shook us up. That knocks the tank out 'till we can get it repaired. Sonofabitches oughta mark their goddamned mines.'"[314]

The attack stalled at its starting point as the tankers searched for a way around the minefield. The infantrymen began to fret. Daylight was near, and they would be clay pigeons while crossing the flat, treeless field between the jump-off point and the twin towns.

The Second Battalion commander sent his foot soldiers into the field alone. The "cans" (infantry jargon for tanks) would have to catch up. Midway across, the tanks rejoined the infantry. The Company G commander appreciated their presence:

"I heard the noisy clatter of tanks behind us, and I breathed an inward sigh of relief. The steel monsters churned past us and took their places in the assault formation—three forward and two back—like waddling ducks going to water."[315]

The tanks and infantry reached the first houses of Heimbach without encountering enemy guns. Catching whiffs of smoke, the Americans found two smoldering homes hit by shell fire during the night. Only civilians occupied the town, which fell without resistance, as did its sister community, Weis. Prodded from their slumber at Weis, twenty half-dressed prisoners were the only Germans clad in Feldgrau.

The sun shone brightly as the tanks and infantry pressed on toward Bendorf-Sayn, and there the cakewalk ended. Suddenly, the harsh, ripping sputter of a German machine gun broke the calm. The racket escalated with the sounds of rifles, machine pistols, and an enemy self-propelled gun. The attacking foot soldiers scrambled into the first houses and advanced no farther. Cannon shells from Company C might have quickly broken the impasse, but, as an infantry sergeant explained to his captain, "The goddamned tankers said they can't come forward."[316] The sergeant knew they feared being ripped apart by the self-propelled gun, but he and his men had no patience.

Minutes later, a young private started the tanks rolling and reported to the captain: "I told them you said they'd better get moving or you'd be down there and shoot hell out of them with a bazooka right smack in their tails."[317]

The tanks pumped shells through every window from which enemy fire emanated. House after house fell to the infantrymen. As resistance faded, they charged toward Sayn Creek, which cut through the town. No sooner did a bridge over the creek come into view than it disappeared in a cloud of smoke and debris. The Germans had waited until the last moment before blasting the ancient structure.

Undaunted, Second Battalion infantrymen waded across the knee-deep water, but the tankers had to wait for engineers to lay a treadway span over the fast-flowing stream.

Most of Bendorf-Sayn lay beyond the wrecked bridge. The infantry trudged ahead, crossing a second ribbon of water, Brex Creek. Another demolition effort partially ruined a bridge over the creek, but the structure had enough strength to support foot troops. The infantry marched on, leaving the armor farther behind. Curious civilians were the only Germans in evidence until the GIs reached the last houses at the far edge of town and gun clatter erupted. American riflemen and machine gunners responded, but their effort to quell the uprising ceased when headquarters ordered a change in plans. The attack shifted in another direction.

Both tank platoons and the Second Battalion pulled out of Bendorf-Sayn and moved toward Stromberg, a village already captured by the First Battalion, 23rd. The tankers used an undamaged bridge to cross near a deserted factory. Wooded high ground beyond the bridge featured a serpentine road that led up to Stromberg. The foot-weary infantry climbed aboard the tanks despite warnings that German armor fled up the road less than fifteen minutes prior. The ease of riding outweighed any concern for being caught by enemy fire.

Company G rode the lead tanks. Its commander positioned himself on the fourth vehicle in the column. "We loaded on the tanks, and the tankers signaled they were ready. Their treads churned up clouds of dust on the dirt road, and we began the ascent."[318]

After several stops to investigate blind corners and probe for mines, the tanks and their entourage reached Stromberg. Exhausted infantrymen from the Second Battalion flopped down in the houses, while the tankers joined the First Battalion and plunged into adjacent villages.

Company C finished the day spread out between three communities on the high ground above Bendorf-Sayn. The company CP and First Platoon stayed at Stromberg, where an enemy self-propelled gun shelled the village during the night. The Third Platoon made its home in neighboring Caan, while the Second Platoon took the choicest spot. Its men claimed Nauort, a prosperous farming community resplendent with fine homes. Crewmen under Sergeants Hascal Combs and Enrique Mariani enjoyed the best accommodations. Each man had a bed with clean linens, maid service, and cognac, but such niceties were always short-lived. During the night, the First Platoon relieved them, and the departing tankers belted out a chorus of profanities.

On March 25, using the Victor Bridge, Company D was the last 741st unit to cross the Rhine, whereupon it banded together with the 9th Infantry. While the 23rd Infantry headed to Stromberg, Company A and the 9th resumed fighting for Bendorf-Sayn. After finishing off the last defenders, the infantry swung northeast.

The 2nd Division redirected its forces in preparation for a breakout from the Remagen bridgehead. The attack had thus far run parallel to the Rhine, but now the division turned its sword tip inland toward the German interior.

Three tank platoons—one light and two medium—coupled with the Second Battalion, 9th, and another light tank platoon joined the Third Battalion. Their objective was Höhr-Grenzhausen. After brushing aside enemy resisters, the four platoons stayed there for the night. In the morning, the remaining tanks and headquarters personnel from Companies A and D joined them.

On March 26, the Medical Detachment welcomed a new leader. Captain Lawrence F. Barker from Saint Louis, Missouri, replaced Captain Godett as battalion surgeon. The departing doctor suffered from combat exhaustion.

"He had two or three close calls," remembered Major Browder. "One time a tank ran up onto the back end of his jeep. Enemy shells also nearly hit him a time or two. It got to the point that he would not even get out of his foxhole to go eat. He put his mess kit on the side of the hole and ate there. We decided to send him back, and we sent him to Paris."[319]

Godett moved to the 152nd Station Hospital, which operated the Medical Department's central blood bank on the European continent. Captain Barker had come from there.[320]

The same day that the two physicians swapped places, the 2nd Division broke loose from the bridgehead and roamed east toward Limburg. The 9th Infantry advanced from Höhr-Grenzhausen to within a kilometer of the Cologne-Frankfurt Autobahn, and the 23rd Infantry also closed on the highway.

The next day, Charley Company tanks crossed the Autobahn and stopped in Wirges. Several men from the company chatted in the street with a young Fräulein when Colonel Lovless roared up in his jeep. Caught fraternizing, the men scattered. In their wake, the colonel collared Ernest R. Pepe, a mere bystander.

"What's your rank, soldier?" Lovless snapped.

"Corporal, sir."

"You *were* a corporal!"[321]

CHAPTER 39 NON-BATTLE CASUALTIES

The halcyon days of National Socialism produced the Reichsautobahn, a highway system constructed to accommodate the Wehrmacht and its transportation needs. The graded, six-lane highways served the invading Allied armies equally well.

On March 28, the 741st motored on the Autobahn for the first time, though part of the battalion kept to secondary roads. Companies A and D navigated roundabout through the backcountry with the 9th Infantry to reach Hadamar, a sizable town directly above Limburg, which had fallen two days earlier to the 9th Armored Division.

Attached to the 23rd Infantry, Company B moved to a marshaling area at tiny Elgendorf. Infantrymen from the Third Battalion loaded up, and the tanks tore ahead toward Limburg, speeding down the Cologne-Frankfurt Autobahn. Displaced persons and former slave laborers lined both sides of the highway, shuffling along toward an uncertain future.

Company C also rallied at Elgendorf. Third Battalion troops climbed aboard the tanks, and together they too roared off toward Limburg. The column wound through Montabaur and crept out of town on Sauertalstrasse, which hugged a towering fortress wall. The narrow downhill road partially collapsed under the lead machine's weight. The unfortunate tank tumbled down a high, stony embankment, overturning twice before landing right-side up. The somersault injured the gunner, Corporal John R. McGowan, as well as the bow gunner, Private First Class Casimir R. Kubian, who fractured an arm and a leg. The tank commander, Second Lieutenant Leonard Henkelman, died afterward from internal injuries. Two riflemen riding on the tank also lost their lives, but the tank itself suffered only scratches.[322]

Henkelman had accepted a battlefield commission in January and became a platoon leader after Covington took charge of the company. Peter Fardella liked the enlisted-man-turned-officer. "He had never put on airs when he received his commission. The only difference about him was the bar on his collar."[323]

Company C and its infantry contingent reached the Autobahn less than a mile beyond the accident site and motored on to their destination.

On March 29, Companies B and C vacated Limburg as Third Army troops flooded into the city along with emaciated prisoners freed from Stalag XIIA. Both tank companies settled nearby in less congested Dietkirchen.

Companies A and D left for Giessen during the afternoon and transported riflemen from the First Battalion, 9th. They found soldiers from the 7th Armored Division picking their way through the streets as random fires smoldered. In the downtown area, many of the tanks parked for the night along Johannes-Strasse, the crews and foot troops sacking out in adjacent buildings.

The barreling advance continued on March 30 through a soft drizzle. General Robertson and two staff officers stood in the rain at a downtown intersection and watched the tanks and infantry depart Giessen. No combat occurred as the tankers followed the 9th Armored Division. Over sixty miles of German soil passed under their grinding tracks. Companies A and D reached Rüdigheim with the 9th Infantry; Companies B and C rolled into Erbenhausen with the 23rd Infantry.

The next day, Able and Dog tore off again, finishing the month in Nieder Warolden and Höringhausen, respectively. Baker and Charley ended up at Wega, where they picked up radio signals from an outfit locked in combat. In the distance, fires and red-tailed tracers lit the darkness.

★★★

April 1, Easter Sunday. The entire 741st remained attached to the 2nd Division and knifed deeper into the Reich.

Company A, still with the 9th Infantry, occupied Elleringhausen but sent one platoon to join a mobile reserve at Nieder Warolden. Company D, also with the 9th, established its CP at Nieder Warolden with one platoon at Ober Warolden. Attached to the 23rd Infantry, Companies B and C reached Volkhardinghausen and discovered a surprise. Rear echelon troops from V Corps headquarters occupied all available billets, forcing the tankers to camp on an open hillside. Headquarters Company and Service Company both reached Mengeringhausen, a large municipality with ample lodging.

Easy sailing continued the next day. Companies A and D moved to Korbach and put up roadblocks on four major roads leading into town. Each barrier included two medium tanks. Companies B and C entered Arolsen, where the former helped establish roadblocks.

The push into Germany halted for two days, time the companies used for rest and vehicle maintenance. During the lull, Second Lieutenant Frank Kurkowski (a recent battlefield commission who replaced Henkelman) made news when he brought in a Luftwaffe flier, whose Messerschmitt crashed in the Service Company area.

Around 6:00 P.M. on April 3, Technician Fifth Grade Barney Gordon from Dog Company and another man stood guard near Korbach, while the rest of their group left for chow. Gordon swigged liquor and quit his post, traipsing off to a farm called Dingeringhausen.

The next afternoon a sergeant from the 804th MP Company visited the farm to investigate a report that an American soldier lay crumpled alongside the road. The policeman found Gordon's lifeless body, a nasty scalp laceration the only visible injury.

The farm owner, Wilhelm von Kleinschmit, told the MP that an intoxicated Gordon had appeared at his door and asked to speak with a soldier named "Smith," who was eating supper inside. Drunk and armed with a carbine, Gordon made everyone nervous, so Smith told him to wait next door. Smith had a motorcycle and promised to take his friend to Korbach.

Gordon staggered away to the adjacent house and climbed stone steps to a landing where a farmhand refused him entry. The tipsy soldier tumbled down the steps as he departed. After pulling himself up, he spotted two approaching vehicles and waved for them to stop. The first truck flew past, but the second slowed to a halt, and Gordon planted a boot on the running board. Reeking of alcohol, he asked for a lift and an argument erupted, whereupon the six-by-six roared away. Gordon flew backward onto the road, where oncoming vehicles ran over him in the dark. During the incident, his friend Smith disappeared into the night on the motorcycle.

The MP sergeant collected Gordon's body and hauled it to a Graves Registration outfit at nearby Bredelar. An officer there positively identified the cadaver and examined it for signs of foul play. With none found, he considered the death accidental.

Eight days later, Gordon's wife received a death notice from the War Department. The same week, Captain Snyder, commander of Company D, mailed her a typewritten condolence letter filled with the usual platitudes. He mentioned nothing about AWOL and intoxication, only revealing the soldier's cause of death.

Mrs. Gordon applied for a death pension from the Veterans Administration. She anticipated monthly pension checks like the annuity checks she received from her late husband's National Service Life Insurance policy, but life insurance was simple.[324] The insurer paid benefits no matter the cause of death, except suicide. VA benefits involved other factors, and Barney Gordon's demise while AWOL resulted in a denied claim.

Gordon's widow appealed the decision several times. Her letters conveyed confusion and frustration. On one occasion she begged, "I have a son to take care of, and I need the money badly." Another time she wrote, "Other children are receiving it, then why not mine?" Her words changed nothing. The VA forever denied benefits and also withheld the precise reason.[325]

Another accident killed a 741st soldier the day after Barney Gordon's death. Technician Fourth Grade Brodie Ritchie died when a deuce-and-a-half truck hit him while walking to the Baker Company kitchen for lunch. Ritchie lost his life in the line of duty, a distinction that spared his family the ordeal endured by Gordon's widow.

CHAPTER 40 TIGERS!

At the 62nd General Hospital, a former German Navy facility outside Paris, Captain Sledge wrote to update his mother on his hand injury:

"The fingers are getting along pretty good now. The thumb is just as good as new, and the forefinger will bend about ¾ of the way toward a clenched fist. So, shouldn't be here much longer. I'm in no hurry to get back to the fighting, but when the time comes to go back, maybe the toughest part will be over."[326]

For Sledge's comrades in Germany, the drive into enemy territory resumed on April 5. The 9th Infantry received orders to strike across the enemy countryside toward Immenhausen. In support, tankers from Companies A and D teamed up with the First and Third Battalions.

An hour before noon, the attacking tanks and infantry began their march. They raced through Freienhagen, Wolfhagen, Obermeiser, and Grebenstein before reaching Immenhausen in midafternoon. No opposition sprang to thwart the armored column.

Recently returned from a Stateside furlough, Lieutenant George Habib stretched the advance even farther. His platoon from Company A operated with a rifle company from the First Battalion, 9th. With infantrymen clinging tight, Habib's tanks ventured into a thick forest flanking the Weser River. Their mission was to clear a river bend above

Hann-Münden, contact troops from the 69th Infantry Division, and perhaps capture a bridge.

All hopeful expectations soon vanished. The attackers found entrenched Germans from SS-Panzerbrigade „Westfalen."[327] The small American force stopped short of the river and burrowed into the forest floor for the night. Several times before morning, the tankers and infantrymen detected the muffled ruckus of SS men prowling in the darkness. Encirclement loomed, but the disquieting noises came to nothing.

On April 5, Companies B and C again joined the 23rd Infantry. Their mutual objective was Hofgeismar. The tankers and infantrymen dashed through empty streets in Ersen, Warburg, and Niedermeiser. In each town, they encountered windows displaying white sheets, bath towels, pillowcases, and other improvised surrender flags.

Hofgeismar presented the opposite, with rifle barrels and machine-gun muzzles protruding from windows. Diehards from the Westfalen brigade intended to make a stand and envisioned glory at the eleventh hour. They reinforced their ranks with eighty teenagers drafted from their homes following a house-to-house canvass by Hofgeismar's military commandant. Clothed in secondhand uniforms and toting Czech rifles, the youngsters cowered in foxholes on the town perimeter. They buried their faces in the dirt and waited for death or capture.

Company B approached Hofgeismar with soldiers from the Second Battalion, 23rd. The lethal chant of mortars and small arms sent the infantrymen springing from atop the armor. They fanned out as the tanks deployed and hammered HE and WP into barns and homes. The buildings burned brightly as the foot soldiers entered town. After exhausting their shells, the tanks withdrew a half-mile where Service Company trucks served up more ammunition.

Again, the tanks approached Hofgeismar, but this time they encountered fire from a hidden Panzer. The German tank scored no hits as the GIs searched in vain for the enemy vehicle. The American tanks shielded themselves from the threat by weaving their way between buildings at the edge of town. The German machine withdrew.

More firepower joined the skirmish as armor from Company C arrived along with troops from the Third Battalion, 23rd. Close combat ensued on the streets and in houses.

The attacking Americans discovered that the defenders possessed more than small arms and mortars. An abundant Panzerfaust supply lay at their hands. One projectile wobbled through the air and crashed against the Company B tank commanded by Sergeant Gordon Crouse. He and his driver, Private First Class Louis A. DesHaies, received minor injuries but remained on duty.

Company C also lost a tank. Sergeant Robert J. Lamberson's vehicle burst into a blazing torch after a Panzerfaust hit. All but one crewman escaped. The flames cremated Private First Class Dock W. Fields, the bow gunner, leaving no identifiable remains. The four survivors escaped with burns on their faces and wrists.

Ushered in by rain and gray clouds, nighttime came early to Hofgeismar. Fires still burned when infantrymen from the Third Battalion, 23rd finally secured the town. The SS fled, and their exodus ended the children's crusade led by the town commandant. The boys who survived landed in the 2nd Division PW cage, many still trembling from the shock of battle.

Early the next morning, four tanks from Company C joined the Second Battalion, 23rd. The remaining Charley Company tanks merged with the Third Battalion, as did Company B and tank destroyers from the 612th TD Battalion. The final push to the Weser River began after sunrise.

Tanks, TDs, and trucks loaded with infantrymen advanced along a road that sloped downhill into Karlsdorf, the first village east of Hofgeismar. Vigilant eyes in the mechanized column suddenly discerned two hulking predators, Königstigers or King Tigers belonging to schwere Panzerabteilung 507, a heavy tank battalion that beefed up the Westfalen brigade. The menacing duo sat on a hill along with a Jagdpanther also from the battalion.

The Tigers fired first.

Frantic truck drivers sped for Karlsdorf, jinking left and right to avoid being hit. All avoided destruction, thanks only to the enemy's unpracticed gunnery. The American armor remained behind to battle the three Panzers.

The matchup turned lopsided at more than one thousand yards. Cannon shells from the American tanks and TDs glanced off the thick-skinned enemy tanks. The Germans fired 88 mm rounds that bored into Sergeant Vernon's tank and crippled it, though no harm resulted to the crew. Staff Sergeant Neidrich's tank also took a hit, and Private Howard Rowe, the driver, emerged with an arm wound. All the Company B tanks dodged mishap.

The exchange ended when 2nd Division howitzers interdicted the hill where the Panzers stood. Meanwhile, a TD ventured close enough to drill the Jagdpanther, destroying it.

The two Tigers slipped away.

Karlsdorf fell to the GI tanks and infantry. Afterward, they branched right and left at a road fork and swept into Udenhausen and Hombressen. Sniper fire dogged the invaders. The advance resumed once the infantry routed enemy sharpshooters from both places.

Between Hombressen and the Weser, violence flared when Company C tanks and troops from the Second Battalion, 23rd drove uphill toward a wooded area. An explosion boomed from the trees, a Panzer cannon's telltale report.

Hidden in the woods, a Tiger crew tracked the Charley Company command tank (without Captain Covington aboard). The first shell splashed a fountain of earth to its right. Soldiers clinging to the tank jumped to the ground with only an instant to spare before another projectile hit the vehicle. The four men inside spilled out, a sprained knee the worst injury. Their tank burst into swirling orange flames. The other tanks turned away, re-entered Hombressen, and offered no further challenge to the German tankmen lording over the area.

Companies A and D wrangled with opposition in late afternoon. At about 5 P.M., Staff Sergeant Ball—just back from furlough—and his crew claimed an obsolete Panzer III from the Westfalen brigade. Company D and a rifle company from First Battalion, 9th Infantry battled an ambush and cleared the enemy from the Gahrenberg Forest. The fight left Staff Sergeant John Cahill with thigh and facial injuries.

Dusk approached as tanks from Companies B and C reached Veckerhagen on the Weser. The little community lay tranquil when the first Americans entered, but that ended when German machine-gun bullets tattooed a supply truck. The fire originated across the river, and tanks in the town banged back at the lone shooter, silencing the gun.

After sunset, armor from both companies in Veckerhagen stood ready to cover a river crossing by the Second Battalion, 23rd. The infantrymen rowed across unmolested in wooden assault boats. The foot soldiers inched their way uphill, while the tankers moved to nearby Vaake and waited for engineers to construct a bridge.

By midday, April 7, armor rolled across the Weser on a floating treadway bridge. Company C crossed first, followed by Companies B, A, and D.

Company B rushed north along the river with the First Battalion, 23rd and skirmished with the enemy south of Glashütte. Over one hundred Germans ended the fight with their hands held high. Two Flakvierlings (quad-barreled 20 mm AA guns) fell into American hands.

The tanks and infantry turned toward Löwenhagen. Two machines from the First Platoon, Company B motored ahead on reconnaissance. Bullets ripped at them as they approached the town, and the tankers answered by setting fire to several farm buildings. Led by Lieutenant Malcolm Reynolds, the entire platoon rendezvoused in Löwenhagen and pushed on to seek out enemy armor reported ahead. (Reynolds had transferred to the platoon from battalion headquarters, replacing Lieutenant Sheppard.)

Beyond Löwenhagen, the tankers spotted two King Tigers in a field ten acres away. The steel monsters appeared unaware that Reynolds's men had them in sight. The Germans slowly backed up toward Imbsen.

Reynolds ordered action.

His gunner, Corporal Boardman, cut loose and so did Sergeant Lucey's gunner.

Boardman's first round streaked past the tank in his gunsight. Though closer, his second shot also missed. His final round hit the tank's frontal armor. "It made a beautiful white flower, many times the size of the tank," Boardman recalled. The crew inside gave no reply other than to continue creeping backward. "It was my experience that enemy tanks didn't fire back if you were hitting them," Boardman explained. Even if a shell caused no damage, the vicious jolt from a projectile impacting at supersonic speed usually rattled the crew.

In the murky twilight, Boardman lost sight of the German titan as it retreated into Imbsen. Within minutes, he heard the whip-crack of exploding ammunition followed by a plume of black smoke. Had the enemy machine succumbed to Boardman's final shot? "I realized that I hadn't knocked it out. They had backed up and set it on fire." Exactly what prompted the Germans to scuttle their own machine puzzled Boardman.[328]

Upon crossing the Weser River at 2:00 P.M., Company C moved east to join infantrymen from the Second Battalion, 23rd. As a team, they shoved off across green fields toward Ellershausen, one of two villages in the area with the same name.

Behind a preparatory barrage, tanks and dismounted infantrymen struck the village. Their spread formation resembled an illustration in a field manual; rarely did combat assume such a textbook appearance. Aloft, three P-47s hurtled down on the village, their wings spitting caliber .50 slugs. The artillery barrage ended when the tanks and infantry reached the first houses.

The enemy reacted with a single machine gun firing from a haystack, but American tracer bullets quickly burned the stack to cinders. Among the prisoners, nearly two hundred belonged to Magenbataillon 281, a "stomach battalion" composed of men with digestive disorders, who received special rations.

Before the attack resumed, enemy shells dropped on Ellershausen. Despite the bombardment, Company C drove toward Varlosen while daylight remained.

The infantry rode atop the tanks, unlike the previous assault. The officer in charge decided the grassland in front of the village warranted a quick crossing. The less time exposed in the open, the better. The infantry riders sat on the sides and rear of each vehicle to give the tank crews unobstructed vision and open fields of fire.

They raced downhill, barreling into Varlosen with the infantry leaping to the ground just before reaching the first houses. Not a single shell descended on them during the speedy journey. Even a tank that dropped behind due to a stalled engine incurred no harm before its motor started again, and the machine chugged into the village. The place fell without a struggle.

The fight that day went much the same for Companies A and D. They stuck to the river for several miles after crossing the Weser with the 9th Infantry. The tanks and foot soldiers eventually turned away from the river near Volkmarshausen. Its defenses quickly collapsed owing to the deft interrogation of a prisoner grabbed at the town entrance and grilled on the spot. He divulged all troop locations, and, within minutes, tank shells dislodged the German defenders. They surrendered without offering resistance.

Company D lodged for the night at neighboring Gimte. Company A and its infantry cohort pushed ahead to Oberscheden before bedding down.

Morning bloomed bright and sunny on April 8. Company A moved out with the 9th Infantry at 7:15 A.M., one platoon with Second Battalion and the others with Third Battalion. Company D also joined Second Battalion.

King Tigers reappeared during the day's advance.[329]

Able Company chanced upon a disabled King Tiger, a tank retriever, and a gasoline truck between Barlissen and Dahlenrode. The First Platoon engaged them at one thousand yards and wrecked all three. Their remains blocked a bridge over Dramme Creek.

Dog Company blasted an ammunition truck near Mariengarten and claimed a King Tiger, a David-versus-Goliath matchup that became legendary in the 741st. David in this story was Private David Neill, a black-haired Irishman born in Belfast, whose family immigrated to California. Busted to private in October by a court-martial, the sometimes impetuous soldier was the gunner in Sergeant Elton L. Parker's light tank.

Spotting German troops escaping from a village, Neill and Parker opened up with high explosives, but their attention suddenly shifted to a nearby hill where a Tiger rattled and clanked into firing position. Hearts racing, the Americans looked in vain for cover. The tank commander had an HE shell chambered, but the 37 mm projectile lacked the punch to penetrate armor.

"Might as well give him something to remember me by," Neill thought, assuming death was moments away. He swiftly laid his gun on the target and fired. The shot exploded on the Tiger and ignited gasoline the enemy had foolishly stored on the engine deck. Neill watched in near disbelief as flames sprang up and turned the enemy goliath into a blazing furnace.

The lucky Irishman received a Bronze Star Medal for his marksmanship.[330]

Outside Reinhausen, Company A hit upon two King Tigers scuttled by their crews. One had its turret blown clean off, the hull still smoking. Another smoldering Tiger stood close-by, one of its tracks broken, tow cables attached, and its barrel bent by an internal explosion.

When Company B neared Imbsen, Lieutenant Reynolds's platoon discovered muddy impressions left by Tiger tracks. Hearts raced when Staff Sergeant Robert Nicol—formerly of Company A—and his crew rounded a corner and spotted the enemy tank straight ahead. The fearful beast sat motionless. Jack Boardman had engaged this King Tiger yesterday before its crew laid waste to their machine. The only shot today was from a Company B tank driver, who snapped a photo with his camera.

The company and First Battalion, 23rd rushed through more farming communities and on toward a fat prize, Göttingen. Dubbed the big "G" by staff officers at division headquarters, this was the largest German city yet to confront the 741st and the 2nd Division.

Only fog slowed the tanks and infantry as they rolled into the metropolitan area. Just past a cemetery, Company B halted at a booby-trapped bridge, which had four aerial bombs hanging from it. Engineers determined that nobody had rigged them for detonation, and Sergeant Nicol's tank crossed first. No defenders guarded the other side, and the advancing column roared toward the city center. Along the way, a German dispatch rider turned a corner on his motorbike and found himself leading the armor and infantry. He quickly pulled aside and waved on his enemies. Unarmed German officers and enlisted men strode into the streets, gawking at the invaders. The city fell with less commotion than most villages in recent days.

The only episode resembling combat developed when a truck stuffed with enemy troops wheeled onto a residential street swarming with Americans. Several GI riflemen lifted their weapons and fired. The German driver mashed down his brake pedal, and his cargo jumped to the road, hands raised. Elsewhere, curious residents surfaced from their cellars to inquire, "*Engländer? Amerikaner?*"[331]

The advance continued eastward and into the Göttinger Forest. The tankers reached the opposite side, where the trees abruptly ended at a precipice overlooking a valley. Checked by topography, the men peered across the picturesque lowland, but something other than beauty attracted their eyes. They spied a German convoy retreating on a country road that ran through a village called Mackenrode. The column included cars, tanks, trucks, bicycles, gun tractors, halftracks, and horse-drawn wagons. This mishmash extended in either direction as far as anyone could see. It was every tank gunner's wildest dream. The targets presented slightly less challenge than plunking mechanical ducks on the midway.

Sergeant Nicol's tank had led the column and was the only one in position to shoot. Narrow spacing between the tree trunks prevented the other tanks from swinging abreast.

Corporal Edwin V. Mikolajczyk, the gunner, hammered away like mad. When his shells ran low, crewmen from other tanks donated their ammunition. Private First Class Thomas J. Quinby, the loader, worked so fast he snatched a brass fire extinguisher and slammed it into the open breech. When it failed to fit, he realized his slipup. The men outside saw a fire extinguisher fly from the turret amid the cascade of spent casings.

The Germans never fired back, unable to locate their assailant.

When the smoke dissipated, Mikolajczyk tallied his score: one tank, two cars, four antiaircraft guns, three halftracks towing field pieces, one 88 mm gun with its halftrack prime mover, and numerous wagons. German corpses and dead horses added to the destruction.

Company C moved out from Varlosen on April 8 with troops from the Second Battalion, 23rd riding the tanks. Village after village fell, artillery pounding each in advance. The only casualties occurred at Varmissen when a German machine-gun crew sprayed infantrymen mounted on a tank. Their lieutenant radioed the Second Battalion CP:

"The sonofabitches shot a machine gun from that wooded hill to the right and caught my fourth tank. Killed five men, and a tank ran over another. Wounded three."[332]

The enemy shooters slipped away.

A blown bridge over the Leine River forced the troop-laden tanks to detour through Göttingen. Peter Fardella recorded his observations.

"We saw a German soldier strolling down the street, spick-and-span and bold as brass, just as if we weren't there. Our driver stepped on the brakes, and we motioned for him to come over. Jerry lost his composure and pointed to a Red Cross armband we hadn't seen on his left arm."[333]

The medical man—probably employed in one of the city's twenty-two hospitals—walked away untouched. His brazenness, and that of many other Germans, resulted from a declaration that Göttingen was an "open city," but that news never reached Fardella and his comrades.

The push continued toward the southern tip of the Göttinger Forest and Klein Lengden, the first village beyond the tip. Only German antiaircraft troops remained in the area. The tanks held back, while dismounted infantrymen hoofed their way toward the village. Artillerymen hit the place with high explosives. After rounding the forest tip, Flakvierlings assailed the foot soldiers. Forced to the ground, the infantry summoned artillery support and help from the tank platoons under Lieutenants Dew and Crisler. Howitzers and tank cannons ousted the antiaircraft men and permitted the infantry to seize Klein Lengden.

As darkness overtook the countryside, the tankers slid backward to a town named Diemarden, retiring there for the night. The infantry kept its fangs in Klein Lengden.

After Göttingen, the first community larger than a village was Duderstadt. The town's capture fell to Company B and its partner in recent weeks, the First Battalion, 23rd. Lieutenant Dudley's platoon led the way with Sergeant Albert H. Kammeyer's tank spearheading the drive, which commenced on April 9, just after sunrise.

Outside Duderstadt, the tank column caught a retreating German convoy. Kammeyer's gunner, Corporal Harrell Gullatt, opened up on the surprised enemy while his driver plowed through the convoy. Panicked Germans tried to disentangle themselves from death and sprinted across adjacent fields. Many failed to save themselves as the tanks behind Kammeyer knocked them to the ground with machine-gun fire. Cannon shells—mostly Gullatt's—accounted for one halftrack equipped with an antiaircraft gun, one 88 limbered and in tow, seven horse-drawn wagons, and four 150 mm artillery pieces.

Sporadic shooting continued as the tanks rolled into Duderstadt, an idyllic town with a medieval fortress wall and streets lined with half-timbered homes, many sporting white flags.

Near the fortress wall, an elderly gentleman stepped outside as Kammeyer and his crew passed through a massive gate tower called the Westerturm. The old man seemed deaf and oblivious to the approaching armor and the shouts from accompanying infantrymen. He stopped with his mouth agape when he saw the American invaders churning toward him.

Gullatt saw him as well as a German officer running hard to flee. The corporal had exhausted the ammunition for his coaxial machine gun, but he wanted to stop the officer. He fired an HE round at him, and shell fragments killed the elderly man. "I wasn't shooting at him, but he got in the way," Gullatt said.[334]

One other civilian died, a forced laborer from the Soviet Union named Natasha. She and her friends rushed out to welcome their liberators only to face machine-gun bullets. Natasha died in a local hospital several days later. The town commandant, Major Roman Link, survived a ricochet that struck his leg when his staff encountered the first tank that turned onto their street. The major had hoped to surrender the town before anyone died.

The prisoners captured in Duderstadt included the Bürgermeister, caught in his office but later released. Several hundred Wehrmacht and Volkssturm troops capitulated to the Americans.

There were other prisoners, too—seven hundred Allied servicemen, including about one hundred Americans. They came from Stalag VIIIA in Upper Silesia, a camp the Germans evacuated on Valentine's Day as the Red Army approached. The internees marched west on foot, a grueling journey that lasted thirty days and left dead prisoners along its route. The exhausted survivors packed a makeshift camp in a brick-making factory outside of Duderstadt. Frostbite, starvation, and festering battle injuries had worn these men to shells and husks of soldiers. Medical Department personnel tended to them, while the combat troops who liberated the prison headed toward Leipzig some one hundred miles away. The advance shifted into high gear.

Tracked vehicles from the 9th Armored Division passed through the 2nd Division all day on April 10 and assumed the lead. For the next three days, 741st tanks and their infantry passengers rolled behind, uprooting enemy holdouts from wooded tracts and bowling over roadblocks, mostly earth-and-timber barriers erected by the Volkssturm. German civilians called them "sixty-one-minute roadblocks." They reasoned the Americans would spend an hour laughing and one minute smashing through.[335]

CHAPTER 41 **FLAK ALLEY**

Friday, April 13. The 9th Armored veered away from Leipzig fifteen miles outside the city. The sidestep maneuver occurred at the Leuna-Merseburg-Schkopau industrial district, home to an array of petrochemical facilities.

Nearly one thousand antiaircraft guns from the 14.Flak-Division protected the area. The heaviest weapons ranged in caliber from 88 mm to 128 mm, but the defenses also included lighter weapons, particularly 20 mm Flakvierlings. The Germans typically planted the guns in pits surrounded by brick walls with dirt piled against them. Outside these gun emplacements, trenches, wooden barracks, and ammunition bunkers added to the infrastructure. In 1940, Flak crews began sprinkling the countryside with wrecked aircraft and slain aviators. The crews now lowered their guns to fight armor and infantry.

The impetus behind the 9th Armored's sidestep maneuver became apparent outside Schotterey, when Company C discovered wrecked vehicles north of the village. Seven tanks from the 19th Tank Battalion had fallen to the antiaircraft guns. The offending weapons stood nearby, also silent and dead. The prospect of battling other such guns sent the 9th on a circuitous path. Armored divisions operated much like old-fashioned cavalry formations, riding and slashing their way deep into enemy territory and breaking free from hot spots, leaving them for infantry units like the 2nd Division.[336]

Schotterey was as far as the 9th Armored ventured toward the industrial zone. Foot soldiers from the 2nd took over and pushed into Bad Lauchstädt without difficulty. Tanks from Company C turned in the direction of the next town, Dörstewitz. Lieutenant Dew's platoon led the charge down a concrete road connecting the two towns, with the lieutenant's tank on point.

Decked with troops from Second Battalion, 23rd, the armor rolled less than halfway to Dörstewitz when explosions ripped limbs from trees lining the road. Shells also burst in adjacent fields and showered the tanks with dirt clods. Antiaircraft crews to the right labored feverishly in their gun pits, but, despite their zealous efforts, they failed to strike a single tank.

An enemy soldier with a Panzerschreck had sharper aim. As Dew's tank neared a culvert beneath the road, the German sprang up with his weapon shouldered. He moved too fast for riflemen on the tank to answer with bullets. The rocket streaked from its launching tube and detonated against the machine's frontal armor.

Two infantrymen died instantly, a third later succumbed to his injuries, and five suffered non-fatal wounds. As for the tank crew, the driver, Sergeant Walter J. Fryer, died in his seat. The veteran sergeant had substituted for Dew's regular driver that morning. Fryer left behind a widow and two-year-old daughter in Bellwood, Pennsylvania. The detonation also punctured Dew's eardrums, and it fractured the skull of his bow gunner, Private First Class Andrew Kostelnik, who required a surgically implanted steel plate to repair the damage.

The lieutenant's hard-hit tank coasted off the road into cropland burgeoning with grain. It rolled to a standstill and smoldered in the bright sunshine. A lifeless infantryman hung upside down from a grab rail on the left sponson, and looted wristwatches lay on the ground, having spilled from musette bags carried by the infantrymen. The three other tanks in the platoon hurried across the culvert and veered off the road.

The Panzerschreck gunner sheltered in the culvert with several comrades, while other enemy soldiers hid nearby in a gully clogged with brush. The infantry attacked the culvert with rifles and hand grenades. The tanks spat machine-gun fire at those in the gully until one German emerged clutching a white flag. The fight ended.

The antiaircraft men lost visual contact with the Americans who reached the culvert, but the big guns continued to roar away. They focused on the rest of Company C and the Second Battalion, forcing tanks and troops back into Bad Lauchstädt.

From the culvert, the road to Dörstewitz looked clear, but the infantrymen with Dew's tanks were too few to risk an attack. The little force gained strength in the afternoon when more riflemen arrived by a circuitous route, but a hitch developed. Dew received orders from Captain Covington to stay put until the infantry alone vanquished the Flak position, twelve 128 mm guns split between two batteries. The infantry leader implored the tankers to join his men, but his appeals brought no movement, at least for now. The foot troops crept forward alone.

They scattered after entering Dörstewitz, corralled several docile prisoners, and captured the village. Six antiaircraft guns responded. Huge shells burst in midair over the red-tile roofs, and the streets reverberated with earsplitting explosions as steel fragments rained down. The mighty guns commanded the ground by day, but that changed after nightfall.

When they fired after dark, 2nd Division artillerymen responded, pumping howitzer shells at the giant muzzle flashes. Afterward, an infantry platoon slithered toward the gun positions in the blackness, only to hear five muffled explosions. The Germans had known an infantry attack was coming, and they demolished several of their guns before abandoning the area. Unaccustomed to ground combat, especially at night, they chose to flee, seeing no profit in dying for the Führer.

During the day, while Company C battled in the Dörstewitz area, Captain Cinquina's Company B drove toward Merseburg. Lieutenant Sheppard's platoon led the way, along with troopers from First Battalion, 23rd. Liberated slave laborers warned them about enemy guns ahead, and the tankers advanced with caution.

Sheppard's tank headed the column as it entered Knapendorf. "My driver, Sergeant Jenkins, was easing our tank along the left side of the road, and suddenly a direct-fire weapon opened up on us, firing straight down the road. Although I couldn't see the gun itself, I knew about where the fire was coming from. I told Jenkins to hold it while we pumped seven or eight rounds in the general direction. The Jerry must have been zeroed in on the center of the road or was in a position where he couldn't traverse right, which was the reason he missed us."[337]

Another gun opened up from the right, and an HE shell wounded infantrymen around Sheppard's tank. Sergeant Angeletti, the tankdozer commander, suffered an eye injury that ended his war. Sergeant Mortzfield's tank threw its left track while backing over a steep bank to escape the shells. The crewmen dismounted, each hauling his personal possessions. The men kept low and hustled across a field to Bündorf, which they had driven through earlier. They congregated in a shed while the enemy poured on more fire.

When the shelling subsided, three other Company B men approached the shed: Corporal Louis A. DesHais, Private First Class Calvin C. Everheart, and Private First Class Philip Bornstein. They chatted with Mortzfield's driver, Technician Fourth Grade George R. Waterman, and inquired if he wanted to help explore a large manor house in Bündorf. The foursome took off toward the eighteenth century estate, where they rummaged for souvenirs before heading back to their company before it moved out, but they were too late.

Abandoned, the men began searching for their outfit. Waterman carried a Luger, Bornstein had a small pistol, "Pluto" Everheart shouldered a carbine, and "Wahoo" DesHais toted a Mauser rifle and a sword he found at the estate. They visited an infantry aid station hoping to find discarded firearms but found only one M1 rifle. Waterman snatched it.

Outside Bündorf, a slave laborer told them about German troops just beyond a wooded area south of Burgstaden. The foursome made the informant lead the way in case it was a trap.

"When we got to the woods," Waterman stated, "I motioned to the other boys to spread out, and in we went. I don't mind saying I was a trifle nervous, not knowing what to expect."

On the other side, he and his buddies found twelve 105 mm Flak guns and ancillary huts. This was Flakstellung Nr. 109. The Americans thought they could roll up the position and fired into the buildings as they marched forward. Waterman spotted a German. "I saw a Kraut make a break for a dugout, and I dropped to one knee, took aim, and picked him off."

The other tankers also worked their trigger fingers while using broken German to urge surrender. "Com Arous," they shouted, "Com Arous."

Defenders in one building responded with rifle fire. A bullet zinged past Waterman, just missing him. He emptied two M1 clips into the structure. On the left, Bornstein and "Pluto" cut down two machine gunners attempting to set up their weapon.

The GIs watched in amazement as Germans streamed out from the buildings, hands over their heads, unwilling to forfeit their lives for a lost cause. The officer in charge, a Luftwaffe Oberleutnant named Eppelsheimer, handed his pistol to Waterman, formally surrendering the garrison. The four Americans captured ninety-six enlisted men, three officers, and five nurses. The tankmen formed their prisoners into three columns and marched them toward the rear. They left behind five dead Germans.

"It was a long walk with many ten-minute breaks," Waterman recounted. "It was just getting dark when we entered the town in which the 23rd Infantry had its headquarters. We were sure glad to see them at last. We marched our prisoners to the town square and turned them over to the MPs. We were congratulated, fed, and given a place to sleep."[338]

Events rolled on while Waterman and his buddies sacked out. The First Battalion, 23rd pushed off before daybreak on April 14 and advanced in a column of companies that included jeeps. They advanced along Lauchstädter-Strasse and passed between a Luftwaffe airfield and Hill 102 before slipping into Merseburg through its Altenburg district.

Unknown to the Americans, the hill was home to Flakstellung Nr. 108 and twelve 105 mm antiaircraft guns divided into east and west batteries.

In the darkness, the Flak crews mistook the passersby for Germans and refrained from firing.

Inside the city, GI machine gunners looking for a firing position shoved open the door to the headquarters of the local Volkssturm battalion. They found sixty unsuspecting occupants just awakening. Everyone surrendered.

Captain Cinquina and his jeep driver, Technician Fourth Grade Lewis H. Lanier, accompanied the First Battalion into the city. The captain received a request for tanks to help solidify the gain. He radioed Company B and directed his men to move in at first light.

The tank column departed Knapendorf around 7:00 A.M., Lieutenant Dudley's Second Platoon and the company's 105 mm assault gun in the vanguard. The lieutenant led the column as it motored along Lauchstädter-Strasse. He and his men were unaware that Flak crewmen at the western battery had spotted them.

After an earsplitting blast, Dudley's machine stopped cold, its engine compartment hit by a high-explosive projectile.

Another shell exploded on the road, missing its target.

Sergeant Mercer commanded the assault gun, and his driver tried to back up, but a round collided with his tank's frontal armor, sending concussive shock waves through the vehicle.

Staff Sergeant Case and his crew maneuvered off the road and returned fire, blasting one Flak gun before a shell struck their tank's engine compartment. Flames soon leaped skyward.

Another shot disabled Sergeant Mazzio's tank.

Sergeant Kammeyer's tank escaped scot-free, finding concealment among nearby trees. His gunner, Harrell Gullatt, recalled that their vehicle presented a less visible target. "Our tank had a Ford motor, and the others had Wrights. The ones with the Wright motors were shooting flames out of their exhaust pipes and were the ones that got hit."[339]

The crewmen in the knocked-out machines clambered out. Several men had cuts and contusions. Kelly Layman emerged from Mazzio's tank with metal fragments in his back and buttocks. Mercer suffered a compound fracture and nerve damage to one arm. Kammeyer and his crew were unable to move their vehicle without losing concealment, so they climbed out and sheltered in a nearby dugout where Mercer's crew joined them.

The tankers from the First and Third Platoons halted after watching the Germans spring their welcome on Dudley's tanks. Suddenly, five jeeps from Company D, 23rd roared up towing trailers stacked with mortar shells. The tankers warned a lieutenant riding in one vehicle about the situation ahead. The lieutenant discounted their words.

"What do you mean? We've got troops in that city over there!"[340]

The jeeps barreled on down the road, and the Flak men ripped into the trucks with antiaircraft shells. The mortarmen fled on foot, except for three soldiers, one of whom was dead.

Near the demolished vehicles, several tank crewmen in the dugout decided to crawl away toward Knapendorf. Mercer needed medical care. After waiting awhile, the others—Gullatt, Kammeyer, Henry Gross Jr., Ivan Schmidt, Clarence Gallow, and Merlin LeBon—did the same. Gullatt led the group. The others followed, ten feet between each man.

As the six tankers crawled near the jeeps, they noticed three enemy soldiers inspecting the wreckage. The six kept still and prayed that their enemies passed. One German turned and leveled his rifle at Schmidt, killing him with a single shot. In response, Gross, Gullatt, and Kammeyer shouted, "*Kamerad! Kamerad!*"[341]

The enemy ordered the three Americans to their feet, while Gallow and LeBon remained silent, just out of sight. The Germans escorted their prisoners at gunpoint to Hill 102.[342]

The captured tankers met the enemy commander, Oberleutnant Jakob, a Luftwaffe officer and former schoolteacher. He asked if anybody in the group spoke German. One man replied "*Ja*." Gross had learned the language from his Austrian-born parents. The Flak men herded the three tankers into a dugout where they met two other prisoners, both from Company D, 23rd, one gravely wounded. They received coffee, canned meat, ersatz bread, and three shots of cognac.

By noontime, Gross persuaded Jakob to allow the prisoners to carry the wounded soldier outside and arrange for American medics to fetch him. Jakob consented but sent along a guard to ensure the uninjured prisoners returned. Before any transfer took place, the man died.[343]

Interned on 102, the American prisoners feared death after 2nd Division howitzers began counter-battery fire to smash the German guns. The open entrance to their dugout posed a problem; nobody could close the door. The explosion from one shell obliterated a range-finding apparatus just outside the door, but no fragments entered the dugout.

Between barrages, the battery commander's teenage daughter darted in to converse with the Americans. In the late afternoon, she tore inside shrieking, "*Schwarze! Schwarze!*"[344] Someone had sighted Negro soldiers, probably the colored platoon from Company F, 23rd. The prisoners immediately played on her racial fears. Once night came, they nervously explained, the black men would creep in undetected and slit every throat they found. An atrocity would take place if her father failed to surrender the hill before sundown. She scurried off to inform papa what horror awaited.

The artillery assault halted in the early evening, and officers from the 23rd dispatched two civilians to the hill with a warning. If Jakob did not surrender within thirty minutes (7:30 P.M.), the shelling would resume until every German was dead. American artillery shells had already killed and wounded many Flak troops. The threat of resumed bombardment weighed on the German commander. He opted to quit but not before the ceasefire elapsed. He hoisted a white flag just as the first rounds landed.

Imperious even in defeat, he turned up his nose at surrendering to his American captives. Capitulation to anyone from the enlisted ranks was beneath him. He insisted on giving up to an American officer and sent two prisoners to find one. While waiting, the commander strode about in the open with a white flag in hand, ordering his troops to exit their dugouts and join him topside. The GI prisoners helped drum up the antiaircraft men.

After an infantry officer arrived, riflemen from the 23rd took charge of the human catch, four officers and 140 enlisted men. They also claimed the twelve 105 mm Flak guns, eight of which were damaged or destroyed, some of them wrecked by their crews.

The liberated Americans ambled away from the hill after spending ten hours in captivity. Gullatt noticed trucks squeezing past Dudley's tank. The abandoned machine looked undamaged. Its armor plate had withstood the shell impact, the round designed to fragment and puncture aircraft aluminum. Gullatt reasoned he could start the tank and drive it off the road. He climbed in, followed the usual start-up procedure, and turned the starter switch. An explosion jolted the tank, and a fire whipped through the machine. Gullatt flew out, unscorched.

Concussion from the antiaircraft shell had broken a fuel line or ruptured a fuel tank. Starting the motor sparked the blast. Gullatt decided to allow maintenance men to inspect the other tanks that looked operable before trying to start them.

Earlier that day, before the Flak guns fell, their fire blocked reinforcements from reaching Merseburg. Cinquina, his driver, and the other US soldiers in town found themselves isolated and short of everything except Germans. The enemy numbered far more than those captured at the headquarters. Over one thousand obstinate Volkssturm defenders battled back, reinforced by scratch troops from the Luftwaffe and Reichsarbeitsdienst. The 2nd Division had never faced militiamen in force, but their heyday ended after the surrender on Hill 102.

Reinforcements arrived in Merseburg, and the two stranded tankmen sped out of town in their jeep. Their company with its nine operable tanks was moving south to join the 38th Infantry, which had returned to the front after weeks in division reserve.

Lanier and Cinquina tailed a truck hauling gasoline to their company. The vehicle poked along, malfunctioning brakes gradually slowing it. The followers grew impatient and passed the laggard, racing down the road in the fading daylight.

Tracer-lit bullets suddenly streaked across their path.

Lanier stomped on the accelerator pedal, but the enemy fire drew closer.

"Hit the ditch!" the captain yelled.

Lanier swerved off the road and plunged their little four-by-four into a roadside ditch. The impact catapulted Cinquina from his seat, but the crash injured neither man.

"The high bank offered us protection from the grass-cutting bullets," Lanier remembered. "We crawled, duck walked, and ran for quite a distance. The captain was a bit on the chubby side, but I couldn't keep up with him."

The pair backtracked from the jeep until they found the gasoline truck. Its brakes had locked up, yet, after sitting idle, they released. The driver suggested locating the nearby bivouac of Company D, 741st. The tankers joined him in the cab and drove away. They stopped to collect fifteen German soldiers, who flagged them down and surrendered. Lanier felt skittish. "There were enough of them to eat us! I'm just glad they were tired of fighting."

After touring the nocturnal countryside, the men found the campsite, but the night was only half finished for Lanier. He journeyed back to retrieve the lost jeep along with two Dog Company noncoms. The trio drove the company halftrack.

They arrived at 1:00 A.M., a "spooky hour" as Lanier recalled. The men aligned the halftrack with the jeep, hooked up a steel cable, and applied power to their winch. The jeep was nearly free when the enemy opened up, firing blindly at noises in the night. Everyone dove into the ditch. After the shooting died down, the men quickly finished the job.

The halftrack backed away, and Lanier tried to start his vehicle but flooded the carburetor. He dashed to meet the noncoms, hoping all three could push the jeep. Lanier saw a flash to the left. He shouted, "Look out!" and a Panzerfaust projectile exploded, another wild shot. The men sprinted away. "I thought we would never get that halftrack turned around," Lanier recounted. "Once we did, we shagged out of there."[345]

The jeep driver's adventure concluded in the morning when he returned with the halftrack and a light tank. The team salvaged the jeep and bagged thirteen prisoners.

The First Battalion, 23rd Infantry and twelve tanks from Company C reached Merseburg by 4:00 A.M. on April 15. They encountered little resistance from the Volkssturm, which had begun retiring from town. The infantry crossed the Saale River in the afternoon on a treadway bridge but abruptly halted at a manmade waterway called the Mittelkanal. Enemy troops poured rifle and machine-gun bullets from slit trenches on the east levee.

The Second Battalion, 23rd found easier going at Schkopau, where it crossed the Saale after dark using two bridges the Germans failed to demolish. Men from the battalion wheeled south to pinch off the canal defenders from behind.

On the outskirts of Leuna, the enemy offered lighter resistance, but small-arms and antiaircraft fire increased as the 9th Infantry and tanks from Companies A and D advanced into the town. Leuna fell after two battalions from the 9th pushed in from opposite directions.

By midnight, the entire 23rd Infantry was over the Mittelkanal with the First and Third Battalions paddling across in wooden assault boats. Ten tanks from Company C supported the crossing with cannon fire, belting shells over the canal for nearly an hour. The tanks later crossed and headed for Tragarth. The armored troops paused there to refuel and restock ammunition while the infantry, aided by artillery, wrestled down two dozen Flak guns.

Prisoners flooded into the division PW stockade at Merseburg. Most considered themselves lucky to be alive and to have avoided Soviet captivity. The few despondent prisoners included the local Nazi party boss, who attempted suicide in the stockade but failed to gulp down enough strychnine. His recovery provoked disappointment among the guards and prisoners.

Interrogators had no luck ferreting out information about the enemy forces defending Leipzig, the fifth-largest metropolis in Germany. Only later would they learn about its defenses and its military commander, Oberst Hans von Poncet.

CHAPTER 42 LEIPZIG

April 16. The approaching GIs heard a spine-chilling howl as air-raid sirens wailed in Leipzig, a prearranged signal that American troops had come within sight of the city.

Hans von Poncet stood ready. He had taken charge the previous week after relieving a general, who intended to declare the city open. The flint-hard colonel had earned a Knight's Cross on the Eastern Front, and desperate situations were nothing new to him.

Poncet mobilized rear-echelon troops, Hitlerjugend boys, and eight Volkssturm battalions. They filled streetcars with rocks and rubble to build roadblocks. The defenders also rigged bridges for demolition and seeded thoroughfares with mines, many electrically detonated from nearby bunkers camouflaged with foliage. Troop strength proved less than hoped, and company commanders found themselves leading platoon-sized units. Uniforms were whatever was handy. Two company commanders donned their First World War tunics, each having the prescribed Volkssturm armband stitched to the left sleeve. Many troops lacked usable firearms after receiving German ammunition for their French rifles. Yet the defenders were far from helpless. An abundant Panzerfaust supply lay at their fingertips.

The HASAG firm in Leipzig developed the antitank weapon and manufactured it by the millions. The company had its headquarters and two production facilities in the city, as well as satellite plants in the surrounding region and facilities outside the area.[346]

Plans to crack open Leipzig developed in a conference held at the 23rd Infantry CP. The chief of staff for the 2nd Infantry Division attended, and he informed Colonel Lovless that General Robertson wished to launch a night attack striking into the western suburbs. The general knew the open ground bordering the city made an ideal killing ground for enemy machine gunners. He reasoned that a nighttime operation would reduce casualties. Lovless thought differently. He believed that his soldiers could move faster and accomplish the same job in daylight. Darkness begets confusion. The 38th Infantry also had a role in the attack, and its commander favored nighttime. Robertson resolved the issue by allowing Lovless to start an hour before the 38th, then hold up at dusk while the 38th continued working through the night.

Before the 23rd Infantry began its push on April 17, Captain Covington reported to the 23rd CP, where he met the regimental operations officer and learned that Company C would only provide fire support during daylight.

At 5:00 P.M., two battalions from the 23rd attacked and flew unimpeded through several villages just outside the city. Both battalions reached the city limits at dusk, and their easy march ended.

Enemy troops opened up on the First Battalion as it approached the Lindenau district. Defenders also contested the Second Battalion, surrounding advance elements at a factory in the Böhlitz-Ehrenberg district. Artillery, not armor, provided support after dark. Barrage after pounding barrage scattered the Germans until a counterattack relieved the trapped infantrymen.

At 2:00 A.M. on April 18, Lieutenant Crisler's platoon set off for Gundorf to connect with the Second Battalion. His tanks joined the fight after sunrise, as did M18 tank destroyers from the 612th TD Battalion. They jarred loose snipers and machine gunners using HE and white phosphorus. This permitted foot soldiers to shake out two districts, Leutzsch and Böhlitz-Ehrenberg. The tanks joined Companies E and F as they mopped up en route to the Hindenburg Bridge, which spanned the Elster Basin, the last barrier before the downtown area.

Just before 2:00 P.M., Company F assaulted the bridge, and within thirty minutes, its men gained the other side. Company E crossed next, then Crisler's tanks. With their final objective won, the infantrymen took up defensive positions. The tanks parted after dark and withdrew across the bridge to refuel and replenish ammunition. The day had passed without mishap for Crisler's men. Far different circumstances befell Lieutenant Kurkowski's platoon.

Well before first light, he and his men left for Lindennaundorf to join the First Battalion. At 5:13 A.M., the combined force drove across open fields and entered Leipzig with the goal of capturing the Zeppelin and Klinger bridges. Potshots failed to daunt the attackers as they sailed into the city along Merseburger-Strasse. Silent buildings and shadowy streets engulfed the tanks and foot soldiers. Each window threatened, every rooftop taunted. Death seemed to leer at them, waiting to tap this or that GI on the shoulder.

The hunt for opposition ended ten blocks into the city. Gunfire echoed from building to building, and tracers ricocheted in furious sparks off the brick streets. Kurkowski split his platoon, ordering one section to Zeppelin Bridge and the other to Klinger Bridge.

The tanks commanded by Sergeant George K. Cuthbert Jr. and Sergeant John Welch navigated through the firing toward Klinger Bridge, the former in the lead. Their vehicles rolled along Zschochersche-Strasse, the crews intending to turn left onto Karl-Heine-Strasse.

Cuthbert, a tall, erudite New Yorker, held a degree from Columbia University and had served as a gunnery instructor at Fort Knox before joining Charley Company as a replacement. The ex-instructor received his sergeant's stripes just that morning and commanded a tank named "Cleopatra." He told his driver to steer their machine alongside a dignified stone building housing a credit union and a photography studio.[347]

As the sergeant's driver edged around the building and started to turn left, the eyes of a Hitlerjugend teenager followed from across the street. Far enough, the boy thought. He let fly with a Panzerfaust, and its fat warhead exploded against the right sponson. The blast produced a metallic jet that pierced the hull at hypersonic speed. In an instant, the jet ignited ammunition stored inside, and Cleopatra gushed flames.

Sergeant Welch's crew spotted the kid's hideout and pounded it with high explosives. He died, and another youthful fighter gave up. Infantrymen found Cuthbert's lanky body outside his tank and covered him with a blanket. There were no survivors among his crew.[348]

Two Signal Corps cameramen recorded the scene. Technician Fifth Grade Richard W. Crampton shot still pictures, and Technician Fourth Grade Jack H. Hutton rolled motion-picture film. William C. Allen from the Associated Press photographed the tank as curious civilians swarmed the area. CBS radio correspondent Edward R. Murrow was also present.[349]

While Cuthbert's machine blazed, shots crackled five hundred yards from Zeppelin Bridge. Enemy resistance flared up at Anger Bridge, a stubby little overpass spanning a narrow tributary river. Kurkowski's other section joined the fight. One at a time, the tanks inched over the bridge, blasted away and backed off.

When Sergeant Norman N. Schneider's tank crossed the river, his gunner fired the cannon and coaxial machine gun into distant trees. After the salvo, Schneider hollered, "Let's back 'er up." His driver, Technician Fifth Grade Verdie M. Guest Jr., also known as "Cowboy," threw the transmission into reverse gear.[350]

Straight ahead, a German launched a Panzerfaust from a coal chute (or basement window). Everyone inside the tank felt the impact and bailed out.

In his haste, Guest left the transmission in gear, and the crewless tank slowly rolled backward on Kuhturm-Strasse. The five tankers eyeballed damage to the left sprocket but saw no penetration and no sign of fire. All but one dashed back into the vehicle as it chugged along. The gunner, Corporal George Hillner, remained behind, crouching on the riverbank, plinking away at the enemy with a Walther P-38 he carried as a sidearm.

Before climbing aboard, Guest noticed two track segments beginning to separate. The enemy projectile had broken one of two steel connectors joining the segments. He steered the machine backward down the street and into Lindenauer marketplace, where he swung onto a side street. As Guest shut down the motor, the segments broke apart, collapsing the entire left track. Within an hour, maintenance men from company headquarters arrived in their halftrack, replaced the sprocket, rejoined the segments, and restored track tension.[351]

Noon approached without success at the bridges, although the Germans inexplicably refrained from blasting them to smithereens. Unknown to the Americans, Poncet honored a request from the police chief and spared the bridges to maintain the gas, water, and electric lines carried through them to the city's west side. The chief also implored him not to fight, but Poncet insisted on a suicidal fight to the last pubescent Hitlerjugend fighter.

Kurkowski's tankers pounded targets around the bridges before the First Battalion launched another assault at 1:30 P.M. Tank shells erased enemy firing positions and enabled the foot troops to charge ahead. Company C, 23rd struck Klinger Bridge and captured it. Riflemen from Company B, 23rd lunged at Zeppelin Bridge and fought their way across by 3:00 P.M.

Enemy opposition collapsed, though firing suddenly erupted in the evening. Task Force Zweibel from the 69th Infantry Division stormed into the city center and sprayed bullets at whatever moved, including Company B, 23rd. The company sheltered in cellars and withheld return fire. Its men suffered no losses before the error became apparent and quiet returned.

After sundown, Kurkowski received word to withdraw his tanks for fuel and ammunition. The tired crewmen replenished their vehicles and then split three cases of champagne that Sergeant Padgett had acquired. No toasts or exuberant backslaps followed, nothing to celebrate the capture of the bridges. The loss of Cuthbert's crew reminded everyone that victory and defeat came at the same price—death. The men drank to forget.

Throughout the battle on April 18, only one platoon from Company B entered combat. Lieutenant Sheppard's tanks helped the 38th Infantry blunt sharp resistance in Markkleeberg on the south side of Leipzig. The tanks plastered several buildings and a roadblock. Resistance folded once GI armor and infantry reached the center of Markkleeberg.

Companies A and D encountered little resistance that day.

Over the next two days, the 23rd Infantry maintained security at the bridges while the 69th Division finished conquering Leipzig. City Hall fell at 9:30 A.M. on April 19, after the mayor and his deputy took poison; seven others, including four women, also committed suicide.

The final enemy bastion stood in a southeastern suburb. The Völkerschlachtdenkmal, or the Battle of the Nations Monument, soared more than three hundred feet tall at the site where the Swedes, Russians, Prussians, and Austrians joined arms in 1813 to defeat Napoleon. Poncet and his last followers holed up in the granite-and-concrete redoubt, where they withstood infantry attacks and artillery shells. The colonel only gave up after nine hours of negotiation and assurances that his men would be released upon surrendering. Those inside the monument emerged with raised hands at 2:00 A.M. on April 20, and so ended the killing in Leipzig.

★★★

With Leipzig quiet, the 2nd and 69th Divisions pushed on and began relieving the 9th Armored Division along the Mulde and Zwickauer Mulde Rivers. The two waterways lay on the "stop line," where the Soviet and Western armies agreed to meet, a plan devised during the Yalta Conference earlier in the year.

Companies A and D reached the Zwickauer Mulde on April 20 at Colditz, a town conquered four days earlier by the 9th Armored. Able Company crossed the river and ventured into Hausdorf, while Dog Company also crossed but stayed in Colditz itself. Headquarters and Service Companies settled well to the rear at Flössberg.

The next day, Companies B and C left Leipzig for two locations on the Mulde River—Grimma and Leisenau. The tanks and accompanying infantrymen secured bridges at both towns.

Company D drove east, scouting for advance Soviet forces. It lent three tanks to a patrol from the First Battalion, 9th Infantry, and two other tanks joined the Second Battalion. The First Battalion expedition met no resistance, but the Second Battalion group collided with enemy fighters at Tautendorf. One shot from a Panzerfaust destroyed Elton Parker's machine, wounding him in the leg and injuring David Neill and Andrew Suster.[352] Both patrols retreated.

Orders from 2nd Division headquarters directed the 741st to conduct "maintenance and rehabilitation" on all its vehicles.[353] Companies A and D pulled back behind the stop line to carry out the program.

Tank crewmen sloshed buckets of water over their vehicles and scrubbed them like housemaids. Road grime disappeared. Crew compartments gleamed. Maintenance personnel descended into engine compartments, tinkering with this and that. The crews field-stripped machine guns, cleaned them with solvent, and displayed the parts on tarpaulins. Hands and arms sponged cannon bores and coated them with oil. Bogie wheels, support rollers, and idler wheels required grease. The tankers tightened tracks until they had just the right amount of slack. The routine continued through every item on the cleaning and maintenance checklist. The colonel visited each company and nodded his approval.

Except for Company D, the 741st had fired its last shot on German soil. The light tanks resumed patrol activity beyond the Zwickauer Mulde with the 9th Infantry. Four tanks left Colditz at noon on April 25. Two machines accompanied jeep-borne infantrymen from the First Battalion, while the other pair escorted soldiers from the Second Battalion. Both patrols hoped to meet Soviet troops but discovered only Germans. The First Battalion patrol killed one and captured another. The other patrol also found the enemy but refrained from firing.

The rest of the 741st undertook occupation duties. The job included one measure of satisfaction for the tankers. The battalion took over a Panzerfaust production facility hidden in a forest a mile from Flössberg. Owned by HASAG, the plant was a subcamp in the Buchenwald system. The tankers found completed weapons, components, and nearly six hundred tons of explosives. The 741st posted guards on April 28 and remained there until a bomb disposal unit from V Corps demolished the factory and its contents the following day.

Near the factory, slave corpses rotted in mass graves. These Jewish prisoners had worked at the plant and died at the hands of their SS taskmasters. Captain James B. Eure, the battalion intelligence officer, used civilians from nearby villages to exhume the bodies and rebury them.

The 741st also started a daily regimen of calisthenics, close-order drill, and training lectures. It was the peacetime army revived. Many tankers hoped the enemy would lob in a few shells and put an end to it, but the Germans made no effort to oblige. Surrounded by the doldrums of garrison life, the Company A commander and his maintenance officer decided to create their own excitement.

Lieutenants McDonough and Klotz contrived a scouting mission to contact the Soviets. During the afternoon of April 29, the two officers took off in a jeep driven by Technician Fifth Grade Edward Teffeteller. "We were just out messing around," he explained years later.[354] Their adventure took them south along the Zwickauer Mulde for more than thirty miles. While passing through Zwickau, they crossed the river and continued south to the Autobahn leading to Chemnitz. They were now east of the stop line.

The trio turned onto the superhighway and raced toward Chemnitz without knowing if Soviet troops held the city. German machine-gun fire dashed their journey after they sped beneath an overpass connecting Lugau and Stollberg. The three men had gambled and lost. The Red Army was still fighting in the Dresden area and had yet to reach Chemnitz.

Unharmed by the fusillade, Teffeteller steered the damaged vehicle off the road. Klotz too avoided injury, but McDonough died in the backseat, blood leaking from a hole in his head. The two survivors tried to flee, but the open landscape offered no escape route. Within seconds, enemy riflemen converged on the scene. Both Americans raised their hands.

Several Germans commandeered the jeep and drove it away with McDonough's body slumped in the back. The enemy marched Klotz and Teffeteller to nearby Lugau and from there drove them to Wistritz in the Sudetenland. There stood Stalag IVC. This oppressive place was now home for the blundering tankers.

By nightfall, their absence rankled Skaggs. Were two of his best officers out swilling vodka with the Russians? He planned a hot reception for them.

Also, promotion orders had come through for McDonough. He was now a captain.

CHAPTER 43 LIBERATORS AGAIN

April 30. Rumors abounded that combat in Europe had ended for the 741st and the 2nd Division. Home or the Pacific lay ahead, but the division operations officer dashed that notion after he requested a billeting detail from the battalion. He explained that all V Corps units would join Third Army for a push into Czechoslovakia to subdue unbeaten enemy units. The move offered at least one benefit: Czechoslovakia was a friendly nation, and this meant the return of something forbidden in Germany—fraternization. The GIs would be free to live in homes with civilians and to mingle with them, women in particular.

Captain Hurdle McDaniel from Headquarters Company reported to the division CP. Primed with maps and marching orders, he and a billeting detail set out for Schönsee on the German-Czech border. The rest of the 741st would begin moving to this little Bavarian town the next day, a two-hundred-mile journey. In preparation, the tank companies pulled back to assembly areas.

May 1, 3:50 P.M.: The battalion convoy commenced movement with Headquarters Company in the lead, trailed by Companies A, B, C, and D. Service Company brought up the rear from where it shuttled gasoline forward and tackled vehicle breakdowns. The extended column slipped south through Zwickau and onto the Autobahn but traveled in the opposite direction taken by McDonough.

The 741st and the 2nd Division followed the highway toward Bayreuth. The road bustled with vehicles heading south to join Third Army. Darkness brought out headlamps, the first time any tanker had seen them since leaving the United States. Lights on the Autobahn meant peacetime was days away. Rather than chance nighttime confusion, the battalion pulled off the road into a wooded area after motoring sixty miles. The men slept as an abrupt weather change settled over them. Snowflakes drifted down.

With fuel tanks topped off, the tankers resumed their trek after sunrise. Icy roads caused delays and accidents, but spring temperatures melted the hazard as the falling snow turned to rain. After passing through Bayreuth, the convoy twisted its way through the foothills of Bavaria. The day's journey covered more than 140 miles before Headquarters Company reached Schönsee just after 9:00 P.M. Captain McDaniel's advance party appropriated billets for the company in nearby Gaisthal. Companies B and C lodged in Schönsee itself, while Companies A and D settled in Weiding. Service Company took up residence in Gaisthal.

Along the border region, 2nd Division troops relieved a regiment from the 97th Infantry Division and two regiments from the 90th Infantry Division. The 2nd also assumed responsibility for the 11.Panzer-Division, a fully intact enemy division surrendering as a whole. Clean-shaven and stern-looking, the Germans approached in a long column of cars, buses, trucks, and armor that stretched to the horizon. Unvanquished in spirit, the men wore defeat like a badge of honor. After passing through 2nd Division lines, each unit unloaded its ammunition and set down its weapons. Stacked rifles, machine guns, and field gear adorned the fields. Many tankers secured that most sought-after prize—a Luger or P-38.

Supreme Headquarters granted permission for a push into Czechoslovakia no deeper than forty miles. Company D sent a platoon to each 2nd Division regiment. The platoon under Lieutenant Rivers departed Schönsee for the 23rd Infantry in the Czech village of Bělá. Lieutenant Payne's platoon left for the 38th Infantry at Waldmünchen, Germany, while Lieutenant Kovachy's platoon went to the 9th Infantry at Neukirchen, another German town. The Company D commander moved his CP to Waldmünchen. Orders arrived before midnight for the medium tank companies to join the regiments at daylight.

General George S. Patton Jr. directed V Corps to attack on May 5 and seize Pilsen. This began at 6:00 A.M. with the 2nd and 97th Divisions spearheading the assault. The 16th Armored Division sat in reserve, ready to gallop through the breach opened by the two infantry divisions.

Artillery fire nullified German riflemen and machine gunners in the 23rd Infantry sector. Armor from Charley Company accompanied the regiment but saw little action. The tank crews busied themselves seizing schnapps and sidearms from enemy prisoners. The 23rd cleared more than a dozen hamlets at the cost of seven wounded and two killed. No organized resistance arose against the 9th Infantry or the 38th Infantry, but the 741st suffered a fatality.

Sergeant Joseph M. Brost from Dog Company died of a fractured skull in the morning after his light tank slid off the road and turned over. Its driver endured angry reproaches for losing control. Tankmen throughout the 741st felt sad for Brost, an Indiana farm boy who, as it turned out, was the last battalion casualty.

During the afternoon, five representatives from the Czech underground reached American lines after driving from Prague. They passed through Pilsen and reported that resistance fighters had the city's German garrison bottled in a large barracks. They also stated that the German commander in Prague wished to capitulate, though not to partisans. The Czech nationalists expected the US Army to race there, accept his surrender, and liberate their capital. No chance. Pilsen was the limit. Prague belonged to the Red Army.

The push to Pilsen continued on May 6. The 2nd Division marched just under twenty miles, dusting away frail opposition. The 16th Armored blitzed ahead and entered the city at 8:00 A.M. Czech partisans guided the Americans to the few spots still under German control.

Elsewhere jubilant citizens, waving Czech flags, flocked to greet the armored columns as they rolled along streets dappled with sweet-scented lilac blossoms. The soldiers grabbed food and drink from outstretched hands—cakes, apples, cookies, and steins foaming with Pilsner beer.

Supreme Headquarters halted offensive operations on May 7 and instructed all combat units to assume a defensive footing. The unconditional surrender of all German military forces would take effect on May 9 at 12:01 A.M.

Before the message filtered down, the 2nd Division strode into Pilsen followed by the 741st. Like the day before, the streets resembled a carnival. Company C halted in the city, while Companies B and D stopped in Doudlevce. Company A eased into Starý Plzenec. Headquarters Company and Service Company encamped at Rosutka.

Germany's formal surrender on May 8 brought the flag-waving Czechs back into the streets. They bombarded the GIs with food and even pursued them for autographs. Boys and girls turned out in brightly colored costumes, while bands paraded around playing polkas and waltzes. Meanwhile, retreating Germans and their camp followers passed in sullen silence. Along the frontline—now the surrender line—enemy soldiers scrambled for US territory to escape the Bolshevik bear in its hour of revenge.

The day brought joy to everyone in the 741st, but festivities among the tank companies paled compared to the merriment in Paris after its liberation. Something was different this time. The great propelling force in their lives had disappeared. A sense of shock overcame many soldiers, like men emerging from darkness into the glare of daylight.

They had reached the dawn of a new world.

EPILOGUE

V-E Day found numerous 741st men far away from the battalion. Most absentees populated hospital wards in Paris, London, and throughout the United States.

Lieutenant Sledge returned to the battalion only the day before from the 62nd General Hospital.[355] His belated homecoming resulted from a metal sliver lodged between his right thumb and forefinger, something missed during his first surgery in March.

After rejoining Able Company, he learned that military police in France had one of his men in custody at the Paris Detention Barracks, a former French military caserne converted to a US Army stockade. Donald Kepplin had surrendered himself on May 5, seven months after sneaking away from his company. Sledge formally leveled charges against Kepplin, and barracks officials transported him by truck to the 741st in Czechoslovakia, where he faced a general court-martial on June 13, 1945.

Kepplin pled guilty to being AWOL, but the court rejected that plea and found him guilty of desertion. The adjudicating officers noted, "He abandoned his organization at a time when it was likely to become engaged in combat with the enemy, and that he returned to the service only when more severe fighting had passed, and the enemy was virtually defeated."[356]

The verdict carried a possible death penalty in wartime, but the court sentenced him to life imprisonment and a dishonorable discharge. The army transferred the convicted soldier to the US Penitentiary at Terre Haute, Indiana, but he served less than three months there before moving to the US Army Disciplinary Barracks at Fort Benjamin Harrison, Indiana, a more appropriate location for a man guilty of a "military offense."

He requested clemency in 1946 and received a reduction in his sentence to five and a half years. Kepplin ultimately returned to his wife and two children in Chicago.

★★★

Other 741st soldiers served prison time but under honorable circumstances. The Germans captured seventeen battalion members on the battlefield, and Allied troops eventually liberated all but one. Thomas Chastain from Company A freed himself.

After the enemy captured him in Normandy, he took an immediate aversion to peacefully sitting behind barbed wire as a flea-bitten "Kriegie" (short for Kriegsgefangener, the German word for prisoner of war). After a brief stay at Stalag XIIA, a transit camp for new prisoners, the Germans sent Chastain to Stalag VIIA, where he immediately began plotting to break free. Assigned to a work detail, he slipped away from his captors while digging potatoes. The escapee traveled at night and slept by day in wooded areas as he fled south into the Austrian Alps. Near the Swiss border, he dozed off against a tree trunk and awoke to see uniformed Germans standing over him. A kid from a nearby village had betrayed him.

The next stop for the recaptured prisoner was Stalag XVIIB at Gneixendorf, Austria. After a spell in solitary confinement, he joined other American prisoners, all of them air force noncoms, in a scheme to tunnel out, one of many such plans hatched at the camp. Their underground handiwork extended several hundred yards.

In January 1945, Chastain was the only one rash enough among his fellow tunnelers to undertake a breakout in freezing weather, especially with the war's end in sight. After slithering through the narrow subterranean passage one night, he emerged from the exit hole and fled across flat, open countryside toward a distant forest. He disappeared into the trees as if he were part of the woodland.

Chastain wandered north into Czechoslovakia and on into Poland, taking charity from villagers. He battled exhaustion, gripping hunger, and frozen feet, but the journey posed more than physical hardship. Red Army soldiers crossed Chastain's path in Poland, and several times he looked down their machine-gun barrels. He once provoked the Reds by refusing to drink alcohol they offered, recoiling in horror when they poured homemade hooch from a jerry can that once held gasoline.

After Poland, Chastain left a trail of boot leather across Ukraine all the way to Odessa on the Black Sea, a repatriation point for Allied prisoners freed by Soviet forces. He boarded an American freighter sent to fetch former prisoners.

Once through the Bosporus and Dardanelles, the ship sailed for Naples where the former tank crewman descended the gangplank and returned to US Army control. After being hospitalized, the army flew him to New York, where he caught another flight to Fort Bragg, North Carolina. Chastain's odyssey ended with a bus ride to his parents' home in Highlands, North Carolina.

Like many things in his life, the escape arose from his rebellious temperament rather than necessity or a sense of duty. Years later he admitted, "I lived better in solitary confinement than I did on the outside."[357]

Frank Klotz and Edward Teffeteller experienced the shortest time in captivity. Their internment lasted twelve days before Soviet troops released them from Stalag IVC. The two tankers commandeered a black Mercedes sedan, used white paint to daub "Allied P.W." on both sides of the vehicle, and took off toward American-held territory. They learned that the 741st had concluded the war in Czechoslovakia, and the duo obtained directions along the road until reaching Pilsen. Here they found a peeved battalion commander, who intended to court-martial them for being AWOL.

Luckily for Klotz and Teffeteller, the War Department provided a means for evading judicial action. Former POWs were eligible for automatic furloughs and priority transport home. Neither man squandered the opportunity to flee before being brought to trial. Their ordeal ended, but the agony had just begun for Roger McDonough's family.

On May 18, McDonough's parents received a telegram at their Long Island City home. It stated the unbelievable—their son was Missing in Action. Twelve days later, the army changed his status to Killed in Action based on details supplied by Klotz. The army eventually provided the parents with a footlocker filled with personal items, but only a kind letter from Klotz explained what transpired on the Autobahn near Stollberg.

In August 1946, after waiting more than a year for information about her son's place of interment, McDonough's mother penned a brief letter to New York Senator James M. Mead. "It is an awful feeling for a mother & dad not to know where their son was buried. I did not write to anyone as I tried to keep patient, waiting from day to day, but my patience is letting out now."[358]

The senator quickly forwarded the letter to the Quartermaster General's office along with a request for an investigation. Colonel William J. McDonald responded. "A burial report for Captain McDonough has not been received. However, an investigation is being conducted by our Graves Registration Service, and every possible effort is being made to locate and identify the remains of Captain McDonough. When a report is received, you will be notified."[359]

By 1947, Cold War politics and Soviet restrictions thwarted the investigation being conducted by the American Graves Registration Command (AGRC). Communist authorities denied all search requests, but if they discovered a body or grave, that was an exception.

Hope for finding the captain faded until June 1948, when the Red Army summoned an AGRC team to Stollberg to collect American remains buried in the town cemetery. The team knew nothing about the McDonough case. Under Soviet escort, the team members wandered to the graveyard's northwest corner, where they discovered five wooden crosses with perennials blooming beneath. Four crosses bore American-looking names while the fifth carried no name, only the enigmatic words "*Ein Amerikanischer Soldat*."[360]

Beneath several feet of clay, the unidentified GI rested in a plain wooden coffin. His decomposed corpse weighed about twenty-five pounds and exhibited a shattered skull and missing left hand. Clothing articles included socks, drawers, a wool undershirt, an M-43 field jacket, OD shirt and trousers, but no boots. The shirt displayed a first lieutenant's bar on the right collar, an Armored Force insignia on the other collar, and a triangular Armored Force patch stitched to the left sleeve. No dog tags or other identification surfaced.

The cemetery caretaker said that civilian "internees impressed into work by the Americans" brought the bodies to him for burial. The AGRC men tried to contact others for additional details, but the Soviet escort party became impatient. One Red Army officer protested, "They're Americans. What more do you want?"[361] The recovery mission ended, and the team transported all five bodies to the US cemetery at Neuville-en-Condroz, Belgium.

An AGRC laboratory known as the Central Identification Point operated at the cemetery. Technicians scrutinized the five bodies and positively identified four. Two of the deceased belonged to the 4th Armored Division. The other two—one being a War Department civilian—died in an ambush on May 4, 1945, while conducting research for the US Strategic Bombing Survey. The fifth body defied identification. The lab designated it X-7626, and workers buried the corpse in the cemetery as an "unknown."

The dead man appeared to be an officer assigned to the bombing survey. Nobody linked X-7626 with McDonough until 1949, when the AGRC compared a dental chart for the unknown to the late captain's army dental charts. Too many differences existed for positive identification. The lab needed something more, and the identification effort stalled until May 1950, when his parents finally found dental records for their son dating back to the 1930s. The elusive records arrived at the cemetery and explained the differences.

The army stenciled the name Roger J. McDonough on the wooden cross marking the grave. His parents received a request: Did they want the body returned to the United States or left in Europe? They opted to make Neuville-en-Condroz his final resting spot.

Along with McDonough's mother and father, Russell Bradsher's kin had to choose whether to repatriate his remains or permanently inter them overseas. Russell drove a tank for Company A and died on July 11, 1944, the sole fatality in his crew that day.

In September 1947, his mother received a "Request for Disposition of Remains" form. She elected to have her son returned for burial at a cemetery near Stokesdale, North Carolina. Then a problem arose, a significant oversight. The AGRC had never positively identified the soldier thought to be Bradsher. This unidentified serviceman, X-110, died on Hill 192, and a collecting team recovered his body in July 1944. Circumstantial evidence connected the case to Bradsher, but the unknown's estimated height was over an inch greater than Bradsher's known height, and pronounced differences existed between their dental charts. In April 1950, the army declared the body unidentifiable.

That same month, Mrs. Bradsher wrote to the Quartermaster General and asked about her son's body. More than two years had passed since she completed the request form. The long-patient mother received a telegram with news that an officer from the Repatriation Branch would visit her home the next day.

Lieutenant Colonel E. M. Brown arrived at Sanford, North Carolina, to meet Russell's mother, sister, and brother. At the Bradsher home, the colonel presented the case history, reviewed tooth charts, and explained the oversight with repeated apologies. The army intended to inter the nameless body as an unknown at a permanent American cemetery in Normandy.

Brown later noted in his travel report that "Mrs. Bradsher received the news without emotion."[362] He never saw the mother's quiet suffering before leaving North Carolina for Washington and his tidy world of form letters, interment records, and administrative procedure.

There was however more to the story, more than Brown and the Bradsher family ever knew. In June 1947, an AGRC investigator named

Wilbert H. Hubbell visited Hill 192 and collected human remains from two Able Company tanks destroyed fifty yards apart in an apple orchard. He quickly identified them as A-741 machines based on markings still visible on their rusted hulls. The AGRC tried to determine who crewed the two vehicles and found Russell Bradsher's name, but which tank was his? The question eluded an answer. The army interred the remains as unknowns X-440 and X-441 at the Blosville temporary cemetery.

Like the Bradshers and McDonoughs, all the other families of war dead coped with loss, some more successfully than others. In a magazine article, DeRonda Elliott described the indelible hurt experienced by her mother after her father, Frank Elliott, died on Omaha Beach.

"My mother never remarried, although she had several opportunities to do so. Heartache and sadness, hard work and worry, punctuated by a few moments of humor in the company of friends and family, characterized the rest of her life. She made the best of her life, but she never could forget her first and lasting love."[363]

DeRonda's mother passed away in 1990 at age seventy, her sadness inconsolable.

Other widows, like Dorothy Miller, remarried and started new families. Her late husband, Victor Miller, served as a platoon leader with Company B, 26th Tank Battalion at Camp Chaffee before joining the 741st as a replacement officer. Dorothy's second husband was Hambleton Baxter Carpenter, the intelligence officer for the 26th and Victor's friend. Dorothy and Baxter raised three children and lived together in South Carolina until her death in 2006. Before her passing, she and Baxter visited Victor's grave at the Henri-Chapelle American Cemetery in Belgium. "Vic is still in our hearts and minds," she wrote afterward.[364]

The cemetery also held the remains of Miller's company commander, Captain Young. In 1947, his widow elected to permanently inter him there. Also that year, she and her three-year-old son attended a ceremony at the National Guard armory in Englewood, New Jersey. Inside the ivy-covered building, an army officer presented the little boy with his father's Bronze Star Medal, which had two oak leaf clusters fixed to its silky ribbon. Dressed in his Sunday suit and bow tie, the youngster gave a sheepish glance toward a newspaper photographer as the officer bent down and pinned on the medal. The reverse side of it bore his dad's name, a man he never knew.[365]

★ ★ ★

Though replete with sad stories like that of Captain Young, the war also brought stories of matrimony and new families.

While in England before D-Day, Sergeant Robert F. Stover from Company D married a nineteen-year-old Wiltshire resident named Joan E. Hollister. She was pregnant when he became a prisoner of war. After returning to freedom, he arranged for his wife and infant son to immigrate to the United States. The young woman and her child sailed to their new home in 1946 aboard an ocean liner filled with war brides. The couple settled in California and raised four children.

Corporal William J. Krunsberg from Company C married a British subject named Edith P. Russell, whom he met in Devon while training there before D-Day. Two years after the war, he paid her airfare to the United States, where the two wed and started a family in Michigan.

First Lieutenant Russel Dotson from Headquarters Company married Bohusiava Puchtova, a Czech woman he met in May 1945. At least three other 741st soldiers married Czechs. Corporal Edward R. Lucas from the Medical Detachment wed Anna Kadlecova, a twenty-two-year-old Pilsen resident.

Anna's first encounter with Americans occurred on May 6, when 16th Armored Division soldiers forced German troops to surrender near her family's apartment. She and her neighbors flooded into the street, offering their liberators hugs, kisses, and flowers. In return, the GIs doled out chocolate bars to the children. The littlest ones looked puzzled. Few Czechs had chocolate during the German occupation, and the youngsters handed Anna the strange bars. She redistributed them among the group and told the kids to taste what was inside the wrapping paper.

That evening, the 16th Armored departed her neighborhood. The next day more Americans arrived—men of the 741st. Anna was home for lunch when several children knocked on her door. Eyes beaming, they begged her to ask the new Americans for chocolate. She pulled on her overcoat, went down to the street, and found 741st vehicles and soldiers, including Ed Lucas. Anna recalled their first encounter: "I was young, and the soldiers wanted to have their pictures taken with girls. Someone took a photo of him and me. … and that's how we met."[366]

Unlike the 16th Armored, the 741st stayed in the area. Anna and Ed continued to see each other and used a Czech-English dictionary to communicate. Ed proposed marriage after little more than a month. Anna accepted. Her parents frowned on the idea, but the couple married in Pilsen on August 22. Four ceremonies made it official: two civil ceremonies, one in Czech and the second in English, and two religious services, one in each language. An army chaplain conducted the English ceremonies.

Shortly after their nuptials, the newlyweds parted when Ed shipped out to the United States. He could not take his wife aboard the troop transport. While waiting to rejoin her husband, Anna studied English with help from a nun in Pilsen. The opportunity to use her new language arrived on July 4, 1946, when she reached New York aboard the USAT *Henry Gibbins*. Her life in America moved west with a rail journey to Seymour, Indiana, where Ed met her at the train station. His sister held a long-delayed wedding reception at her home. The couple resided thereafter at a farmhouse in Brownstown, Indiana, Ed's home since he was six weeks old. They lived there until his death in 1994. The wartime marriage produced a son, daughter, and five grandchildren.

Lieutenant Colonel Skaggs relinquished command of the 741st on May 28 and reported to Camp Beale, California. The battalion executive officer, Major Jack Browder, temporarily served as the commanding officer. Under his leadership, the battalion vacated the Pilsen area and relocated forty-two miles southwest to Starý Klíčov. The soldiers moved into a tent city pitched in the grassy meadows outside of town.

Ten days later, the 741st parted from the 2nd Infantry Division. Their yearlong relationship ended on June 16. The battalion joined the 8th Armored Division, which moved from Germany to relieve the 2nd Division in Czechoslovakia.[367]

Browder stepped aside on July 3 when Lieutenant Colonel Preston R. Bishop became the new commander.[368] His arrival and the battalion's attachment to the 8th Armored registered as peripheral events for most 741st troops. Recreation time meant far more. Almost everyone received seven-day furloughs to Paris or the French Riviera.[369]

The soldiers also had great interest in their Adjusted Service Rating scores, a point system used to allocate honorable discharges to officers and enlisted men. This demobilization program awarded one point for each month of military service since September 1940, one for each month overseas, five for each battle star awarded, and five for each medal received. Fathers with minor children earned extra points. Anyone with eighty-five points met discharge eligibility.

Medals seemed frivolous in combat, weighing little against the fight for survival, but suddenly awards and decorations had tangible worth. Some men reached eighty-five points when the 8th Armored bestowed dozens of Bronze Star Medals and several Silver Star Medals.

Sergeant Mitchell Adamek from Company D acquired points in an unusual way. The Detroit native received credit for twenty-one months of service overseas with the Polish Armed Forces in the West, military units loyal to the Polish government exiled in England. His ASR score ballooned to 106.

Based on points, 741st men headed home for discharge less than a month after VE-Day. At first, men left on an individual basis. Later, outbound soldiers transferred to units returning Stateside. Numerous 741st men joined the 35th Infantry Division in July.

The largest exodus occurred on August 4, when about three hundred men moved to the 6th Armored Division, including the last "old timers" who had formed the battalion at Fort Meade in 1942. Besides the 6th Armored, other 741st troops joined the 4th Armored Division.

In September, the requisite points decreased to eighty, and more men departed for home. Lieutenant Colonel Bishop waved goodbye on September 2 before joining the 146th Engineer Combat Battalion along with a group of 741st enlisted men. Bishop's position as battalion commander passed to Major Browder.

Eight days after he assumed leadership, higher headquarters alerted the 741st for movement to the United States. After this announcement, nearly five hundred "high-point men" joined the 741st from three artillery battalions and various rear-echelon units. "Low-point men" transferred out. Browder's battalion was a tank unit in name only. Unfamiliar faces filled its ranks, many men never having set foot inside an armored vehicle.

The trip home commenced on September 13, when 741st troops began leaving Czechoslovakia. Over the next four days, the battalion moved by rail and truck to an assembly area near Reims, France. Camps in the area bore American city names. The 741st settled at Camp Washington DC, where the battalion waited for orders to an embarkation port once shipping space became available.

The call to port soon came. On October 1, the 741st started a 615-mile rail journey to Camp Calas outside Marseilles, France. A two-week wait ensued there before the battalion traveled into Marseilles on October 15. That morning, the men boarded the USAT *George Washington*. The men adorned it with a giant banner that read: "741 TK. BN. First of V Corps to hit Beach—Last to go Home."

Late that afternoon, the vessel steamed out to sea. The ocean voyage lasted more than a week. The *George Washington* reached New York on October 26 where trains whisked the men to Camp Kilmer, New Jersey. The final curtain fell at 11:59 P.M. on October 27 when the battalion furled its colors, officially inactivated.

The 741st no longer existed.

The erstwhile tankmen traveled to separation centers across the country to complete their discharge paperwork and accept their final pay. Each veteran departed with a "ruptured duck" insignia sewn above his right breast pocket, an emblem identifying him as a civilian.

Many battalion veterans described their release from active duty as the happiest day they ever experienced, bar none. They eagerly donned civilian clothes and resumed lives put on hold, but the door to the past did not close for many. The war burned on in their minds.

After returning home, Herb Covington found himself hesitant to pull off the highway while driving the family car, fearing the shoulders contained mines. "It's okay, the war's over," his wife reminded him.[370] His hypervigilance gradually faded over time.

Phil Zodda ruminated on mental snapshots, memories of this field and that hill. He never put his ghosts to flight. Darkness crept into his head, and he struggled with clinical depression. More than once his wife found him balled up, unable to separate past from present.

His friend Phil Fitts drank to numb the hurt, but his alcohol elixir only offered a temporary reprieve and threatened to destroy his life. He mustered the strength to walk away from that self-destructive path before it was too late.

The war occasionally crashed back on Jack Boardman. He awakened one Saturday morning and broke down in tears. "I was crying because we had lost the war. It was weird because I knew full well we had not lost."[371] His emotional outpouring sprang from the fearful moment on Omaha Beach when the invasion fleet disappeared from view. The men on the beach thought they had failed. Later during the 1970s, Boardman recognized an unexpected change. He realized that D-Day had stopped penetrating his daily thoughts. Until then, haunting recollections had occurred *every* day since June 6, 1944.

Memories of Omaha Beach never stopped tormenting Bruce E. Elliott, who served with Company A and made sergeant before leaving the army in September 1945. On the thirty-eighth anniversary of D-Day, his wife heard a gunshot in their bedroom. The sixty-two-year-old veteran had killed himself, his misery outweighing his will to live.

Former platoon leader Alfred Steineker returned home to Louisville with a prosthetic leg. He worked for the Veterans Administration before moving to Montgomery, Alabama, in search of mild winters that were easier on his leg. Steineker married there in 1948 and attended the University of Alabama, where he earned an accounting degree. The disabled veteran fathered two sons and two daughters and became Alabama's chief examiner of public accounts. All the while, memories of the war led to periodic bouts of melancholy. He finally escaped the battlefield in 1975, shooting himself to death at age fifty-three. No suicide note was found.

Michael Kark's spinal-cord injury left him in a wheelchair. The army retired him at 100 percent disability after 308 days of hospitalization. He added to his government pension by working as a mechanical engineer for over thirty years before retiring. Kark died a widower in 1992, leaving a son, daughter, and two stepchildren.

Emory Hopper, the sole survivor from Bolick Smulik's tankdozer, spent 592 days hospitalized with severe burns on his face and left hand. He lost fingers, endured painful skin grafts, and slowly recovered at the 129th General Hospital in England and later Northington General Hospital in Alabama. Hopper carried the war with him in the most visible way possible, and his devoted wife remained at his side throughout. He had met Audrey in 1942 while at Fort Meade, and they married later that year. The couple raised five children and lived together until he died at age sixty-three from lung cancer.

Another burn victim and sole survivor, Charles Buchanan, settled in Tucson, Arizona, and spent forty-two years as a conductor for the Southern Pacific Railroad. After retiring with a railroad pension and VA disability pension, he worked as a school crossing guard three days a week. He passed away in 2003 at age eighty-one, after battling diabetes for years. His wife, Clara, died three months after him.

More than twelve hundred men served in Europe with the 741st. Many veterans lived their lives at or near the place they called home during the war, while others relocated in search of greater opportunities. Ed Lucas farmed on his Indiana property and labored in a paper mill. Val Fister worked as a draftsman for Goodyear Aircraft Corporation in Akron, Ohio, and commanded a VFW post. Phil Fitts owned a Ford dealership in Pennsylvania. Nelson Hall worked as a radio announcer, advertising agency executive, and finished his working years at First Security Bank in Salt Lake City. Jack Boardman loved technology and tinkering. He concluded his working years as a technician for a Connecticut-based company that built infrared microscopes.

Joe Dew also loved technology. The former platoon leader earned a bachelor's degree from Iowa State College and accepted a position with General Motors in Michigan. He climbed the corporate ladder to become superintendent of production for Chevrolet's pressed-metal unit, but inventing was his passion. Dew personally held four patents that earned millions for GM, and he continued working on innovations after retiring from the auto industry, one project aimed at improving wind turbines. Family life for the inventor revolved around his wife, Evelyn, and their son and daughter.

Russel Dotson pursued possibly the most unusual postwar profession. An experienced high diver in college, he founded an entertainment team that performed comedy diving. He appeared at amusement parks around the country until 1975 when a stunt went wrong. Dotson hit concrete instead of water and died in front of nine hundred spectators at Hershey Park in Pennsylvania.

Bill King, the former battalion operations officer, retired as a lieutenant colonel in the Army Reserves and became director of student activities at Clarkson College of Technology in Potsdam, New York. In 1961, he returned to his alma mater, the New York Military Academy, as its assistant director of admissions before moving on to the McQuade Foundation, a nonprofit organization that provided children's services. King had four kids, including a son who also graduated from the academy and served as a US Army officer.

King's successor as operations officer, Cecil Thomas, built a career in the North Carolina Department of Agriculture, rising to director of research stations. He finally married at age fifty. Cecil and his wife, Frances, enjoyed ten years together before emphysema and heart failure ended his life in 1972. The couple had no children.

Edward Sledge, who became the Company A commander after Thomas moved up to battalion, returned to Alabama and rose to vice president at First National Bank in Mobile and later became president of an investment firm. He received a brief mention in his brother's acclaimed memoir about the meat-grinder battles for Peleliu and Okinawa. Edward's suicide in 1985 shocked many of the men who had served under him, yet they knew all too well the torment behind such an act.

Jack Browder stayed in the army as a reserve officer for three years after the war. He returned to Oklahoma and worked for Coca-Cola before purchasing a 7-Up bottling plant. He operated the business for a dozen years before becoming a stockbroker and real-estate agent.

Bob Skaggs accepted a Regular Army commission after the war and rose to full colonel before calling it quits in January 1958 after more than thirty years of service. Later that year, his second wife, Anna Mae, died prematurely at age fifty-three. The colonel lived the remainder of his life in Fort Lauderdale, Florida, where he sold boats and outboard motors. He remarried three more times, and, after two divorces, his fifth wife, Evelyn, outlived him. He died in 1985.

Like Skaggs, several enlisted men chose the military for a career. One was Thomas Chastain, but his service ended in 1962 as a Specialist Fourth Class. "I knocked home-sweet-home out of a company commander for calling me an SOB," he explained. "I was court-martialed and kicked out with an unsatisfactory discharge."[372] Chastain became a long-haul trucker after returning to civilian life in North Carolina. He eventually retired to Florida.

Cecil L. Greenwall also planned to stay in uniform. The former tankman from Company A served as a mess sergeant with the 38th Infantry Regiment during the Korean War. Despite being a cook, he eventually found himself in combat, again.

In February 1951, Chinese Communist forces surrounded Greenwall's unit at a vital bridge. Wave upon wave of enemy soldiers poured into the battle, and every able-bodied GI shouldered a weapon. After repeated attempts, the embattled Americans managed to withdraw from the cauldron. The harrowing struggle resulted in over fourteen hundred casualties, including Cecil Greenwall. The army listed him as Missing in Action for five months until finally clarifying his status—KIA 12 Feb 51.

Robert J. Ewald, Virgil L. Swander, and Herman Sendelbach organized the first postwar reunion for Able Company veterans. The three former tankers lived in Tiffin, Ohio, and served together throughout the European campaign, Ewald and Swander in the same tank crew. The first get-together took place on Labor Day weekend 1950 at the Ewald farm outside Tiffin. Eighteen men attended, including Cecil Thomas, one of the company's former commanders. Gatherings occurred each September until those still living and able to travel dwindled. The annual meetings ended in 1998 after only three men showed up for a reunion in Louisville, Kentucky.

For more than a decade after Company A began meeting, veterans from the other companies held no reunions. The impetus for change came one Sunday in 1962, when George Ganger visited Leonard "Dick" Trimpe at his home in Seymour, Indiana. Their conversation left Dick with a thought, "Why don't we have a reunion?"[373]

In July 1963, Trimpe and his wife hosted the first battalion reunion at Shields Park in their hometown.[374] Six veterans and their wives attended. The following year, several others joined the group, including Jim and Helen Correll. The Corrells hosted several reunions including the 1970 gathering, which drew 260 people, a record never surpassed. Attendance declined little by little and dipped below one hundred for the first time in 1995 at Salt Lake City.

Though reunions shrank in size, their purpose remained constant. Each former tanker sought to maintain his link to the Seven-Forty-First fraternity. Their shared combat experience forever connected them, and a sense of family also evolved. Wives developed friendships and helped organize the reunions along with their husbands. Children also became acquainted, especially during the 1960s when they lived at home with their parents. In later years, several wives continued attending after losing their husbands.

The men who appeared year after year at reunions were the postwar keepers of the flame, but they constituted a minority among 741st veterans. The majority seldom or never attended. Some lacked the money to travel, while others lost touch with the battalion after the war and never regained contact. There were also men with no desire to attend, even if a reunion took place next door. These veterans kept a tight lid on the past, the memories too painful. Reunion absentees also included men who had become alienated from their fellow veterans. One was a former Baker Company lieutenant, who deliberately bashed his knee against a tank, injuring himself enough to warrant evacuation. He never surfaced at a reunion.[375] Another no-show was Donald Kepplin, the deserter from Able Company.

Ralph Woodward attended reunions and spoke at local schools in Wisconsin after he retired. Almost every time, the boys and girls reacted with the same question. "What did you do when you had to go to the bathroom?" He answered, "Well, you didn't say, 'Cease fire. I've gotta go!'" He explained that sometimes he used a shell casing as a chamber pot, but helmets and empty ammo cans worked, too. Other times, in combat, "You just went in your shorts and threw them out the hatch later." That statement always knocked the students backward like a wave.[376]

Annual reunions kept wartime endeavors alive among those who attended, however, for many years, few outsiders knew about the 741st and its history, but events in 1979 sparked new interest in the battalion. That year, the Saint-Laurent mayor notified the US Army that fishermen had snagged sunken tanks with their nets in waters off Omaha Beach. Perhaps these barnacle-encrusted relics contained skeletal remains.

Representatives from the US Army Memorial Affairs Activity-Europe traveled from Germany to meet the mayor. The parties laid the groundwork for a search. An army diver subsequently plunged beneath the waves, but, owing to rough and murky water, failed to locate a single tank. His search lasted less than two days before the team leader suspended the fruitless effort. The US Total Army Personnel Command in Alexandria, Virginia, later halted it altogether by stating that "funds are not available to finance the salvage of the tanks. The army's responsibility begins when the tanks are out of the water."[377]

Gossip about the diving effort spread along the Norman coast. Stories circulated that an American consortium planned to raise the tanks and sell them to collectors in the United States, but there was an obstacle to such a scheme (if it ever existed). The French government had sold recovery rights for the D-Day coastline to salvage entrepreneurs Jean Demota and Jacques Lemonchois. They owned exclusive rights, in alternating blocks, for the coastal waters extending twelve miles from shore. Any recovery operation had to involve one or both men.

Lemonchois possessed more than twenty years of diving experience in the area and had pinpointed ten DD tanks from the 741st and another eight lost by the Canadian 6th Armored Regiment near Juno Beach. Demota's salvage company had recovered a Canadian DD years earlier, and now Lemonchois moved to grab two from the 741st.[378]

He selected a pair lost eighty feet down. The quest to raise them using a seaborne crane began in July 1980, but a problem developed with the first one. Corrosion had eaten through the four lifting hooks on the tank's upper hull, and the recovery team found it impossible to weave pickup straps through the hooks. The only alternative was to loop the straps through the open turret hatches. When the crane brought tension on the straps, the turret separated from the hull.

Lemonchois and his men hoisted the turret from the water and sailed for the other tank. They found its hooks intact. The team pulled the thirty-five-ton armored amphibian to the surface and hauled it to nearby Port-en-Bessin. French authorities welded its hatches shut until an ordnance disposal team arrived to extract ammunition stowed inside. Workers also sifted through the silt and muck accumulated in the crew compartment. They discovered no human remains, though numerous GI artifacts came to light. Once declared safe, Lemonchois moved the tank to acreage about a kilometer outside town. At this rural site, he opened the Musée des Épaves sous-marines du Débarquement in 1982 and placed the DD tank on display along with other artifacts from the deep. Two years later, he raised another DD tank off Omaha Beach. Lemonchois again found no skeletal remains inside.

In July 1986, D-Day survivors from the 741st returned to Normandy to see the two tanks and the museum. The veterans included Lynn Hurte, Paul Ragan, Enrico Valente, Jack Boardman, Joseph DePhillips, and Richard Maddock. They wondered how many of their comrades still lay inside tanks on the seabed. The next year, Phil Fitts organized another trip to Europe for more 741st veterans. They also visited the salvaged tanks.

In an interview with the editor of *After the Battle* magazine, Lemonchois asserted that he knew about "other Shermans and some shipwrecks in which skeletons or bones still remain."[379] He considered those dark vaults inviolable. Others felt differently.

André Georgin of Lisieux, France, wrote to President Ronald Reagan in 1987 suggesting that human remains might rest in an American tank sunk off the Norman coast. Ever since the abortive 1979 search, the army refused to reopen the investigation, but that changed with Georgin's letter and pressure from the Reagan administration.

Lieutenant Colonel Lawrence J. Gomez, mortuary officer for the 21st Support Command, and two other servicemen drove to Lisieux in September 1987. The trio traveled from Germany to meet Georgin and to verify his claim. It turned out that Georgin had no firsthand knowledge, just hearsay information from sport scuba divers. Georgin introduced Gomez and his colleagues to the divers.

The Americans included a dive officer from the Corps of Engineers, who accompanied the French divers underwater. They descended roughly one hundred feet to a tank with all its hatches firmly closed. The French divers used a nautical map to point out other tanks—ones with open hatches—which they had searched on prior dives.

The focus shifted to casualty analysis across the Atlantic. Spurious accounts held that scores of 741st men drowned while trapped inside their DD tanks. Gomez needed an accurate count. He consulted with an employee at the US Army Military Personnel Center, who identified five missing men associated with the sunken tanks. Wartime casualty files housed at the Washington National Records Center sharpened the picture. Naval personnel fished two of the men from the water in June 1944, established positive identification, and interred their bodies at sea. That narrowed the list to three names—Alberto Lopez, Hoover Fugate, and Raymond Dahlman.

Gomez also traveled to Fort Eustis, Virginia, home of the US Army Dive School. He sought expertise and operational support for the upcoming mission, a joint army-navy project. Delays beset the mission, mostly due to clearance requirements demanded by the French Navy and the French government. Diplomats at the US Embassy in Paris helped break the bureaucratic logjam, opening the way for the search to begin in November 1987.

Thirteen divers assembled for the operation—frogmen from the 18th Engineer Brigade in Germany and the US Navy EOD Dive Detachment in England. Before going down, the team consulted Lemonchois, who briefed them on the dangers below, especially unexploded ordnance.

Over ten days, the team combed two previously undisturbed tanks but found nothing.

In September 1988, Gomez provided a brief report to Phil Fitts, who had written an inquiry on behalf of himself and his fellow 741st veterans. The colonel closed by saying, "I can assure you, your association would have been proud of the valiant efforts of the combined services dive team. They worked long and hard, over difficult and adverse weather conditions to find conclusive evidence that there were remains inside those tanks. With no trace of evidence found, the mission was completed and formally closed."[380]

There ended the hunt for Lopez, Fugate, and Dahlman. The army relinquished them to the sea, though nobody ever located all twenty-seven sunken tanks.

The search for Russell Bradsher also reached a dead end until the author of this book discovered new information. Photographs and motion-picture film emerged showing two Able Company tanks destroyed on Hill 192, and interviews with veterans confirmed which one Fair commanded and which one belonged to Hecox. The author also found AGRC files regarding two sets of remains recovered in 1947 by Wilbert Hubbell. These included charred bone fragments collected near the driver's seat of Hecox's machine. Bradsher drove that vehicle, and nobody else died with him, but the AGRC never determined who crewed the tank. Today, the remains, X-441, rest at the Normandy American Cemetery, Plot C, Row 23, Grave 9—A COMRADE IN ARMS KNOWN BUT TO GOD.

Hubbell also culled remains from Fair's shattered tank, but the AGRC never connected the wreck to Fair and his crew. The remains, X-440, belonged with other bones hastily pulled from the same tank during the war, but that consolidation never occurred. The army segregated the wartime recoveries between a grave for Raymond Perkins and a group burial for Fair, Nixon, and Pasqualini. Today, X-440 lies in a separate grave at the Normandy American Cemetery.

DNA analysis may someday permit identification of X-440 and X-441, assuming genetic material survived the flames.

Other men like Captain Jimmie Thornton occupy unrecorded graves. While visiting the Luxembourg American Cemetery decades after the war, Jack Boardman found the captain's name on the Wall of the Missing. The aging veteran stared at the letters neatly carved in stone, his mind turning over the mystery of Thornton's death, a conundrum with no answer.

The missing officer had dared all dangers on the hard road to victory, a triumph he never lived to see.

MAPS

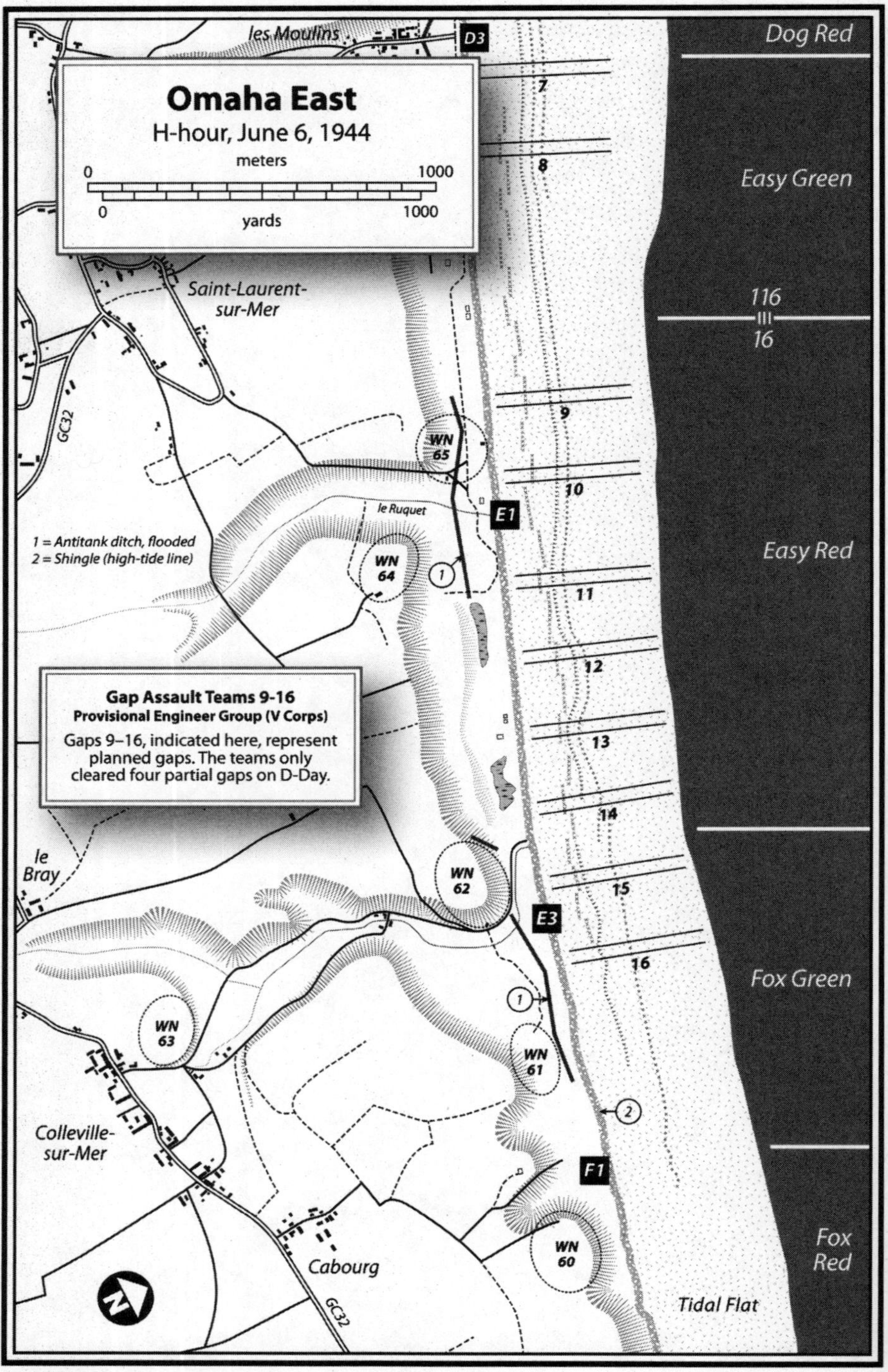

Omaha East

H-hour, June 6, 1944

meters
0 1000
0 1000
yards

les Moulins
D3
Dog Red
7
8
Easy Green
Saint-Laurent-sur-Mer
116
16
GC32
9
WN 65
10
le Ruquet
E1
1 = Antitank ditch, flooded
2 = Shingle (high-tide line)
WN 64
1
Easy Red
11
12

Gap Assault Teams 9-16
Provisional Engineer Group (V Corps)
Gaps 9–16, indicated here, represent planned gaps. The teams only cleared four partial gaps on D-Day.

13
14
le Bray
WN 62
15
E3
16
Fox Green
1
WN 63
WN 61
2
Colleville-sur-Mer
F1
Fox Red
WN 60
Cabourg
GC32
Tidal Flat

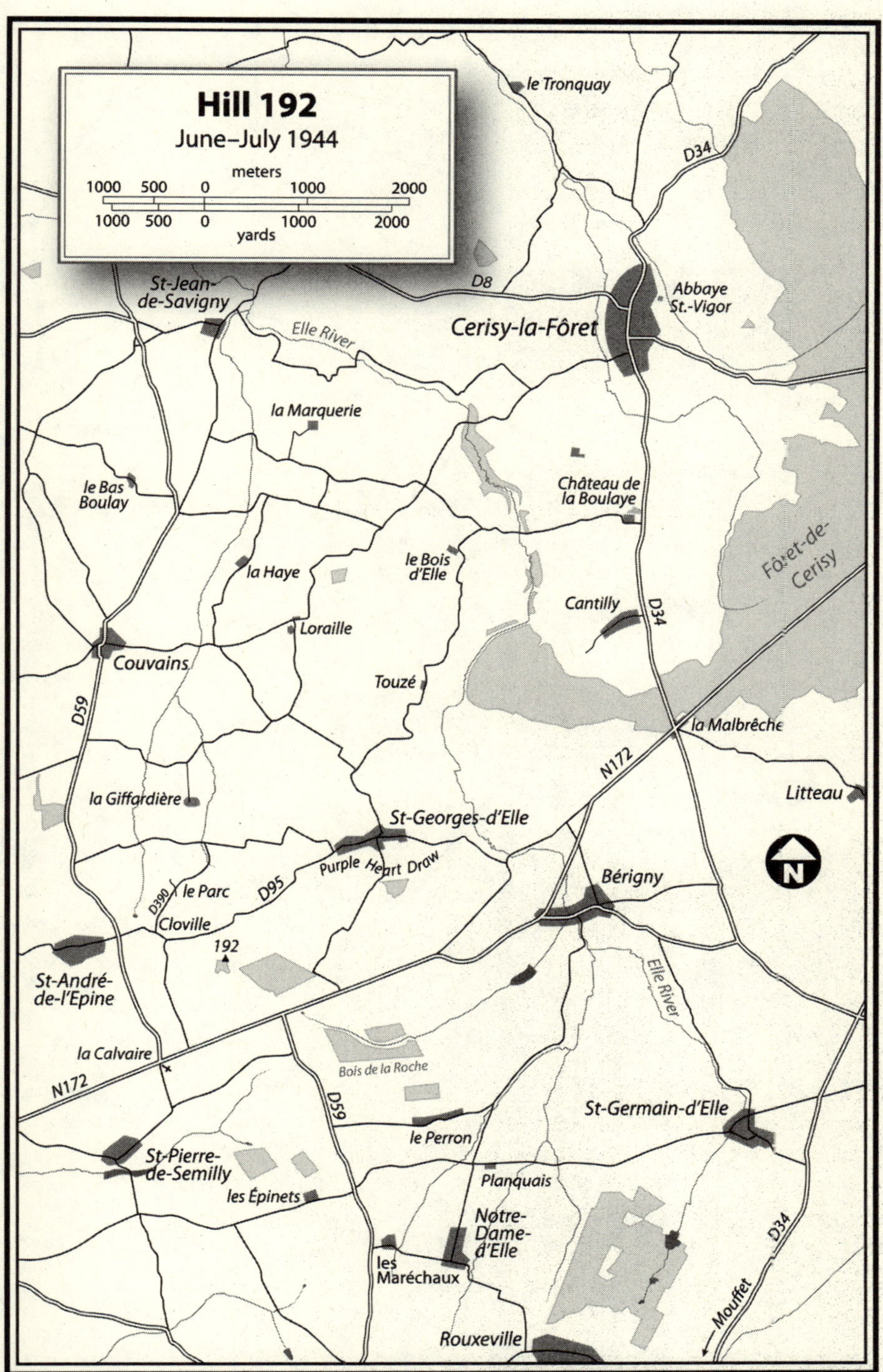
Hill 192
June–July 1944
meters
1000 500 0 1000 2000
1000 500 0 1000 2000
yards
le Tronquay
D34
D8
Abbaye St.-Vigor
St-Jean-de-Savigny
Cerisy-la-Fôret
Elle River
la Marquerie
le Bas Boulay
Château de la Boulaye
Fôret-de-Cerisy
la Haye
le Bois d'Elle
Cantilly
Loraille
D34
Couvains
Touzé
D59
la Malbrêche
N172
la Giffardière
Litteau
St-Georges-d'Elle
Purple Heart Draw
D390
le Parc
D95
Bérigny
Cloville
192
St-André-de-l'Epine
Elle River
la Calvaire
N172
Bois de la Roche
D59
St-Germain-d'Elle
le Perron
St-Pierre-de-Semilly
Planquais
les Épinets
Notre-Dame-d'Elle
D34
les Maréchaux
Mouffet
Rouxeville

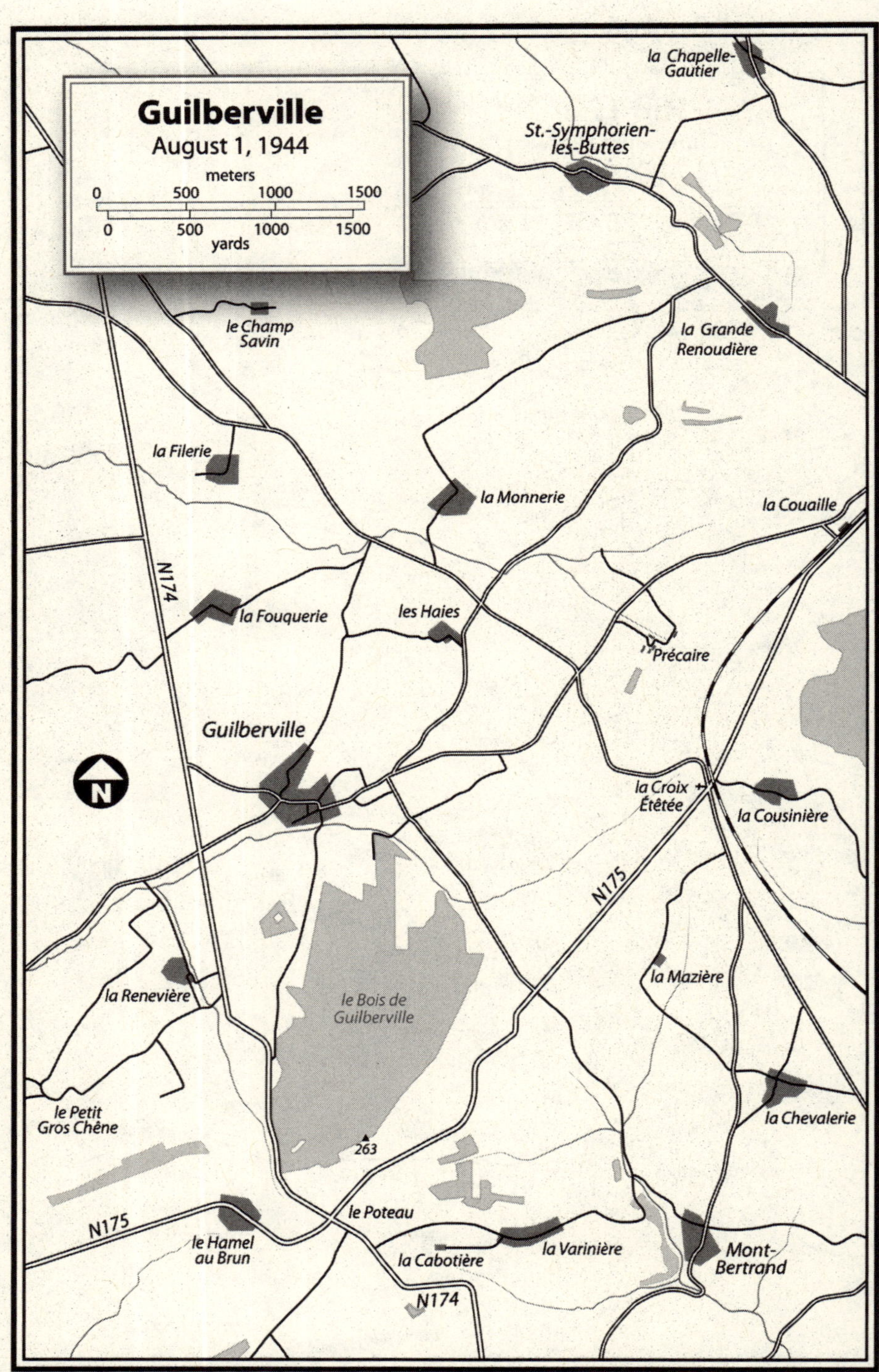
Guilberville
August 1, 1944
meters
0 500 1000 1500
0 500 1000 1500
yards
la Chapelle-Gautier
St.-Symphorien-les-Buttes
le Champ Savin
la Grande Renoudière
la Filerie
la Monnerie
la Couaille
N174
la Fouquerie
les Haies
Précaire
Guilberville
N
la Croix Étêtée
la Cousinière
N175
la Mazière
la Renevière
le Bois de Guilberville
le Petit Gros Chêne
la Chevalerie
263
le Poteau
N175
le Hamel au Brun
la Varinière
Mont-Bertrand
la Cabotière
N174

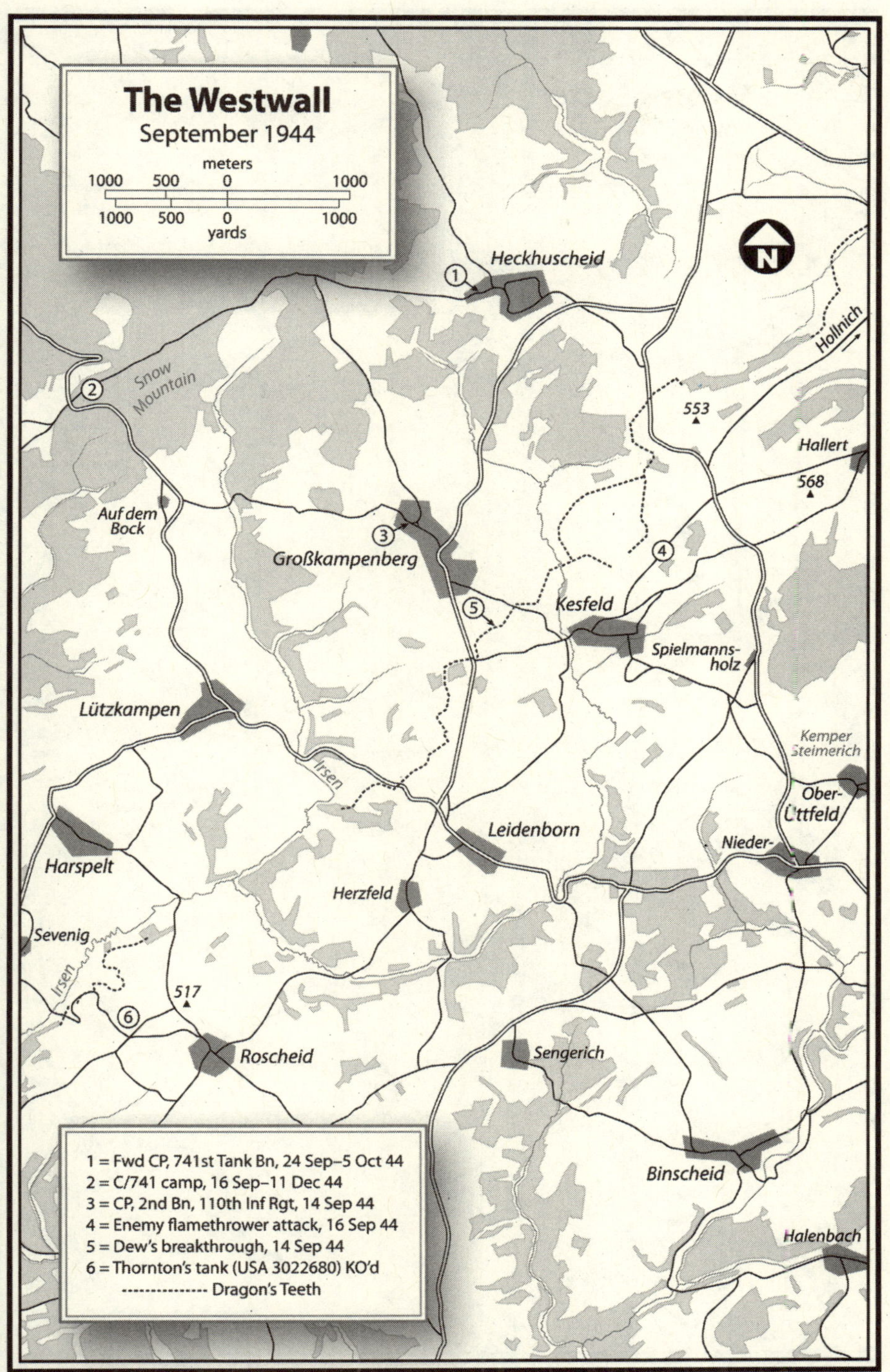
The Westwall
September 1944
meters
1000 500 0 1000
1000 500 0 1000
yards
Heckhuscheid
Snow Mountain
Hollnich
553
Hallert
568
Auf dem Bock
Großkampenberg
Kesfeld
Spielmanns-holz
Lützkampen
Irsen
Kemper Steimerich
Ober-
Üttfeld
Leidenborn
Nieder-
Harspelt
Herzfeld
Sevenig
Irsen
517
Roscheid
Sengerich
Binscheid
Halenbach
1 = Fwd CP, 741st Tank Bn, 24 Sep–5 Oct 44
2 = C/741 camp, 16 Sep–11 Dec 44
3 = CP, 2nd Bn, 110th Inf Rgt, 14 Sep 44
4 = Enemy flamethrower attack, 16 Sep 44
5 = Dew's breakthrough, 14 Sep 44
6 = Thornton's tank (USA 3022680) KO'd
--------------- Dragon's Teeth

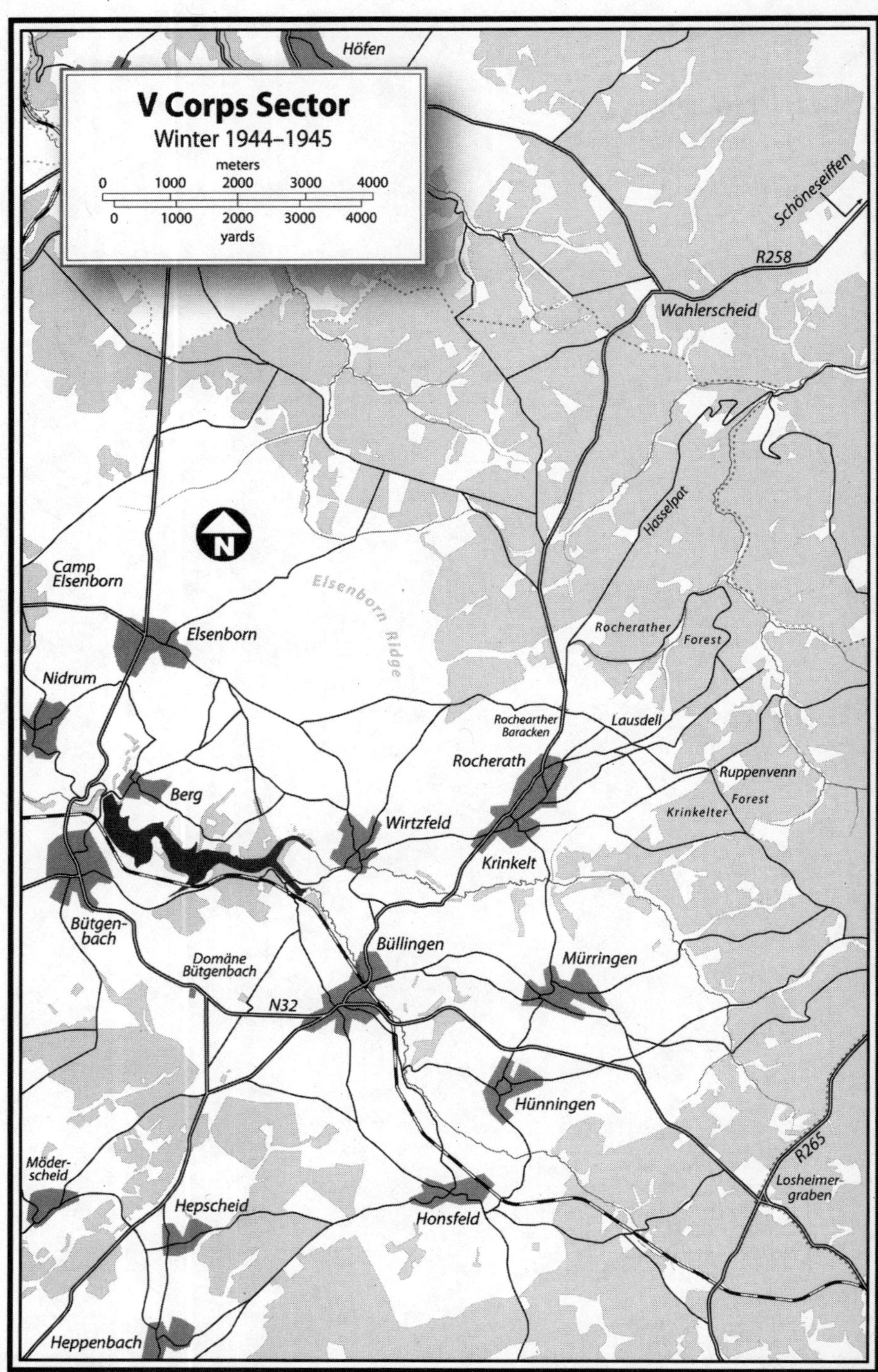
V Corps Sector
Winter 1944–1945
meters
0 1000 2000 3000 4000
0 1000 2000 3000 4000
yards
Höfen
Schöneseiffen
R258
Wahlerscheid
Hasselpat
Camp Elsenborn
Elsenborn
Elsenborn Ridge
Rocherather Forest
Nidrum
Rochearther Baracken
Lausdell
Rocherath
Ruppenvenn
Krinkelter Forest
Berg
Wirtzfeld
Krinkelt
Bütgenbach
Büllingen
Domäne Bütgenbach
Mürringen
N32
Hünningen
R265
Möderscheid
Losheimergraben
Hepscheid
Honsfeld
Heppenbach

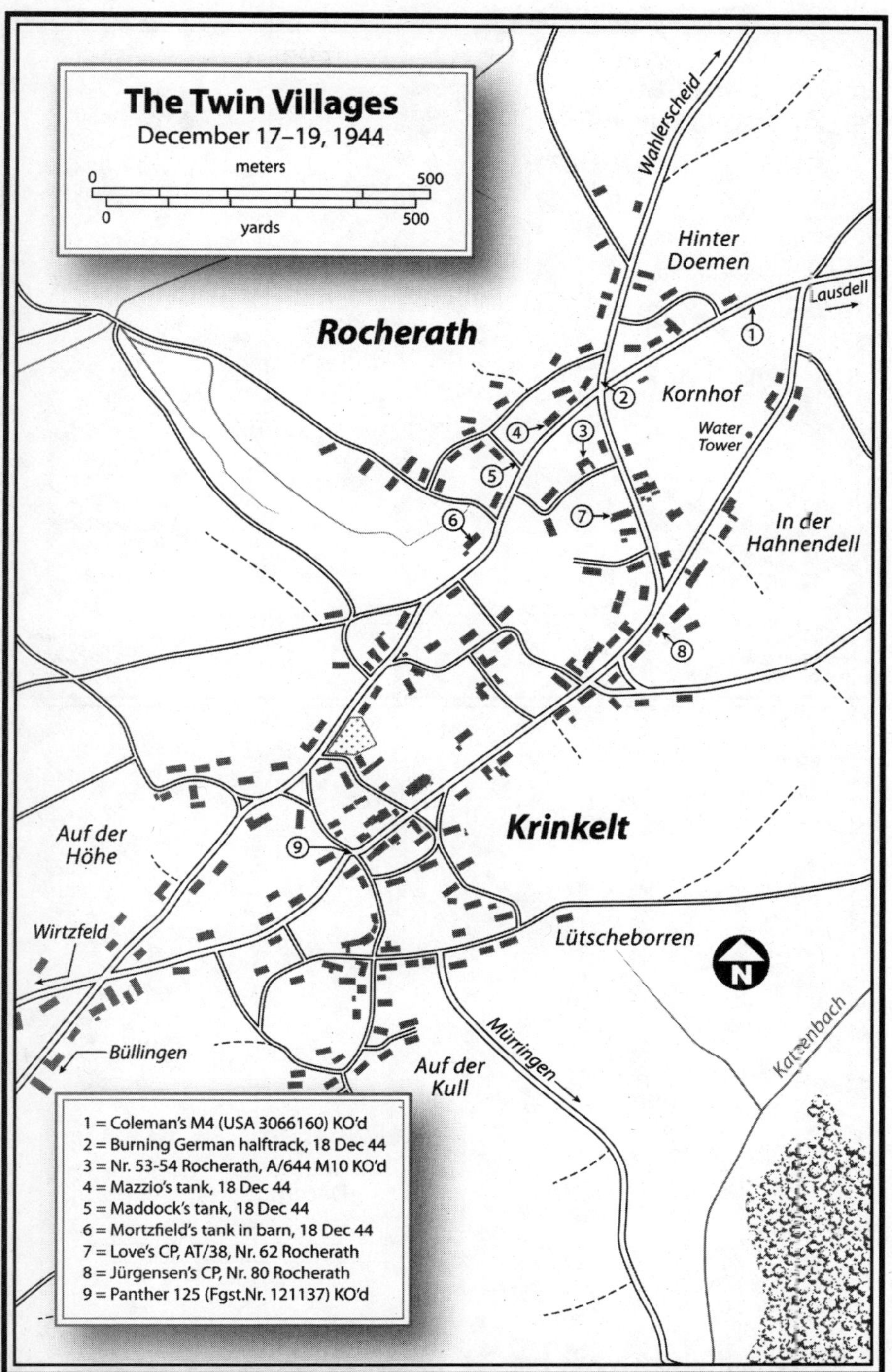
The Twin Villages
December 17–19, 1944
meters
0
500
0
yards
500
Wahlerscheid
Hinter Doemen
Lausdell
Rocherath
Kornhof
Water Tower
In der Hahnendell
Krinkelt
Auf der Höhe
Wirtzfeld
Lütscheborren
N
Büllingen
Auf der Kull
Mürringen
1 = Coleman's M4 (USA 3066160) KO'd
2 = Burning German halftrack, 18 Dec 44
3 = Nr. 53-54 Rocherath, A/644 M10 KO'd
4 = Mazzio's tank, 18 Dec 44
5 = Maddock's tank, 18 Dec 44
6 = Mortzfield's tank in barn, 18 Dec 44
7 = Love's CP, AT/38, Nr. 62 Rocherath
8 = Jürgensen's CP, Nr. 80 Rocherath
9 = Panther 125 (Fgst.Nr. 121137) KO'd

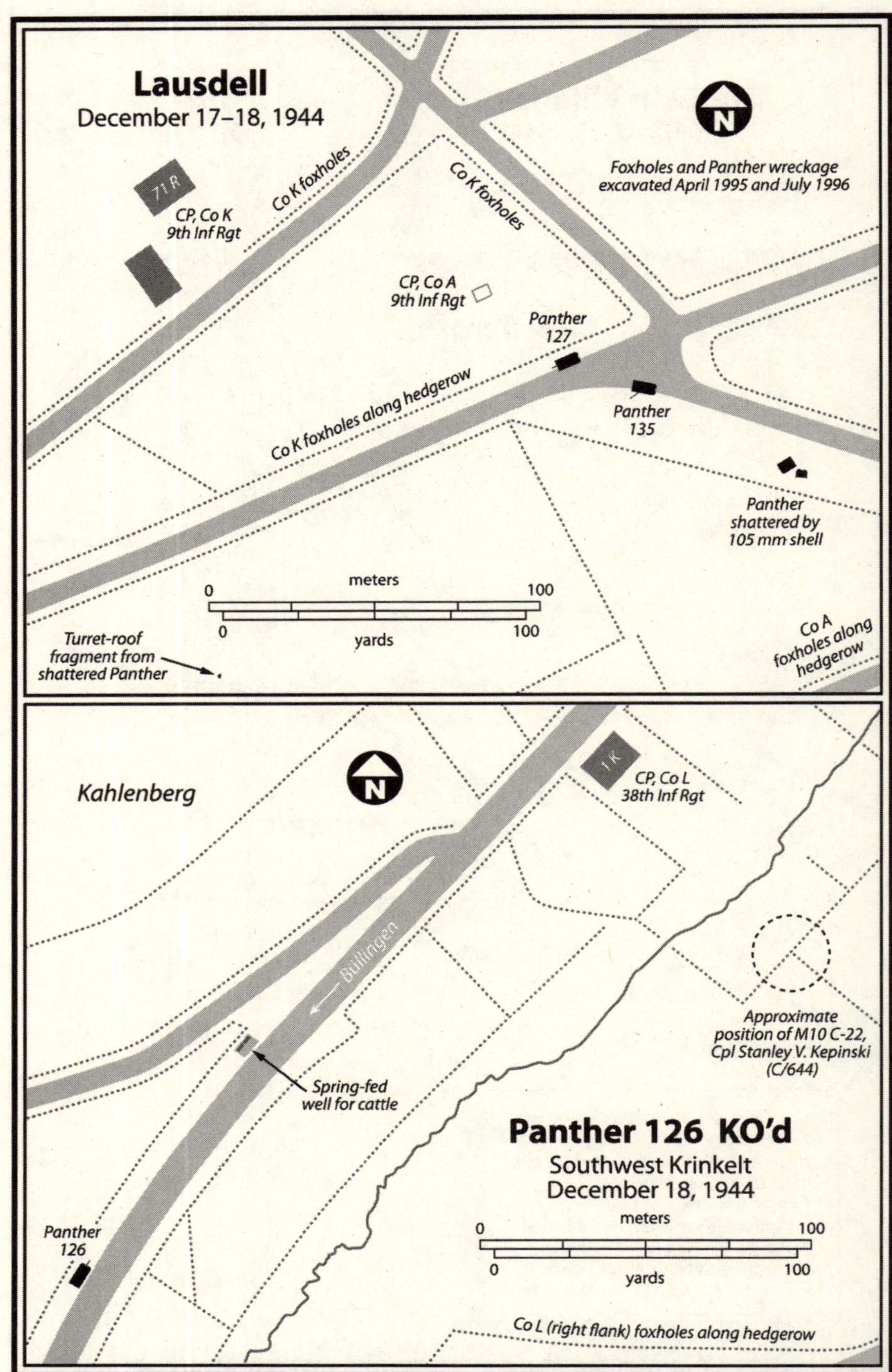
Lausdell
December 17–18, 1944
71 R
CP, Co K
9th Inf Rgt
Co K foxholes
Co K foxholes
Foxholes and Panther wreckage
excavated April 1995 and July 1996
CP, Co A
9th Inf Rgt
Panther
127
Co K foxholes along hedgerow
Panther
135
Panther
shattered by
105 mm shell
meters
0
100
0
yards
100
Turret-roof
fragment from
shattered Panther
Co A
foxholes along
hedgerow
Kahlenberg
1 K
CP, Co L
38th Inf Rgt
Büllingen
Spring-fed
well for cattle
Approximate
position of M10 C-22,
Cpl Stanley V. Kepinski
(C/644)
Panther 126 KO'd
Southwest Krinkelt
December 18, 1944
meters
0
100
0
yards
100
Panther
126
Co L (right flank) foxholes along hedgerow

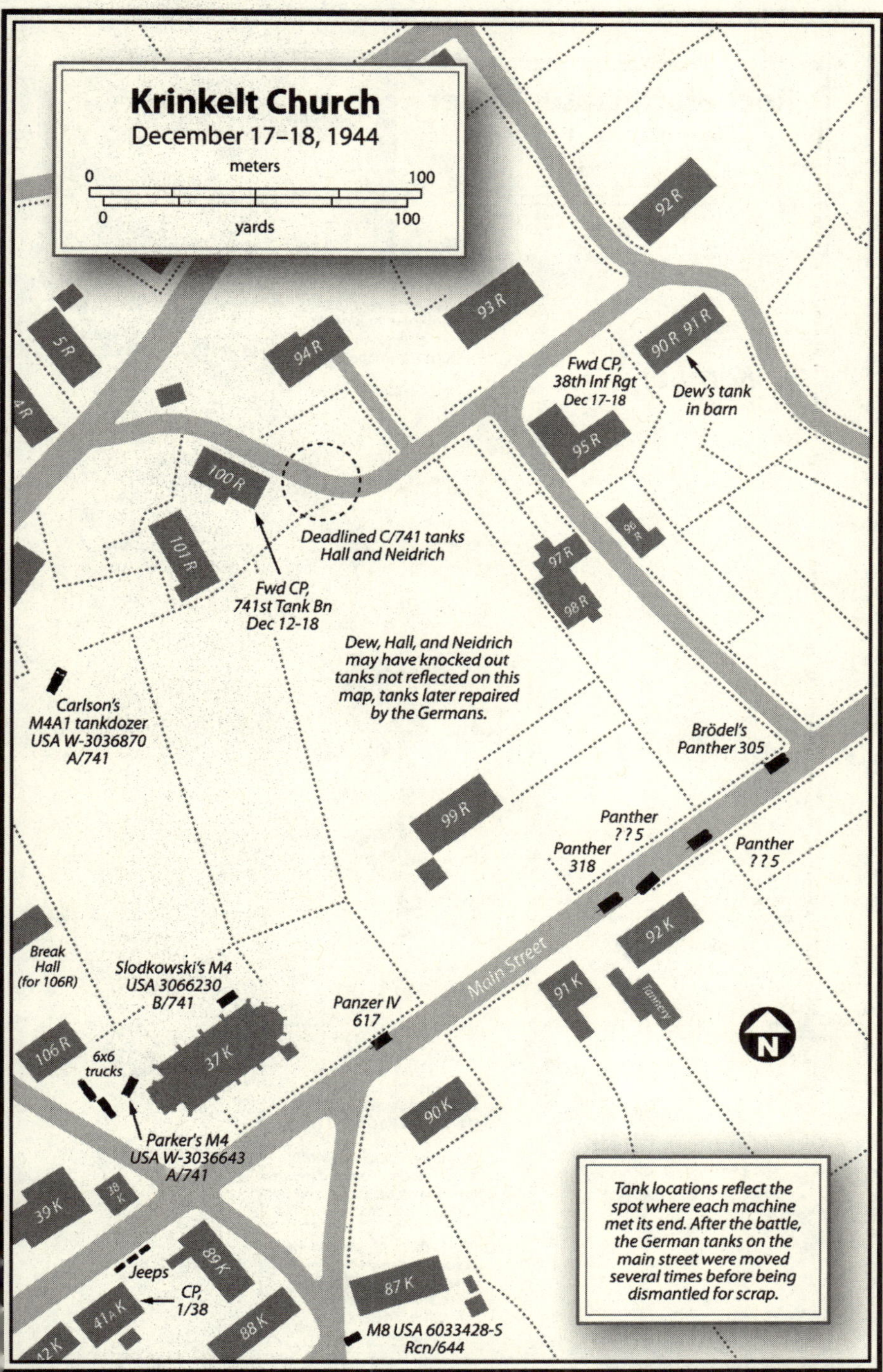
Krinkelt Church
December 17–18, 1944
meters
0
100
0
yards
100
92 R
93 R
94 R
5 R
4 R
90 R 91 R
Fwd CP,
38th Inf Rgt
Dec 17-18
Dew's tank
in barn
95 R
100 R
101 R
Deadlined C/741 tanks
Hall and Neidrich
96 R
97 R
98 R
Fwd CP,
741st Tank Bn
Dec 12-18
Dew, Hall, and Neidrich
may have knocked out
tanks not reflected on this
map, tanks later repaired
by the Germans.
Carlson's
M4A1 tankdozer
USA W-3036870
A/741
Brödel's
Panther 305
99 R
Panther
??5
Panther
318
Panther
??5
92 K
Main Street
91 K
Tannery
Break
Hall
(for 106R)
Slodkowski's M4
USA 3066230
B/741
Panzer IV
617
106 R
6x6
trucks
37 K
Parker's M4
USA W-3036643
A/741
90 K
39 K
38 K
89 K
Jeeps
CP,
1/38
41A K
87 K
88 K
42 K
M8 USA 6033428-S
Rcn/644
Tank locations reflect the
spot where each machine
met its end. After the battle,
the German tanks on the
main street were moved
several times before being
dismantled for scrap.

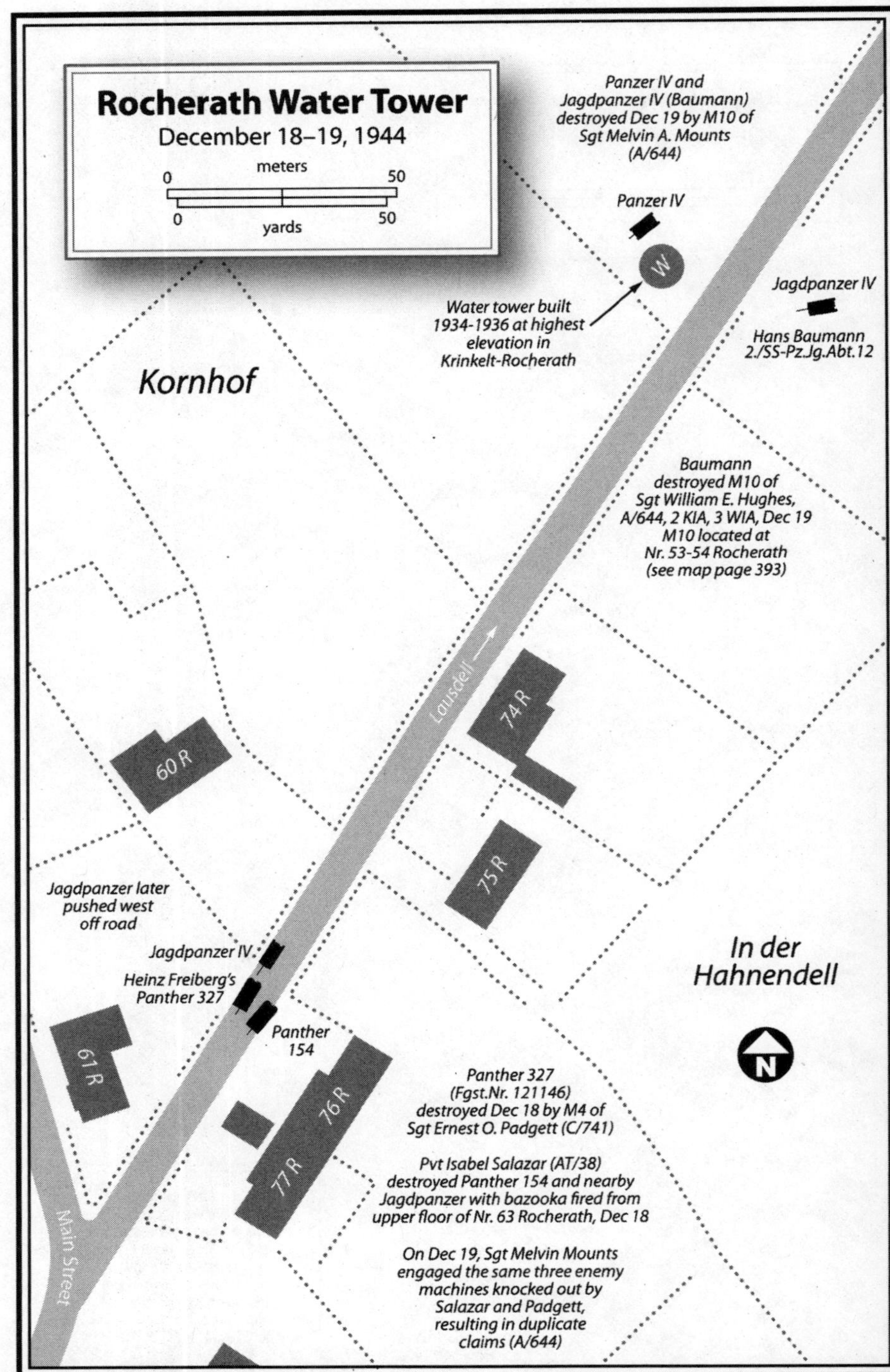
Rocherath Water Tower
December 18–19, 1944
meters
0
50
0
yards
50
Panzer IV and
Jagdpanzer IV (Baumann)
destroyed Dec 19 by M10 of
Sgt Melvin A. Mounts
(A/644)
Panzer IV
W
Jagdpanzer IV
Water tower built
1934-1936 at highest
elevation in
Krinkelt-Rocherath
Hans Baumann
2./SS-Pz.Jg.Abt.12
Kornhof
Baumann
destroyed M10 of
Sgt William E. Hughes,
A/644, 2 KIA, 3 WIA, Dec 19
M10 located at
Nr. 53-54 Rocherath
(see map page 393)
Lausdell
74 R
60 R
75 R
Jagdpanzer later
pushed west
off road
Jagdpanzer IV
Heinz Freiberg's
Panther 327
Panther
154
In der
Hahnendell
N
61 R
76 R
77 R
Panther 327
(Fgst.Nr. 121146)
destroyed Dec 18 by M4 of
Sgt Ernest O. Padgett (C/741)
Pvt Isabel Salazar (AT/38)
destroyed Panther 154 and nearby
Jagdpanzer with bazooka fired from
upper floor of Nr. 63 Rocherath, Dec 18
On Dec 19, Sgt Melvin Mounts
engaged the same three enemy
machines knocked out by
Salazar and Padgett,
resulting in duplicate
claims (A/644)
Main Street

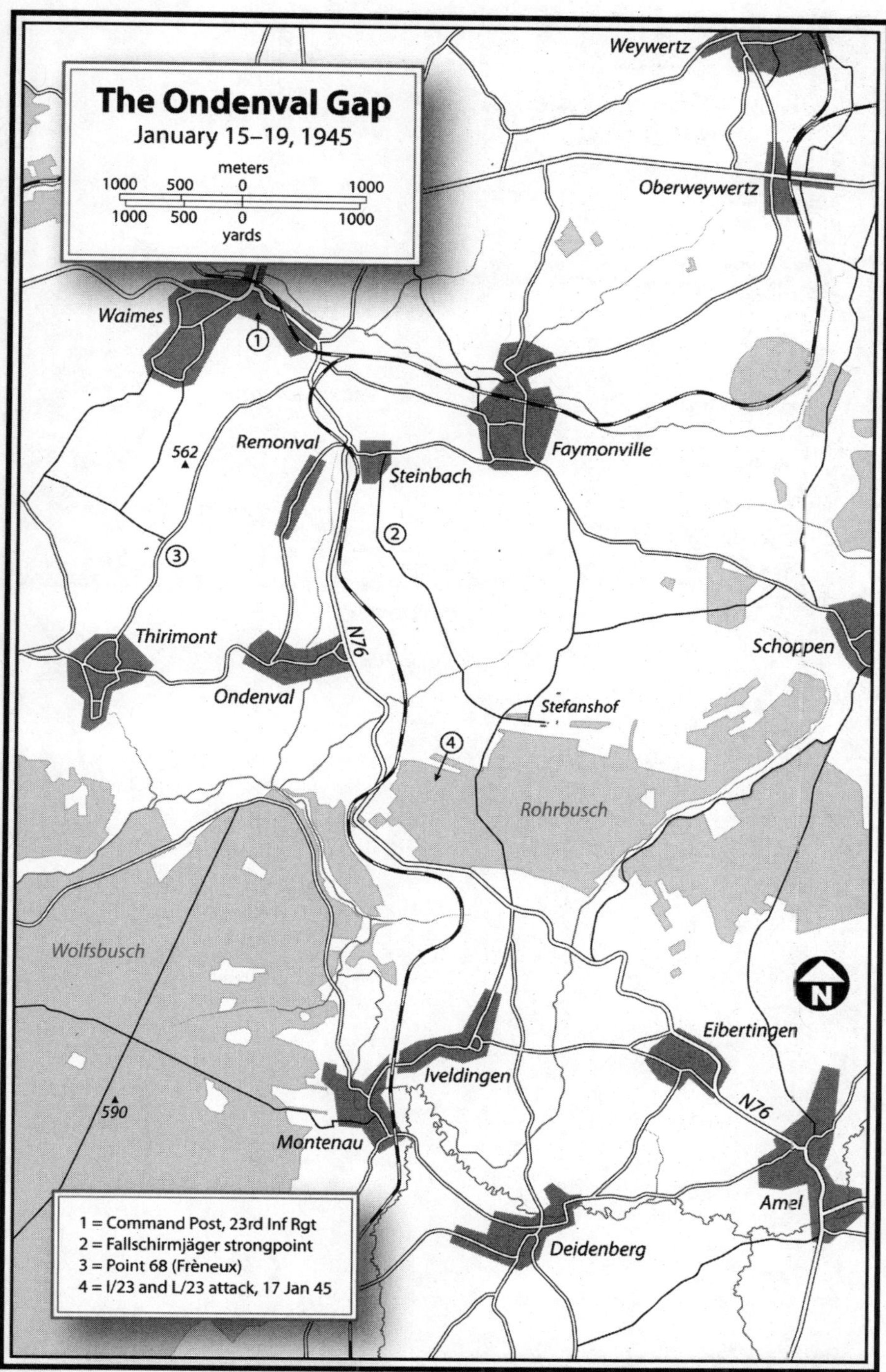
The Ondenval Gap
January 15–19, 1945
meters
1000 500 0 1000
1000 500 0 1000
yards
Weywertz
Oberweywertz
Waimes
562
Remonval
Steinbach
Faymonville
Thirimont
Ondenval
N76
Schoppen
Stefanshof
Rohrbusch
Wolfsbusch
Eibertingen
Iveldingen
N76
590
Montenau
Amel
Deidenberg
1 = Command Post, 23rd Inf Rgt
2 = Fallschirmjäger strongpoint
3 = Point 68 (Frèneux)
4 = I/23 and L/23 attack, 17 Jan 45

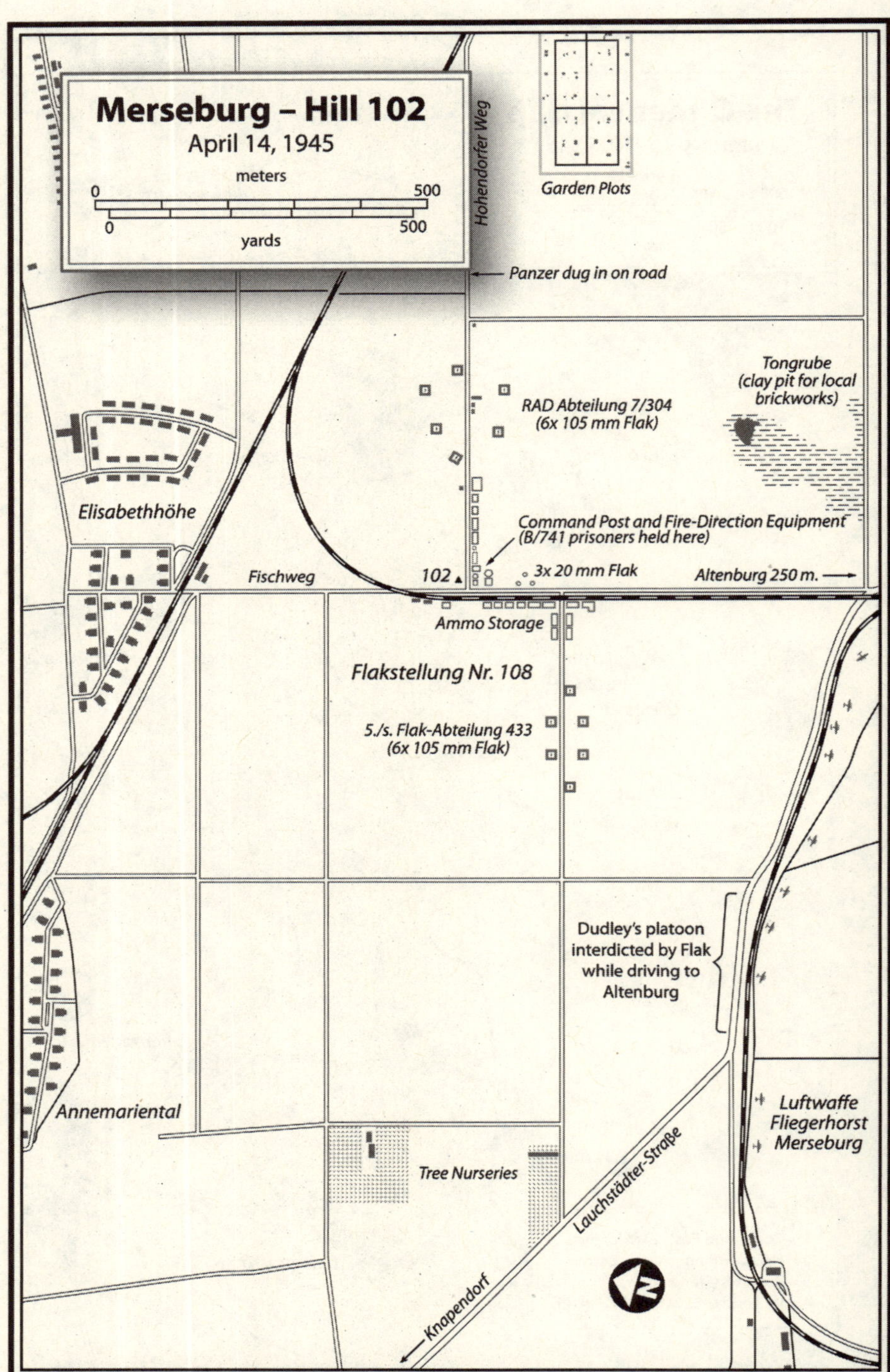

Merseburg – Hill 102
April 14, 1945
meters
0
500
0
500
yards
Hohendorfer Weg
Garden Plots
Panzer dug in on road
Tongrube
(clay pit for local
brickworks)
RAD Abteilung 7/304
(6x 105 mm Flak)
Elisabethhöhe
Command Post and Fire-Direction Equipment
(B/741 prisoners held here)
Fischweg
102
3x 20 mm Flak
Altenburg 250 m.
Ammo Storage
Flakstellung Nr. 108
5./s. Flak-Abteilung 433
(6x 105 mm Flak)
Dudley's platoon
interdicted by Flak
while driving to
Altenburg
Annemariental
Luftwaffe
Fliegerhorst
Merseburg
Tree Nurseries
Lauchstädter-Straße
Knapendorf
N

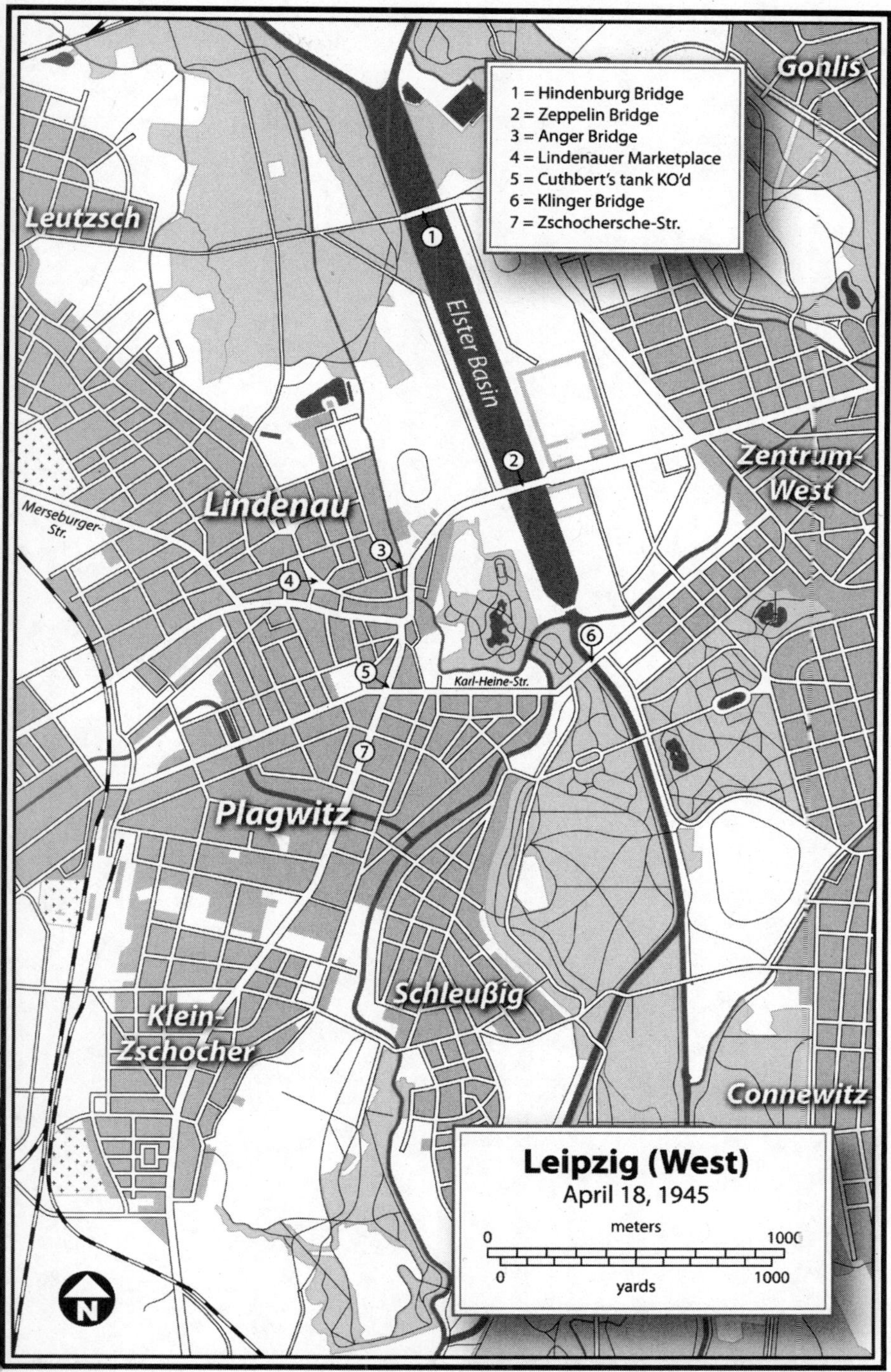
1 = Hindenburg Bridge
2 = Zeppelin Bridge
3 = Anger Bridge
4 = Lindenauer Marketplace
5 = Cuthbert's tank KO'd
6 = Klinger Bridge
7 = Zschochersche-Str.
Gohlis
Leutzsch
Elster Basin
Zentrum-West
Lindenau
Merseburger-Str.
Karl-Heine-Str.
Plagwitz
Schleußig
Klein-Zschocher
Connewitz
Leipzig (West)
April 18, 1945
meters
0
1000
0
yards
1000
N

Appendix A

Honor Roll

During the Second World War, 118 members of the 741st Tank Battalion died in Europe. The author compiled the following roster from AGO records, ABMC records, Quartermaster records, battle casualty reports, company morning reports, National Cemetery registers, and microfiche Ref. No. 601-07, Rosters WW-II Dead (All Services) World War II. Due to space constraints, each soldier's place of permanent interment appears separately at the end of this appendix.

Name	Rank	Co	Status	Death	Temporary Cemetery
Aldrich, Robert H.	PFC	B	KIA	18 Dec 44	Henri-Chapelle #1, NNN-05-087
Bay, Bertrand S.	PFC	HQ	DOW	14 Jul 44	La Cambe, J-04-072
Beck, Merwin C.	Sgt	A	KIA	20 Sep 44	Body not recovered or identified
Becker, Elmer W.	T/Sgt	C	KIA	22 Dec 44	Henri-Chapelle #1, MM-09-175
Blankenship, Clarence W.	Sgt	A	DOW	25 Dec 44	Henri-Chapelle #1, OO-06-101
Bradsher, Russell M.	T/5	A	KIA	11 Jul 44	Body not recovered or identified
Brost, Joseph M.	Sgt	D	KIA	5 May 45	St. Avold, CC-02-030
Brown, Efford L.	PFC	D	KIA	22 Sep 44	Fosse, K-09-168
Campbell, Jesse T.	Pvt	B	KIA	6 Jun 44	St. Andre, L-01-004
Camplin, Raymond C.	Cpl	HQ	DOW	21 Dec 44	Henri-Chapelle #1, RR-10-189
Cerletti, Lawrence E.	Sgt	B	KIA	14 Sep 44	St. Avold, ZZZ-11-128
Chapman, Merrill E.	T/5	B	KIA	6 Jun 44	St. Laurent #1, N-03-050
Charnick, Joseph A.	PFC	C	KIA	6 Jun 44	Brookwood, T-02-021
Clements, Harold R.	T/4	C	KIA	14 Sep 44	Fosse, B-05-095
Coe, Ralph K.	Sgt	C	KIA	6 Jun 44	Blosville, DD-08-141
Coleman, George R.	1/Lt	B	KIA	18 Dec 44	Henri-Chapelle #1, LLL-08-141
Colvin, Grady E.	Pvt	B	KIA	6 Jun 44	Buried at Sea, 26 Jun 44
Cuny, Jess V.	PFC	B	KIA	24 Mar 45	Henri-Chapelle #1, KKKK-10-194
Cuthbert, George K. Jr.	Sgt	C	KIA	18 Apr 45	Margraten, OO-03-55
Dahlman, Raymond	Pvt	C	KIA	6 Jun 44	Body not recovered or identified

Davis, Stanley G.	Pvt	C	KIA	6 Jun 44	St. Laurent #1, H-10-188
Dedmon, Guy A.	Cpl	C	KIA	17 Dec 44	Henri-Chapelle #1, MMM-07-125
Domenighini, Elmo	T/5	B	KIA	6 Jun 44	Brookwood, T-01-011
Dyer, Harry C.	Sgt	C	KIA	7 Jun 44	Brookwood, T-01-031
Edwards, Arthur G.	PFC	B	KIA	6 Jun 44	St. Laurent #1, A-04-065
Elliott, Frank M.	Cpl	A	KIA	6 Jun 44	St. Laurent #1, A-06-118
Fair, Thomas R.	S/Sgt	A	KIA	11 Jul 44	St. Laurent #1, S-04-078 *
Fields, Dock W.	PFC	C	KIA	5 Apr 45	Body not recovered or identified
Figg, Robert J.	Pvt	A	KIA	6 Jun 44	St. Laurent #1, D-04-078
Ford, Donald E.	T/4	B	KIA	6 Jun 44	Buried at Sea, ca. 27 Jun 44
Freed, Robert L.	Pvt	A	KIA	26 Jul 44	St. Laurent #1, W-04-072
Froehlich, John	T/5	HQ	DOW	21 Dec 44	Henri-Chapelle #1, RR-10-193
Fry, Robert W.	Cpl	B	KIA	26 Jul 44	St. Laurent #1, S-03-056
Fryer, Walter J.	Sgt	C	KIA	13 Apr 45	Margraten, NN-10-230
Fugate, Hoover	T/5	B	KIA	6 Jun 44	Body not recovered or identified
Garber, Roy A.	Pvt	B	KIA	18 Dec 44	Henri-Chapelle #1, LLL-08-146
Glatt, William E.	PFC	C	KIA	18 Apr 45	Margraten, SS-03-072
Gordon, Barney	T/5	D	DNB	3 Apr 45	Margraten, XX-04-094
Gray, Earnest L.	2/Lt	A	KIA	19 Jun 44	La Cambe, D-02-035
Hall, Curtis	Cpl	C	KIA	31 Jan 45	Henri-Chapelle #1, MMM-04-068
Hammans, Robert R.	PFC	C	KIA	6 Jun 44	Brookwood, T-02-013
Henkelman, Leonard J.	2/Lt	C	DOI	28 Mar 45	Margraten, TT-04-083
Hornberger, Herbert J.	T/4	C	KIA	6 Jun 44	St. Andre, H-02-038
Hough, William R.	Pvt	D	DOW	8 Sep 44	Solers, B-05-092
Ireson, Charles R.	Pvt	A	KIA	18 Dec 44	Henri-Chapelle #1, MMM-04-064
Jacob, Robert P.	Pvt	B	KIA	6 Jun 44	St. Laurent #1, U-09-170
Jacobs, Edwin H.	2/Lt	C	KIA	27 Jul 44	St. Laurent #1, W-08-157
James, Thurston O.	PFC	A	KIA	1 Aug 44	Marigny, N-05-091
Jones, Walter R.	T/4	B	KIA	6 Jun 44	St. Laurent #1, H-02-037
Kaminesky, Joseph B.	Pvt	B	KIA	6 Jun 44	Brookwood, S-02-025
Keller, Cleo B.	PFC	B	KIA	24 Mar 45	Henri-Chapelle #1, KKKK-10-186
Kettles, Warren E.	Pvt	B	FOD	14 Sep 44	Hamm, R-02-049
Kinkade, Arthur E.	T/4	B	FOD	14 Sep 44	Hamm, R-02-050

Kitchen, Woodrow W.	Pvt	C	KIA	27 Jul 44	St. Laurent #1, W-08-153
Kovar, Kleophas E.	Cpl	C	KIA	14 Sep 44	Foy, J-12-291
Kunkle, Ralph E.	Pvt	B	KIA	6 Jun 44	Brookwood, T-01-003
Laird, Florus W.	2/Lt	B	KIA	6 Jun 44	Buried at Sea, 8 Jun 44
Lajewski, Henry C.	Pvt	A	KIA	26 Jul 44	St. Laurent #1, T-06-113
Lanier, Alvin	Pvt	C	KIA	17 Dec 44	Henri-Chapelle #1, MMM-C7-127
Lawson, Sloan W.	PFC	B	KIA	6 Jun 44	St. Laurent #1, H-07-127
Lee, George W. Jr.	Pvt	C	KIA	27 Jul 44	St. Laurent #1, W-10-194
Letso, Michael	Cpl	B	KIA	14 Sep 44	St. Avold, ZZZ-11-129
Levan, Edwin A.	Cpl	C	KIA	6 Jun 44	Brookwood, T-01-005
Lombardo, Charles	PFC	C	KIA	18 Apr 45	Margraten, SS-03-070
Longmire, James D.	T/4	B	KIA	6 Jun 44	St. Laurent #1, G-02-021
Lopez, Alberto	Cpl	C	FOD	6 Jun 44	Body not recovered or identified
Magures, Leroy	T/4	C	KIA	6 Jun 44	Brookwood, T-01-030
Manganaro, Frank	Cpl	B	KIA	18 Dec 44	Henri-Chapelle #1, LLL-08-145
McDonough, Roger J.	Capt	A	KIA	29 Apr 45	Neuville, GG-05-107
McKinney, Francis	Pvt	C	KIA	6 Jun 44	Brookwood, T-01-026
McKnight, John	Pvt	C	KIA	27 Jul 44	St. Laurent #1, W-06-115
Meads, Ernest M.	T/5	A	KIA	7 Mar 45	Henri-Chapelle #1, AAAA-09-172
Meyers, Calvin B.	2/Lt	A	KIA	26 Jul 44	St. Laurent #1, U-05-081
Middleton, Elmer	Sgt	B	KIA	13 Aug 44	Gorron #1, D-04-062
Miller, Kenneth J.	Cpl	C	KIA	11 Jul 44	St. Laurent #1, R-10-199
Miller, Victor L. Jr.	1/Lt	C	KIA	17 Dec 44	Henri-Chapelle #1, MMM-07-126
Mizer, Lavander D.	Cpl	A	DOW	15 Jun 44	Brookwood, T-08-017
Momany, Donald O.	T/5	HQ	DOW	19 Oct 44	Cambridge, Q-10-024
Morris, Clarence	T/5	C	KIA	6 Jun 44	Champigneul, H-04-083
Moskalski, Robert V. I.	PFC	B	KIA	6 Jun 44	La Cambe, BE-10-185
Myers, William J.	T/5	A	KIA	1 Jan 45	Henri-Chapelle #1, VV-02-027
Nation, Stanley V.	Sgt	D	KIA	30 Sep 44	Henri-Chapelle #1, G-02-034
Nickel, Kenneth W.	Cpl	C	KIA	18 Apr 45	Margraten, SS-03-064
Nixon, Willis E.	T/4	A	KIA	11 Jul 44	St. Laurent #1, S-04-078
Oglesbee, Vemma K.	Pvt	B	KIA	11 Jul 44	St. Laurent #1, R-10-183
O'Riley, Harold C.	Cpl	A	KIA	20 Sep 44	Body not recovered or identified

Paschal, Preston C.	T/5	C	KIA	6 Jun 44	St. Andre, H-06-120
Pasqualini, Frank	Cpl	A	KIA	11 Jul 44	St. Laurent #1, S-04-078 *
Perkins, Raymond L.	Pvt	A	KIA	11 Jul 44	St. Laurent #1, N-09-163
Popovic, Samuel	PFC	A	KIA	26 Jul 44	St. Laurent #1, T-06-101
Price, Warren H.	Pvt	A	KIA	1 Aug 44	Marigny, N-05-091
Reed, Ernest S.	T/4	B	KIA	18 Dec 44	Henri-Chapelle #1, NNN-05-091
Reeves, Charles R.	Pvt	A	KIA	1 Aug 44	Marigny, N-05-091
Rhode, Herbert F.	Pvt	HQ	KIA	21 Dec 44	Henri-Chapelle #1, SS-10-197
Rich, Valjean H.	PFC	B	FOD	26 Jul 44	Body not recovered or identified
Rieman, Thomas H.	Sgt	A	KIA	17 Jun 44	St. Laurent #1, H-08-159
Ritchie, Brodie	T/4	B	DNB	4 Apr 45	Margraten, WW-04-078
Rohrbaugh, Clyde	Cpl	A	KIA	4 Aug 44	Marigny, Q-05-092
Ross, Pearl	T/5	C	KIA	27 Jul 44	St. Laurent #1, V-05-096
Rued, Evald J.	T/4	C	KIA	11 Jul 44	St. Laurent #1, R-10-198
Rush, Fields V.	T/5	C	KIA	6 Jun 44	St. Laurent #1, M-08-146
Schild, Fred A.	Sgt	B	KIA	14 Sep 44	Fosse, B-08-148
Schmidt, Ivan E.	T/4	B	KIA	14 Apr 45	Margraten, NN-04-086
Sertell, John R.	S/Sgt	C	KIA	6 Jun 44	St. Laurent #1, B-06-113
Shirk, Clifford	1/Lt	D	KIA	11 Mar 45	Henri-Chapelle #1, BBBB-06-108
Skiba, Walter J.	S/Sgt	A	KIA	6 Jun 44	St. Laurent #1, A-09-175
Sloan, Ray I.	Pvt	A	KIA	29 Sep 44	Body not recovered or identified
Smulik, Bolick	T/4	A	KIA	1 Aug 44	Marigny, N-05-091
Tarnow, Duane E.	T/4	C	DOI	28 Jul 44	St. Laurent #1, U-07-140
Testa, William J.	PFC	B	KIA	6 Jun 44	Brookwood, T-02-006
Thompson, Dale E.	T/4	C	KIA	6 Jun 44	La Cambe, AF-08-159
Thornton, James G. Jr.	Capt	B	FOD	14 Sep 44	Body not recovered or identified
Unterseher, John	Sgt	HQ	DOW	21 Dec 44	Henri-Chapelle #1, RR-10-190
Waldkoetter, Walter J.	T/4	A	KIA	6 Jun 44	St. Laurent #1, F-07-123
Whited, James D.	Pvt	A	KIA	6 Jun 44	St. Laurent #1, A-06-110
Wilson, George R.	PFC	C	KIA	18 Apr 45	Margraten, SS-03-069
Young, Charles R.	Capt	C	DOW	3 Feb 45	Henri-Chapelle #1, JJJ-05-098
Zoeller, Adolph Jr.	Cpl	C	KIA	6 Jun 44	St. Laurent #1, S-08-141

* Group burial for S-04-078 (X-105), S-05-082 (X-109), S-05-088 (X-108)

Name	Permanent Cemetery
Aldrich, Robert H.	Henri-Chapelle American Cemetery, E-3-10
Bay, Bertrand S.	Idaho - Morris Hill Cemetery, Boise
Beck, Merwin C.	Henri-Chapelle American Cemetery, Wall of the Missing
Becker, Elmer W.	Indiana - Emanuel Lutheran Cemetery, Allen County
Blankenship, Clarence W.	Ohio - Oak Dale Cemetery, Urbana
Bradsher, Russell M.	Normandy American Cemetery, Wall of the Missing
Brost, Joseph M.	Indiana - St. Mary's Cemetery, Dunnington
Brown, Efford L.	Kentucky - Zachary Taylor National Cemetery, C-24
Campbell, Jesse T.	Ardennes American Cemetery, C-30-8
Camplin, Raymond C.	Illinois - Rock Island National Cemetery, D-3
Cerletti, Lawrence E.	California - Golden Gate National Cemetery, O-1292
Chapman, Merrill E.	Virginia - Arlington National Cemetery, 12-2468
Charnick, Joseph A.	Pennsylvania - St. Michaels Church Cemetery, Centre County
Clements, Harold R.	Pennsylvania - Franklin Cemetery, Dunbar
Coe, Ralph K.	Ohio - McPherson Cemetery, Clyde
Coleman, George R.	Illinois - Bluff City Cemetery, Elgin
Colvin, Grady E.	Normandy American Cemetery, Wall of the Missing
Cuny, Jess V.	South Dakota - Mountain View Cemetery, Rapid City
Cuthbert, George K. Jr.	New York - Long Island National Cemetery, J-13682
Dahlman, Raymond	Normandy American Cemetery, Wall of the Missing
Davis, Stanley G.	Wisconsin - Wanderer's Rest Cemetery, Milwaukee
Dedmon, Guy A.	Henri-Chapelle American Cemetery, C-15-42
Domenighini, Elmo	California - Willows District Cemetery, Willows
Dyer, Harry C.	Alabama - Mt. Carmel Cemetery, West Blocton
Edwards, Arthur G.	North Carolina - French Broad Cemetery, Alexander
Elliott, Frank M.	Normandy American Cemetery, I-16-3
Fair, Thomas R.	Virginia - Richmond National Cemetery, 1C-5501
Fields, Dock W.	Henri-Chapelle American Cemetery, Wall of the Missing
Figg, Robert J.	Virginia - Arlington National Cemetery, 12-3144
Ford, Donald E.	Normandy American Cemetery, Wall of the Missing
Freed, Robert L.	Normandy American Cemetery, J-3-11
Froehlich, John	Henri-Chapelle American Cemetery, G-9-54

Fry, Robert W.	Normandy American Cemetery, I-14-7
Fryer, Walter J.	Netherlands American Cemetery, B-2-9
Fugate, Hoover	Normandy American Cemetery, Wall of the Missing
Garber, Roy A.	New York - Long Island National Cemetery, H-11930
Glatt, William E.	Netherlands American Cemetery, J-8-10
Gordon, Barney	Netherlands American Cemetery, M-15-4
Gray, Earnest L.	Minnesota - Ft. Snelling National Cemetery, B-334
Hall, Curtis	Kentucky - Dry Creek Cemetery, Dry Creek
Hammans, Robert R.	Cambridge American Cemetery, C-6-53
Henkelman, Leonard J.	Netherlands American Cemetery, C-20-15
Hornberger, Herbert J.	Indiana - St. John Lutheran Cemetery, Sunman
Hough, William R.	Texas - Ore City Cemetery, Ore City
Ireson, Charles R.	Virginia - Maplewood Cemetery, Tazewell
Jacob, Robert P.	Indiana - Calvary Cemetery, Terre Haute
Jacobs, Edwin H.	Normandy American Cemetery, I-14-25
James, Thurston O.	Tennessee - Knoxville National Cemetery, X-58A
Jones, Walter R.	Tennessee - Hopewell Cemetery, Medina
Kaminesky, Joseph B.	Ohio - Calvary Cemetery, Toledo
Keller, Cleo B.	Henri-Chapelle American Cemetery, F-1-47
Kettles, Warren E.	Georgia - West Hill Cemetery, Dalton
Kinkade, Arthur E.	Pennsylvania - Beaver Falls Cemetery, Beaver Falls
Kitchen, Woodrow W.	Kentucky - Bowling Cemetery, Carter County
Kovar, Kleophas E.	Texas - Plum Catholic Cemetery, Plum
Kunkle, Ralph E.	Pennsylvania - Willow Street Mennonite Cemetery, Lancaster Co.
Laird, Florus W.	Normandy American Cemetery, Wall of the Missing
Lajewski, Henry C.	New Jersey - Beverly National Cemetery, F-1812
Lanier, Alvin	Henri-Chapelle American Cemetery, C-6-56
Lawson, Sloan W.	Normandy American Cemetery, H-26-8
Lee, George W. Jr.	Normandy American Cemetery, I-23-19
Letso, Michael	Lorraine American Cemetery, E-8-2
Levan, Edwin A.	Illinois - Chapel Hill Cemetery, Dixon
Lombardo, Charles	New Jersey - Beverly National Cemetery, G-90

Longmire, James D.	Normandy American Cemetery, I-7-4
Lopez, Alberto	Normandy American Cemetery, Wall of the Missing
Magures, Leroy	Texas - Kemp Cemetery, Kemp
Manganaro, Frank	New York - Long Island National Cemetery, H-10767
McDonough, Roger J.	Ardennes American Cemetery, B-33-50
McKinney, Francis	North Carolina - Raleigh National Cemetery, 24-1351
McKnight, John	Normandy American Cemetery, J-13-3
Meads, Ernest M.	Henri-Chapelle American Cemetery, D-10-1
Meyers, Calvin B.	District of Columbia - Congressional Cemetery, R25/38
Middleton, Elmer	Virginia - Ely Cemetery, Jonesville
Miller, Kenneth J.	New Jersey - George Washington Memorial Park, Paramus
Miller, Victor L. Jr.	Henri-Chapelle American Cemetery, G-5-41
Mizer, Lavander D.	Oklahoma - Woodlawn Cemetery, Claremore
Momany, Donald O.	Cambridge American Cemetery, C-6-53
Morris, Clarence	Epinal American Cemetery, B-14-38
Moskalski, Robert V. I.	Virginia - Richmond National Cemetery, 2C-5510
Myers, William J.	Pennsylvania - Northwood Cemetery, Philadelphia
Nation, Stanley V.	Henri-Chapelle American Cemetery, B-4-50
Nickel, Kenneth W.	Netherlands American Cemetery, N-22-11
Nixon, Willis E.	Virginia - Richmond National Cemetery, 1C-5501
Oglesbee, Vemma K.	Normandy American Cemetery, I-16-27
O'Riley, Harold C.	Henri-Chapelle American Cemetery, Wall of the Missing
Paschal, Preston C.	Texas - Covington Cemetery, Covington
Pasqualini, Frank	Virginia - Richmond National Cemetery, 1C-5502
Perkins, Raymond L.	Normandy American Cemetery, G-15-4
Popovic, Samuel	Normandy American Cemetery, G-1-28
Price, Warren H.	Tennessee - Knoxville National Cemetery, X-58A
Reed, Ernest S.	Tennessee - Reed Cemetery, Savannah
Reeves, Charles R.	Tennessee - Knoxville National Cemetery, X-58A
Rhode, Herbert F.	New Jersey - Beverly National Cemetery, F-1597
Rich, Valjean H.	Brittany American Cemetery, Wall of the Missing
Rieman, Thomas H.	Normandy American Cemetery, I-10-35
Ritchie, Brodie	Kentucky - John S. Combs Cemetery, Vest

Rohrbaugh, Clyde	Brittany American Cemetery, D-6-4
Ross, Pearl	Kentucky - Ross Cemetery, Rush
Rued, Evald J.	Normandy American Cemetery, G-10-8
Rush, Fields V.	Normandy American Cemetery, J-17-18
Schild, Fred A.	Kentucky - Highland Cemetery, Covington
Schmidt, Ivan E.	Netherlands American Cemetery, J-1-22
Sertell, John R.	Normandy American Cemetery, H-20-23
Shirk, Clifford	Iowa - Grant City Cemetery, Grant City
Skiba, Walter J.	Ohio - Calvary Cemetery, Toledo
Sloan, Ray I.	Henri-Chapelle American Cemetery, Wall of the Missing
Smulik, Bolick	Tennessee - Knoxville National Cemetery, X-58A
Tarnow, Duane E.	Normandy American Cemetery, G-28-18
Testa, William J.	New Jersey - Holy Cross Cemetery, North Arlington
Thompson, Dale E.	Indiana - Crown Hill Cemetery, Salem
Thornton, James G. Jr.	Luxembourg American Cemetery, Wall of the Missing
Unterseher, John	North Dakota - Peace Cemetery, Dunn Center
Waldkoetter, Walter J.	Normandy American Cemetery, J-11-7
Whited, James D.	Normandy American Cemetery, J-6-22
Wilson, George R.	Netherlands American Cemetery, A-16-1
Young, Charles R.	Henri-Chapelle American Cemetery, C-16-57
Zoeller, Adolph Jr.	Illinois - Maplewood Cemetery, Marion

Coe's body recovered at Dannes-Camiers, France, 135 miles from Omaha Beach

Paschal's body recovered at St. Quentin-en-Tourment, France, 120 miles from Omaha

X-440 (Normandy American Cemetery, A-5-34) unidentified remains from Fair's tank

X-441 (Normandy American Cemetery, C-23-9) unidentified remains from Hecox's tank

Appendix B

Distinguished Unit Citations

The 741st Tank Battalion is cited for outstanding performance of duty in action. In the invasion of France, the 741st Tank Battalion rendered outstanding performance of duty in action against the enemy on the Colleville-sur-Mer beach on 6–7 June 1944. Armored units of the battalion led the assault wave in landing in the face of extensive and elaborate enemy minefields which had not been cleared or gapped and under the withering artillery, antitank, machine gun, and mortar fire from organized and fortified enemy positions. Despite extremely heavy losses the attack was pushed against the unexpectedly stiff enemy resistance, thereby enabling the infantry to hold the beach. When planned exits from the beach were unusable because of enemy installations of road blocks, minefields, and fortifications, the 741st Tank Battalion forced new exits from the beach against the enemy positions and by such action enabled a beach line to be secured and held, insuring success of the landing and the beachhead. Only through the courageous performance of the officers and enlisted men was the objective secured. In the midst of unceasing enemy fire and often in the face of certain death, duties were performed unhesitatingly and with utter disregard for personal safety. The courage and devotion to duty shown by members of the 741st Tank Battalion are worthy of emulation and reflect the highest traditions of the Army of the United States. (General Orders No. 85, War Department, 3 November 1944) *First announced in General Orders No. 42, Headquarters 1st United States Infantry Division, 19 July 1944.*

Companies A, B, and C of the 741st Tank Battalion are cited for outstanding performance of duty against the enemy from 17 to 19 December 1944. During a violent enemy counterattack in the Krinkelt-Rocherath area, Belgium, personnel inexperienced in combat such as cooks, clerks and drivers, manned deadlined tanks and with superb courage met the headlong plunge of the enemy and inflicted severe casualties upon him. With utter disregard for their personal safety, the officers and men of these three gallant companies faced devastating hostile tank, antitank and self-propelled artillery fire and fought on tenaciously against overwhelming enemy forces. Again and again the infuriated enemy threw armor and infantry against the dauntless defenders but for three days and nights these assaults were turned back by the unwavering fortitude of the inspired men. When it became necessary to withdraw to a more tenable defensive position the tankmen covered the withdrawal and were the last to leave the scene of battle. During the bitter three-day engagement, they had destroyed twenty-seven enemy tanks, five armored vehicles and two trucks. Their indomitable fighting spirit and unflinching devotion to duty are in keeping with the highest traditions of the Armed Forces. (General Orders No. 157, Headquarters Third United States Army, 1 July 1945)

Appendix C

Individual Decorations

Distinguished Service Cross

Second Lieutenant Gaetano R. Barcellona, O-1017167, Infantry, United States Army, for extraordinary heroism in action against the enemy on 6 June 1944, in France. Second Lieutenant Barcellona, commanding a tank platoon, landed with the assault force in the invasion on the coast of France. Though on landing, his tank received several hits, he continued it in operation and fearlessly exposed himself to severe enemy rifle, machine gun and artillery fire in order to select enemy targets and direct the fire of his unit. When his ammunition supply became exhausted, Second Lieutenant Barcellona, completely disregarding his own safety, dismounted from his tank and moved across the fire swept beach to secure additional ammunition. He then returned through the heavy fire to his tank and continued to carry on the fight. The valor, aggressiveness and leadership exhibited by Second Lieutenant Barcellona reflects great credit on himself and is in keeping with the highest traditions of the Armed Forces. Entered military service from Texas. (General Orders No. 37, Headquarters First United States Army, 21 July 1944) *Barcellona was born in San Antonio, Texas, on November 8, 1915. He died from lung cancer on February 8, 1993.*

First Lieutenant Joseph H. Dew (then Second Lieutenant), O-1017451, 741st Tank Battalion, United States Army, for extraordinary heroism in action against the enemy on 15 September 1944, in Germany. First Lieutenant Dew's tank platoon was supporting an infantry company and a group of engineers in reducing a defensive installation of the Siegfried Line. Intense artillery, mortar and small arms fire pinned the foot troops to the ground, preventing them from blowing a gap through the dragons teeth and retarding the advance of the tanks. Realizing the heavy defensive fire was preventing demolition attempts by the engineers, First Lieutenant Dew, with utter disregard for his own safety, climbed from his tank turret. Standing in a completely exposed position atop his tank, he sighted along the tube of the gun, assisted his gunner in aiming the weapon and, after repeated attempts, successfully blasted a path through the dragons teeth sufficient for the passage of his tanks. Although foot troops were still unable to advance due to the intensity of the hostile fire, First Lieutenant Dew heroically led his tanks through the breach and personally directed them in destroying two antitank guns and forcing the withdrawal of others. The personal bravery, tenacity of purpose and conspicuous leadership displayed by First Lieutenant Dew exemplify the highest traditions of the Armed Forces. Entered military service from Iowa. (General Orders No. 7, Headquarters First United States Army, 10 January 1945) *Dew was born on January 1, 1920 and died on July 9, 2003. Captain Charles Young recommended him for the DSC but gave an erroneous date. The action occurred on September 14, 1944.*

Sergeant George A. Habib, 36230655, Infantry, United States Army, for extraordinary heroism in action against the enemy on 6 June 1944, in France. Sergeant Habib was in command of a tank dozer in the initial assault on the coast of France with the mission of removing enemy beach obstacles. Shortly after landing, both tracks of the tank dozer were thrown. Although under heavy and sustained small arms and observed artillery fire, Sergeant Habib continually exposed himself in directing his crew in the repair of the thrown tracks. After successfully repairing the tank dozer, Sergeant Habib fearlessly directed the removal of beach obstacles. The resourcefulness, aggressive leadership and courage displayed by Sergeant Habib reflects great credit on himself and is in keeping with the highest traditions of the Armed Forces. (General Orders No. 37, Headquarters First United States Army, 21 July 1944) *Habib was born on February 7, 1914 and died on December 4, 1979.*

Lieutenant Colonel Robert N. Skaggs, O-260683, Armored Force, United States Army, for extraordinary heroism in action against the enemy on 6 June 1944, in France. Three companies of Lieutenant Colonel Skaggs' tank battalion were landed with the assault echelon in the invasion on the coast of France. Lieutenant Colonel Skaggs landed shortly thereafter, finding the assault echelon held to the fire swept beach and considerably disorganized. Lieutenant Colonel Skaggs immediately initiated steps to rally his tanks and get them into action. Constantly exposing himself to intense enemy small arms, machine gun and artillery fire, he moved along the beach organizing and directing the fire of his tanks. Notwithstanding the difficulties encountered and the constant enemy fire, he continued his efforts throughout the entire day of the assault. The personal bravery and outstanding leadership displayed by Lieutenant Colonel Skaggs reflects great credit on himself and is in keeping with the highest traditions of the Armed Forces. (General Orders No. 37, Headquarters First United States Army, 21 July 1944) *Skaggs was born on December 2, 1907 and died on November 23, 1985.*

Technician Fourth Grade Bolick Smulik, 6890110, Infantry, United States Army, for extraordinary heroism in action against the enemy on 6 June 1944, in France. Technician Fourth Grade Smulik landed with the assault force in the invasion on the coast of France. Shortly after landing, the tank dozer which he was operating was destroyed by enemy fire. Completely disregarding his own safety, Technician Fourth Grade Smulik left his destroyed tank dozer, crossed the fire swept beach and mounted an abandoned bull dozer. Though completely exposed to the heavy enemy fire, he placed the bull dozer in operation and proceeded to assist in preparing a lane through the beach obstacles which permitted the beaching of landing craft. The outstanding courage, determination and initiative of Technician Fourth Grade Smulik reflects great credit on himself and is in keeping with the highest traditions of the Armed Forces. (General Orders No. 37, Headquarters First United States Army, 21 July 1944) *Smulik died on August 1, 1944 and is interred at Knoxville National Cemetery, Tennessee. He rests in a common grave with three members of his crew who died with him.*

Private First Class Lawrence G. Sweeney, 33266510, 741st Tank Battalion, United States Army, for extraordinary heroism in action against the enemy on 6 June 1944, in France. Upon landing with the initial assault wave in the invasion of France, Private First Class Sweeney's tank received several direct hits. Courageously, he remained within the tank and fired at the enemy until his ammunition supply was exhausted. Dismounting, he fearlessly ran across the fire-swept beach, secured additional ammunition and returned to continue his devastating fire. A short while later, he was assigned the hazardous mission of reconnoitering an exit from the beach. Crawling through intense artillery, mortar and small arms fire, he reached his objective and observed that heavy fire from the area had pinned the assaulting infantrymen to the ground. With great valor, he voluntarily mounted an abandoned tank, fired at hostile strongpoints and neutralized several emplacements. On another dangerous assignment, Private First Class Sweeney, with complete disregard for his own safety, preceded several tanks in an assault and, as heavy caliber shells fell dangerously near him, coolly and calmly removed wounded men from the path of his tanks. By his great courage, selfless devotion to duty and marked valor, Private First Class Sweeney exemplified the highest traditions of the Armed Forces. Entered military service from Pennsylvania. (General Orders No. 19, Headquarters First United States Army, 31 January 1945) *Sweeney was born on January 14, 1917 and died on August 9, 1975.*

Silver Star Medal

Annunziata, Anthony E.	Cpl	C	6 Jun 44	GO 2A, 20 Jun 44, V Corps
Ball, Lloyd C.	S/Sgt	A	20 Sep 44	GO 70, 5 Nov 44, V Corps
Barner, John H.	T/4	A	16 Jun 44	GO 17, 16 Feb 45, 2nd ID
Barner, John H.	T/5	A	6 Jun 44	GO 78, 3 Dec 44, V Corps
Beetson, Clarence	Sgt	SV	6 Jun 44	GO 66, 21 Aug 44, 1st ID
Brown, Jerry B.	2/Lt	C	11 Jul 44	GO 54, 18 Jul 44, 2nd ID
Camplin, Raymond C.	Cpl	HQ	6 Jun 44	GO 66, 21 Aug 44, 1st ID
Carlson, Thomas	Sgt	A	6 Jun 44	GO 15, 29 Jun 44, V Corps
Covington, John H. Jr.	1/Lt	C	15 Jan 45	GO 23, 3 Mar 45, 2nd ID
Crouse, Gordon D.	Sgt	B	5 Apr 45	GO 78, 31 Jul 45, 8th AD
Daum, Maurice E.	T/4	HQ	6 Jun 44	GO 66, 21 Aug 44, 1st ID
Dudley, Robert L.	1/Lt	B	18 Dec 44	GO 13, 6 Feb 45, 2nd ID
Fox, Michael W.	T/4	C	17 Dec 44	GO 13, 6 Feb 45, 2nd ID
Henkelman, Leonard J.	2/Lt	C		

Hurte, Lynn E.	T/4	C	6 Jun 44	GO 2A, 20 Jun 44, V Corps
King, William M.	Capt	HQ	6 Jun 44	GO 46, 22 Aug 44, V Corps
Klotz, Frank A.	1/Lt	A	6 Jun 44	GO 66, 23 Oct 44, V Corps
Lamattina, Joseph D.	T/5	HQ	11 Jul 44	GO 78, 31 Jul 45, 8th AD
Lee, William	T/5	C	6 Jun 44	GO 66, 21 Aug 44, 1st ID
Lombardo, Charles	Pvt	C	6 Jun 44	GO 2A, 20 Jun 44, V Corps
Long, Floyd D.	Pvt	C		GO 112, 24 Sep 44, 2nd ID
Mains, Arthur W.	Pvt	C	6 Jun 44	GO 2A, 20 Jun 44, V Corps
Malpass, Paul J.	T/5	MD	14 Apr 45	GO 95, 7 Sep 45, 8th AD
Mariani, Enrique	Sgt	C	27 Jul 44	GO 78, 31 Jul 45, 8th AD
McClintock, William D.	T/Sgt	HQ	6 Jun 44	GO 21, 18 Dec 78, DA
McGowan, John R.	Cpl	C	14 Sep 44	GO 78, 31 Jul 45, 8th AD
Momany, Donald O.	T/5	HQ	6 Jun 44	GO 78, 31 Jul 45, 8th AD
O'Brien, Malcolm C.	T/Sgt	HQ	6 Jun 44	GO 78, 31 Jul 45, 8th AD
Parker, Theodore	S/Sgt	A	18 Dec 44	GO 13, 6 Feb 45, 2nd ID
Ragan, Paul W.	S/Sgt	B	6 Jun 44	GO 21, 18 Dec 78, DA
Rohrbaugh, Clyde	Cpl	A		GO 86, 2 Aug 44, 2nd ID
Sheppard, Turner G.	S/Sgt	B	6 Jun 44	GO 59, 3 Oct 44, V Corps
Skaggs, Robert N.	Lt Col	HQ	17–19 Dec 44	GO 17, 16 Feb 45, 2nd ID
Skiba, Walter J.	S/Sgt	A	6 Jun 44	GO 38, 1 Aug 44, V Corps
Sledge, Edward S. II	1/Lt	A	6 Jun 44	GO 27, 6 Jul 44, V Corps
Trimpe, Leonard H.	Sgt	SV	6 Jun 44	GO 27, 6 Jul 44, V Corps
Williams, Kenneth V.	Sgt	B	6 Jun 44	GO 78, 31 Jul 45, 8th AD
Wiltrout, Gerald E.	Sgt	D	8 Sep 44	GO 62, 3 Oct 44, 28th ID

American Battle Monuments Commission records indicate that Henkelman received the SSM, but no provenance has been found for that award.

Soldier's Medal

Kowalski, Frank A.	Pvt	HQ	4 Jul 43	GO 54, 7 Sep 43, WD
Mercer, Clyde W.	Sgt	B	6 Jun 44	GO 90, 6 Aug 44, 2nd ID

Bronze Star Medal

Aaron, Roy M.	2/Lt	6–17 Jun 44	GO 55, 19 Jul 44, 2nd ID
Adamek, Mitchell F.	Pvt	20 Dec 44	GO 12, 1 Feb 45, 2nd ID
Adametz, Joseph M.	T/5	16–20 Jul 44	GO 84, 31 Jul 44, 2nd ID
Anderson, Andrew G.	PFC	11 Jul 44	GO 71, 24 Jul 44, 2nd ID
Anderson, Charles D.	Pvt	6 Jun 44	GO 24, 2 Jul 44, V Corps
Anderson, Charles D.	Sgt	6 Jun 44 – 8 May 45	GO 77, 30 Jul 45, 8th AD
Ashby, Grover D.	Cpl	7–17 Jun 44	GO 55, 19 Jul 44, 2nd ID
Asker, Earl H.	1/Lt	6–17 Jun 44	GO 55, 19 Jul 44, 2nd ID
Ates, Barnette	M/Sgt	17 Jun 44 – 22 Jul 44	GO 84, 31 Jul 44, 2nd ID
Austin, James H.	Pvt	16–20 Jul 44	GO 84, 31 Jul 44, 2nd ID
Ayers, Edward L.	Sgt	19–20 Sep 44	GO 69, 31 Oct 44, V Corps
Azzato, Joseph	T/4	6–17 Jun 44	GO 55, 19 Jul 44, 2nd ID
Ball, Lloyd C.	Sgt	6 Jun 44	GO 24, 2 Jul 44, V Corps
Ball, Lloyd C.	S/Sgt	4 Aug 44	GO 95, 15 Aug 44, 2nd ID
Barcellona, Gaetano R.	1/Lt	18 Dec 44	GO 31, 22 Mar 45, 2nd ID
Barker, Arthur W.	Pvt	6 Jun 44	GO 24, 2 Jul 44, V Corps
Barker, Lawrence F.	Capt	23 Mar 45 – 23 Apr 45	GO 53, 9 May 45, 2nd ID
Basile, Joseph	Cpl	19–20 Sep 44	GO 69, 31 Oct 44, V Corps
Becker, Elmer W.	T/Sgt	6–17 Jun 44	GO 55, 19 Jul 44, 2nd ID
Berardinelli, Charles F.	S/Sgt	11 Jul 44	GO 71, 24 Jul 44, 2nd ID
Bigelow, Raymond W.	PFC	17 Jun 44 – 8 May 45	GO 77, 30 Jul 45, 8th AD
Blahnik, Frank W.	S/Sgt	20 Jun 44 – 1 Aug 44	GO 95, 15 Aug 44, 2nd ID
Blankenship, Clarence W.	T/5	6 Jun 44	GO 15, 29 Jun 44, V Corps
Boardman, Jack B.	Cpl	26 Jul 44 – 5 Aug 44	GO 95, 15 Aug 44, 2nd ID
Bottron, Raymond A.	PFC	17 Jun 44 – 8 May 45	GO 77, 30 Jul 45, 8th AD
Boyd, Henry D.	T/5	17 Jun 44 – 8 May 45	GO 77, 30 Jul 45, 8th AD
Bradsher, Russell M.	T/5	11 Jul 44	GO 77, 25 Jul 44, 2nd ID
Brooks, Thomas A.	1/Lt	2 Oct 44 – 8 May 45	GO 77, 30 Jul 45, 8th AD
Browder, Jack H.	Maj	17 Jun 44 – 22 Jul 44	GO 84, 31 Jul 44, 2nd ID
Brown, James E.	PFC	6 Jun 44 – 22 Jul 44	GO 84, 31 Jul 44, 2nd ID
Browning, Virgil L.	T/4	11 Jul 44	GO 78, 31 Jul 45, 8th AD

Buchanan, Charles R.	PFC	11 Jul 44	GO 71, 24 Jul 44, 2nd ID
Buckholtz, Robert C.	T/4	11 Jul 44 – 5 Aug 44	GO 95, 15 Aug 44, 2nd ID
Buster, James F.	T/5	26 Apr 45	GO 74, 11 Jun 45, 2nd ID
Butterfield, Robert J.	T/4	6 Jun 44 – 8 May 45	GO 77, 30 Jul 45, 8th AD
Cahill, John	S/Sgt	5–6 Apr 45	GO 67, 30 May 45, 2nd ID
Call, John W.	Sgt	6 Jun 44	GO 29, 9 Jul 44, V Corps
Cambron, George R.	Pvt	17–19 Dec 44	GO 12, 1 Feb 45, 2nd ID
Campbell, William W.	1/Lt	17 Dec 44	GO 74, 11 Jun 45, 2nd ID
Cansdale, Edwin J. Jr.	Pvt	19–20 Sep 44	GO 69, 31 Oct 44, V Corps
Case, Millard I. Jr.	Sgt	11 Jul 44	GO 71, 24 Jul 44, 2nd ID
Cernoch, Merton A.	Cpl	19 Dec 44	GO 58, 15 May 45, 2nd ID
Chastain, Edgar J.	T/5	6 Jun 44 – 22 Jul 44	GO 84, 31 Jul 44, 2nd ID
Cinquina, Vincent A.	1/Lt	27 Jul 44	GO 95, 15 Aug 44, 2nd ID
Cinquina, Vincent A.	Capt	17–19 Dec 44	GO 22, 1 Mar 45, 2nd ID
Clement, George J.	T/4	18 Dec 44	GO 12, 1 Feb 45, 2nd ID
Coaker, Robert D.	Sgt	6 Jun 44	GO 24, 2 Jul 44, V Corps
Cola, John F.	Sgt	26–27 Jul 44	GO 95, 15 Aug 44, 2nd ID
Combs, Hascal	T/4	17–19 Dec 44	GO 12, 1 Feb 45, 2nd ID
Compton, Robert W.	S/Sgt	6 Jun 44	GO 15, 29 Jun 44, V Corps
Compton, Robert W.	S/Sgt	21 Dec 44	GO 12, 1 Feb 45, 2nd ID
Conley, Victor G.	Sgt	6 Jun 44 – 12 Aug 44, 1 Mar – 8 May 45	GO 77, 30 Jul 45, 8th AD
Conner, Gordon R.	PFC	17–19 Dec 44	GO 12, 1 Feb 45, 2nd ID
Conrad, John J.	Sgt	7–17 Jun 44	GO 55, 19 Jul 44, 2nd ID
Correll, James A.	Sgt	11 Jul 44 – 5 Aug 44	GO 95, 15 Aug 44, 2nd ID
Covington, John H. Jr.	2/Lt	27 Jul 44	GO 95, 15 Aug 44, 2nd ID
Crisler, Norman K.	2/Lt	6 Jun 44 – 8 May 45	GO 75, 14 Jun 45, 2nd ID
Cron, Melvin S.	Sgt	17 Jan 45 – 8 May 45	GO 77, 30 Jul 45, 8th AD
Crouse, Gordon D.	Sgt	6 Jun 44 – 8 May 45	GO 79, 23 Jun 45, 2nd ID
Dallas, Everon H.	Sgt	17 Jun 44 – 8 May 45	GO 77, 30 Jul 45, 8th AD
D'Ambrose, Alphonse	Sgt	5–6 Apr 45	GO 74, 11 Jun 45, 2nd ID
Dart, John W.	T/4	6 Jun 44 – 8 May 45	GO 79, 23 Jun 45, 2nd ID
DeLiso, Anthony	T/4	19 Jun 44 – 5 Aug 44	GO 95, 15 Aug 44, 2nd ID

DeNota, Peter	PFC	11–26 Jul 44	GO 95, 15 Aug 44, 2nd ID
Detz, Joseph F.	PFC	20 Jun 44 – 1 Aug 44	GO 95, 15 Aug 44, 2nd ID
Dew, Joseph H.	1/Lt	17–19 Dec 44	GO 12, 1 Feb 45, 2nd ID
Dewitt, Leonard A.	WOJG	16–20 Jul 44	GO 84, 31 Jul 44, 2nd ID
Dickson, Ray A.	Sgt	11 Jul 44	GO 71, 24 Jul 44, 2nd ID
DiClementi, Louis J. Jr.	PFC	17 Jun 44 – 8 May 45	GO 77, 30 Jul 45, 8th AD
Dilling, Charlie G. Jr.	T/4	17 Jun 44 – 22 Jul 44	GO 84, 31 Jul 44, 2nd ID
Donohue, Lawrence J.	Sgt	11–26 Jul 44	GO 95, 15 Aug 44, 2nd ID
Dorsett, Richard J.	Cpl	17–19 Dec 44	GO 12, 1 Feb 45, 2nd ID
Downing, Joel H.	Pvt	26 Apr 45	GO 74, 11 Jun 45, 2nd ID
Duffe, William J.	PFC	17 Jun 44 – 10 Aug 44	GO 95, 15 Aug 44, 2nd ID
Dwinchick, Russell R.	PFC	5 Apr 45	GO 78, 31 Jul 45, 8th AD
Eddleblute, Lester T.	PFC	7–17 Jun 44	GO 55, 19 Jul 44, 2nd ID
Elliott, Bruce E.	Sgt	6 Jun 44 – 4 Aug 44, 29 Oct 44 – 8 May 45	GO 77, 30 Jul 45, 8th AD
Ellis, Billie H.	T/5	17 Jun 44 – 22 Jul 44	GO 84, 31 Jul 44, 2nd ID
Erwin, Troy E.	T/4	11 Jul 44	GO 90, 6 Aug 44, 2nd ID
Erwin, Troy E.	T/4	17–19 Dec 44	GO 12, 1 Feb 45, 2nd ID
Esposito, Francis G.	Sgt	17 Jun 44 – 8 May 45	GO 77, 30 Jul 45, 8th AD
Eure, James B.	Capt	19 Feb 45 – 8 May 45	GO 77, 30 Jul 45, 8th AD
Evans, Willis B.	T/5	11 Jul 44	GO 71, 24 Jul 44, 2nd ID
Everts, Isaac T.	T/5	6 Jun 44 – 10 Aug 44	GO 95, 15 Aug 44, 2nd ID
Ewald, Robert J.	T/4	11–26 Jul 44	GO 95, 15 Aug 44, 2nd ID
Fair, Thomas R.	S/Sgt	11 Jul 44	GO 77, 25 Jul 44, 2nd ID
Fannin, Ponder M.	S/Sgt	17 Jun 44 – 15 Apr 45	GO 79, 23 Jun 45, 2nd ID
Fardella, Peter J.	T/4	17 Jun 44 – 8 May 45	GO 77, 30 Jul 45, 8th AD
Ferro, Anthony A.	T/5	17 Jun 44 – 8 May 45	GO 77, 30 Jul 45, 8th AD
Fiorito, Frank J.	Pvt	17–19 Dec 44	GO 12, 1 Feb 45, 2nd ID
Fiscus, William J.	T/4	20 Jun 44	GO 78, 31 Jul 45, 8th AD
Fitzsimmons, John R.	T/5	6 Jun 44 – 8 May 45	GO 77, 30 Jul 45, 8th AD
Fodness, Arnold V.	T/4	25 Jul 44	GO 145, 15 Dec 44, 2nd ID
Fouty, Glenn A.	T/4	16–20 Jul 44	GO 84, 31 Jul 44, 2nd ID
Franks, Stanley E.	T/Sgt	11 Jul 44	GO 71, 24 Jul 44, 2nd ID

Froehlich, John	T/5	6 Jun 44	GO 24, 2 Jul 44, V Corps
Fry, Robert W.	Cpl	26 Jul 44	GO 95, 15 Aug 44, 2nd ID
Funk, Willard R.	T/5	6 Jun 44 – 8 May 45	GO 77, 30 Jul 45, 8th AD
Gagnon, Michael J.	T/5	7–17 Jun 44	GO 55, 19 Jul 44, 2nd ID
Galliher, Hanson B.	T/4	4 Feb 45	GO 31, 22 Mar 45, 2nd ID
Gaustad, Clayton H.	PFC	26 Jul 44	GO 95, 15 Aug 44, 2nd ID
Geddes, George R. Jr.	Sgt	6 Jun 44	GO 15, 29 Jun 44, V Corps
George, Willie E.	S/Sgt	19 Jun 44 – 5 Aug 44	GO 95, 15 Aug 44, 2nd ID
Gillaspie, John W.	T/4	16–20 Jul 44	GO 84, 31 Jul 44, 2nd ID
Girnius, Alexander J.	T/5	6 Jun 44 – 8 May 45	GO 77, 30 Jul 45, 8th AD
Godwin, Archie B.	PFC	17 Jun 44 – 10 Aug 44	GO 95, 15 Aug 44, 2nd ID
Goetz, Everett D.	Sgt	2 Feb 45	GO 31, 22 Mar 45, 2nd ID
Gorut, William L.	Pvt	11 Jul 44	GO 84, 31 Jul 44, 2nd ID
Greenwall, Cecil L.	T/5	19–20 Sep 44	GO 69, 31 Oct 44, V Corps
Gross, Henry Jr.	Cpl	14 Apr 45	GO 78, 31 Jul 45, 8th AD
Grossi, Antonio	S/Sgt	26 Jul 44 – 5 Aug 44	GO 95, 15 Aug 44, 2nd ID
Grzegorczyk, John S.	Pvt	11 Jul 44	GO 84, 31 Jul 44, 2nd ID
Guest, Verdie M. Jr.	T/5	18 Apr 45	GO 78, 31 Jul 45, 8th AD
Guiney, John J.	T/4	6 Jun 44 – 8 May 45	GO 77, 30 Jul 45, 8th AD
Gullatt, Harrell E.	Sgt	6 Jun 44 – 8 May 45	GO 79, 23 Jun 45, 2nd ID
Habib, George A.	S/Sgt	19 Sep 44	GO 68, 27 Oct 44, V Corps
Hall, Curtis	Cpl	17–19 Dec 44	GO 12, 1 Feb 45, 2nd ID
Hall, Joe B.	T/5	6 Jun 44 – 8 May 45	GO 77, 30 Jul 45, 8th AD
Hamilton, Pearl L.	M/Sgt	15–20 Jan 45	GO 35, 4 Apr 45, 2nd ID
Hamilton, R. C.	PFC	6 Jun 44 – 8 May 45	GO 63, 24 May 45, 2nd ID
Hecox, Donald W.	Sgt	11 Jul 44	GO 71, 24 Jul 44, 2nd ID
Helmberger, Marvin M.	T/Sgt	17–19 Dec 44	GO 12, 1 Feb 45, 2nd ID
Helmberger, Marvin M.	T/Sgt	6–17 Jun 44	GO 55, 19 Jul 44, 2nd ID
Hillner, George	Cpl	27 Jul 44	GO 95, 15 Aug 44, 2nd ID
Hixson, Donald G.	T/Sgt	16–20 Jul 44	GO 84, 31 Jul 44, 2nd ID
Hoffer, Steve J.	Sgt	6 Jun 44 – 8 May 45	GO 77, 30 Jul 45, 8th AD
Holm, Eilert H.	Cpl	6 Jun 44 – 8 May 45	GO 77, 30 Jul 45, 8th AD
Hopper, Emory T.	T/5	7–17 Jun 44	GO 55, 19 Jul 44, 2nd ID

Hopper, Wardell C. N.	Cpl	26 Jul 44 – 5 Aug 44	GO 95, 15 Aug 44, 2nd ID
Ireson, Charles R.	Pvt	11 Jul 44	GO 71, 24 Jul 44, 2nd ID
Jacobs, Edwin H.	2/Lt	26–27 Jul 44	GO 95, 15 Aug 44, 2nd ID
Jenkins, Lee O.	T/4	6 Jun 44 – 8 May 45	GO 77, 30 Jul 45, 8th AD
Johns, Gordon T.	1/Sgt	19 Jun 44 – 5 Aug 44	GO 95, 15 Aug 44, 2nd ID
Johnson, Gilbert A.	Sgt	26 Jul 44	GO 95, 15 Aug 44, 2nd ID
Johnson, Herman A.	Cpl	17–19 Dec 44	GO 12, 1 Feb 45, 2nd ID
Johnson, Vernon E.	S/Sgt	17 Jun 44 – 8 May 45	GO 77, 30 Jul 45, 8th AD
Jones, Charles W.	T/5	16–20 Jul 44	GO 84, 31 Jul 44, 2nd ID
Joven, Albert O. G.	Capt	17 Jun 44 – 5 May 45	GO 63, 24 May 45, 2nd ID
Kaminski, Larry F.	Sgt	7 Mar 45	GO 35, 4 Apr 45, 2nd ID
Kark, Michael	S/Sgt	6 Jun 44	GO 95, 15 Aug 44, 2nd ID
Kelley, Ralph G.	PFC	6 Jun 44 – 8 May 45	GO 77, 30 Jul 45, 8th AD
Kieffer, Glenn W.	T/5	19 Jun 44 – 5 Aug 44	GO 95, 15 Aug 44, 2nd ID
King, William M.	Capt	6 Jun 44 – 10 Aug 44	GO 95, 15 Aug 44, 2nd ID
Kinkade, Arthur E.	T/5	26 Jul 44 – 5 Aug 44	GO 95, 15 Aug 44, 2nd ID
Kirchhoff, Clarence C.	T/5	6 Jun 44 – 22 Jul 44	GO 84, 31 Jul 44, 2nd ID
Knapp, Harry	T/5	6–17 Jun 44, 22 Oct 44 – 8 May 45	GO 77, 30 Jul 45, 8th AD
Kovachy, George M. Jr.	1/Lt	14–16 Jun 44, 11 Nov 44 – 8 May 45	GO 77, 30 Jul 45, 8th AD
Kraft, Raleigh E.	1/Sgt	9 Jan 45 – 8 May 45	GO 79, 23 Jun 45, 2nd ID
Kroeger, Anthony E.	Cpl	6 Jun 44 – 8 May 45	GO 77, 30 Jul 45, 8th AD
Kulick, Harry	Cpl	19 Sep 44	GO 68, 27 Oct 44, V Corps
Kult, Henry P.	Sgt	19–20 Sep 44	GO 69, 31 Oct 44, V Corps
Kurkowski, Frank	Sgt	4 Aug 44	GO 95, 15 Aug 44, 2nd ID
Lance, Arthur L.	T/4	6 Jun 44 – 22 Jul 44	GO 84, 31 Jul 44, 2nd ID
Landrove, Rafael	T/4	6 Jun 44	GO 78, 31 Jul 45, 8th AD
Lanier, Lewis H.	T/4	17–19 Dec 44	GO 12, 1 Feb 45, 2nd ID
Larsen, James B.	Sgt	18 Dec 44	GO 12, 1 Feb 45, 2nd ID
Latshaw, Wilbur T.	1/Lt	17 Jun 44 – 26 Apr 45	GO 77, 30 Jul 45, 8th AD
Le Bon, Merlin D.	Cpl	6 Jun 44 – 8 May 45	GO 77, 30 Jul 45, 8th AD
Lee, William	T/5	27 Jul 44	GO 95, 15 Aug 44, 2nd ID

Levinsky, Maurice H.	T/5	11 Jul 44 – 5 Aug 44	GO 95, 15 Aug 44, 2nd ID
Lillegard, Floyd H.	T/5	5 Feb 45	GO 74, 11 Jun 45, 2nd ID
Lippman, Harold E.	Capt	20 Jun 44 – 1 Aug 44	GO 95, 15 Aug 44, 2nd ID
Lucas, Edward R.	Cpl	17 Dec 44	GO 74, 11 Jun 45, 2nd ID
Lucey, Donald P.	Sgt	11 Jul 44	GO 84, 31 Jul 44, 2nd ID
Maddock, Richard L.	Sgt	6 Jun 44	GO 80, 11 Dec 44, V Corps
Mangold, Leonard	PFC	27 Jul 44	GO 78, 31 Jul 45, 8th AD
Markby, Raymond E.	T/4	6–17 Jun 44	GO 55, 19 Jul 44, 2nd ID
Martin, Marshall	T/4	6 Jun 44 – 8 May 45	GO 77, 30 Jul 45, 8th AD
May, Willie M.	T/5	6 Jun 44 – 8 May 45	GO 77, 30 Jul 45, 8th AD
Mayo, Thomas L.	PFC	17 Dec 44	GO 74, 11 Jun 45, 2nd ID
Mazzio, Edward G.	Sgt	6 Jun 44 – 8 May 45	GO 79, 23 Jun 45, 2nd ID
McCarthy, Thomas V.	T/5	17 Jun 44 – 22 Jul 44	GO 84, 31 Jul 44, 2nd ID
McCormick, Robert A.	Cpl	26 Jul 44 – 5 Aug 44	GO 95, 15 Aug 44, 2nd ID
McCroskey, Marshall L.	Cpl	6 Jun 44	GO 36, 24 Jul 44, V Corps
McDaniel, Hurdle E. Jr.	Capt	17 Jun 44 – 8 May 45	GO 77, 30 Jul 45, 8th AD
McDonough, Roger J.	1/Lt	16 Sep 44	GO 22, 1 Mar 45, 2nd ID
McIntosh, Lovie E.	T/5	6 Jun 44 – 8 May 45	GO 77, 30 Jul 45, 8th AD
McKellar, Francis P.	T/4	17–19 Dec 44	GO 12, 1 Feb 45, 2nd ID
McKenzie, Kenneth C.	T/4	6 Jun 44 – 8 May 45	GO 77, 30 Jul 45, 8th AD
McLendon, Robert E.	1/Sgt	20 Jun 44 – 1 Aug 44	GO 95, 15 Aug 44, 2nd ID
McMullen, Robert B.	Sgt	6–17 Jun 44	GO 55, 19 Jul 44, 2nd ID
Merkert, William S.	T/4	6 Jun 44 – 8 May 45	GO 77, 30 Jul 45, 8th AD
Metzger, Donald R.	Sgt	31 Jan 45	GO 78, 31 Jul 45, 8th AD
Meyers, Calvin B.	2/Lt	11 Jul 44	GO 71, 24 Jul 44, 2nd ID
Michael, Virgil	Cpl	6 Jun 44 – 18 Dec 44, 30 Jan 45 – 8 May 45	GO 77, 30 Jul 45, 8th AD
Middleton, Elmer	Sgt	11 Jul 44	GO 71, 24 Jul 44, 2nd ID
Morgando, Tony	T/4	2 Feb 45	GO 78, 31 Jul 45, 8th AD
Morton, Lytle C.	T/4	6 Jun 44 – 22 Jul 44	GO 84, 31 Jul 44, 2nd ID
Mortzfield, Wilbert C.	Sgt	11 Jul 44	GO 71, 24 Jul 44, 2nd ID
Muncy, Paul D.	Sgt	17 Jun 44 – 7 May 45	GO 82, 25 Jun 45, 2nd ID
Myers, Allen V.	T/5	6 Apr 45	GO 74, 11 Jun 45, 2nd ID

Myers, William J.	T/5	7–13 Jul 44	GO 84, 31 Jul 44, 2nd ID
Myler, John R.	1/Sgt	17 Jun 44 – 5 Aug 44	GO 95, 15 Aug 44, 2nd ID
Neidrich, Frederick A.	Sgt	17–19 Dec 44	GO 12, 1 Feb 45, 2nd ID
Neill, David	Pvt	8 Apr 45	GO 67, 30 May 45, 2nd ID
Nicol, Robert A.	S/Sgt	6 Jun 44	GO 15, 29 Jun 44, V Corps
Nixon, Willis E.	T/4	11 Jul 44	GO 77, 25 Jul 44, 2nd ID
Nobles, Roy C.	T/4	11–26 Jul 44	GO 95, 15 Aug 44, 2nd ID
Nolan, James D.	T/4	16 Jun 44 – 8 May 45	GO 71, 6 Jun 45, 2nd ID
Norton, John C.	T/5	6 Jun 44 – 22 Jul 44	GO 84, 31 Jul 44, 2nd ID
Oette, Emil A.	T/4	27 Mar 45	GO 67, 30 May 45, 2nd ID
O'Hara, Edward	T/5	19 Jun 44 – 5 Aug 44	GO 95, 15 Aug 44, 2nd ID
O'Herron, Lawrence J. Jr.	Cpl	6 Jun 44 – 8 May 45	GO 77, 30 Jul 45, 8th AD
O'Shaughnessy, Patrick J.	2/Lt	18 Jun 44 – 5 Aug 44	GO 95, 15 Aug 44, 2nd ID
Padgett, Ernest O.	S/Sgt	17 Dec 44	GO 78, 31 Jul 45, 8th AD
Padgett, Robert B.	T/4	17 Jun 44 – 8 May 45	GO 77, 30 Jul 45, 8th AD
Pappas, Pete	T/5	17 Jun 44 – 1 Aug 44	GO 95, 15 Aug 44, 2nd ID
Pardini, Robert J.	T/4	11 Jul 44	GO 87, 3 Aug 44, 2nd ID
Park, William E.	1/Lt	17 Jun 44 – 10 Aug 44	GO 95, 15 Aug 44, 2nd ID
Parker, Elton L.	Sgt	17 Jun 44 – 21 Apr 45	GO 82, 25 Jun 45, 2nd ID
Parker, Theodore	Sgt	19–20 Sep 44	GO 69, 31 Oct 44, V Corps
Pasqualini, Frank	Cpl	11 Jul 44	GO 77, 25 Jul 44, 2nd ID
Payne, Cecil J.	1/Lt	17 Jun 44 – 31 Mar 45	GO 42, 15 Apr 45, 2nd ID
Pepe, Ernest R.	Pvt	27 Jul 44	GO 95, 15 Aug 44, 2nd ID
Perkins, Raymond A.	Pvt	11 Jul 44	GO 77, 25 Jul 44, 2nd ID
Petoskey, Roland M.	PFC	6 Jun 44 – 22 Jul 44	GO 84, 31 Jul 44, 2nd ID
Philip, Andy	T/4	16 Jun 44 – 8 May 45	GO 79, 23 Jun 45, 2nd ID
Phillips, Howard V.	Sgt	17 Jun 44 – 8 May 45	GO 77, 30 Jul 45, 8th AD
Pollard, Kenneth E.	Sgt	17 Jun 44 – 7 May 45	GO 79, 23 Jun 45, 2nd ID
Powell, Kenneth V.	T/5	19 Jun 44 – 8 May 45	GO 77, 30 Jul 45, 8th AD
Prager, Herbert W.	PFC	26 Jul 44	GO 104, 12 Sep 44, 2nd ID
Ragan, Paul W.	S/Sgt	11 Jul 44	GO 71, 24 Jul 44, 2nd ID
Ragan, Paul W.	S/Sgt	26 Jul 44	GO 95, 15 Aug 44, 2nd ID
Rains, Lee C.	Pvt	17–19 Dec 44	GO 12, 1 Feb 45, 2nd ID

Ray, Herman G.	Cpl	5 Apr 45	GO 78, 31 Jul 45, 8th AD
Ray, Jack	T/5	27 Jul 44	GO 95, 15 Aug 44, 2nd ID
Resar, John E.	T/4	6 Jun 44 – 8 May 45	GO 77, 30 Jul 45, 8th AD
Reynolds, Malcolm A.	2/Lt	6 Jun 44 – 12 May 45	GO 67, 30 May 45, 2nd ID
Rifkin, Albert	T/5	1 Aug 44 – 1 Apr 45	GO 46, 22 Apr 45, 2nd ID
Riley, Howard E.	Cpl	18 Dec 44	GO 78, 31 Jul 45, 8th AD
Ristau, George J.	T/4	17 Jun 44 – 8 May 45	GO 77, 30 Jul 45, 8th AD
Rivera, Robert O.	Sgt	17 Jun 44 – 7 May 45	GO 79, 23 Jun 45, 2nd ID
Rivers, Henry F.	2/Lt	17 Jun 44 – 12 Dec 44, 5 Feb 45 – 8 May 45	GO 77, 30 Jul 45, 8th AD
Rowe, Howard C.	Pvt	17–19 Dec 44	GO 12, 1 Feb 45, 2nd ID
Rudhman, Carl A.	T/5	6 Jun 44 – 22 Jul 44	GO 84, 31 Jul 44, 2nd ID
Rusnak, George	T/4	11 Jul 44 – 5 Aug 44	GO 95, 15 Aug 44, 2nd ID
Santoro, Salvatore W.	S/Sgt	27 Sep 44	GO 78, 31 Jul 45, 8th AD
Schuh, Shirley R.	T/5	17–19 Dec 44	GO 12, 1 Feb 45, 2nd ID
Schuh, Shirley R.	T/5	20 Jun 44 – 1 Aug 44	GO 95, 15 Aug 44, 2nd ID
Semmont, William J.	T/5	6–17 Jun 44	GO 55, 19 Jul 44, 2nd ID
Sergent, Carl	Pvt	6 Jun 44	GO 24, 2 Jul 44, V Corps
Shapiro, Arthur	T/4	17–19 Dec 44	GO 12, 1 Feb 45, 2nd ID
Sharpless, Howard F.	1/Lt	6–17 Jun 44	GO 55, 19 Jul 44, 2nd ID
Short, Estle I.	Sgt	6 Feb 45	GO 31, 22 Mar 45, 2nd ID
Simon, Walter	S/Sgt	17 Jun 44 – 7 May 45	GO 79, 23 Jun 45, 2nd ID
Simoni, Lillio	Sgt	17 Jun 44 – 10 Apr 45	GO 48, 24 Apr 45, 2nd ID
Skaggs, Robert N.	Lt Col	13–21 Dec 44	GO 20, 20 Feb 45, 2nd ID
Sledge, Edward S. II	Capt	15 Dec 44	GO 12, 1 Feb 45, 2nd ID
Smith, George M.	Cpl	6 Jun 44 – 8 May 45	GO 77, 30 Jul 45, 8th AD
Smith, Lyle W.	T/5	16–20 Jul 44	GO 84, 31 Jul 44, 2nd ID
Smith, William	T/Sgt	17 Jun 44 – 8 May 45	GO 77, 30 Jul 45, 8th AD
Smulick, Bolick	T/4	1 Aug 44	GO 95, 15 Aug 44, 2nd ID
Snike, Alfred	Cpl	6 Jun 44 – 8 May 45	GO 77, 30 Jul 45, 8th AD
Snyder, Thomas N.	Capt	17 Jun 44 – 8 May 45	GO 77, 30 Jul 45, 8th AD
Sorensen, Stuart H.	T/5	16–20 Jul 44	GO 84, 31 Jul 44, 2nd ID
Steineker, Alfred W. Jr.	2/Lt	20 Sep 44	GO 68, 27 Oct 44, V Corps

Sterling, June P.	PFC	6 Jun 44 – 8 May 45	GO 77, 30 Jul 45, 8th AD
Sterrett, John D.	Capt	6–17 Jun 44	GO 55, 19 Jul 44, 2nd ID
Stokan, Thomas	T/Sgt	7–17 Jun 44	GO 55, 19 Jul 44, 2nd ID
Stookey, Gerald E.	T/4	15 Apr 45	GO 74, 11 Jun 45, 2nd ID
Suydam, John G.	S/Sgt	6 Jun 44 – 8 May 45	GO 79, 23 Jun 45, 2nd ID
Svrcek, Martin	PFC	20 Jun 44 – 1 Aug 44	GO 95, 15 Aug 44, 2nd ID
Swander, Virgil L.	Cpl	11–26 Jul 44	GO 95, 15 Aug 44, 2nd ID
Sweeney, Lawrence G.	PFC	17 Jun 44	GO 95, 15 Aug 44, 2nd ID
Teffeteller, Edward	T/5	6 Jun 44 – 8 May 45	GO 77, 30 Jul 45, 8th AD
Thiele, Edward J.	Pvt	19–20 Sep 44	GO 69, 31 Oct 44, V Corps
Thomas, Cecil D.	Capt	6–17 Jun 44	GO 55, 19 Jul 44, 2nd ID
Thomas, Francis G.	T/4	6 Jun 44 – 8 May 45	GO 77, 30 Jul 45, 8th AD
Thornton, James G. Jr.	Capt	26 Jul 44	GO 95, 15 Aug 44, 2nd ID
Thornton, James G. Jr.	Capt	6 Jun 44 – 9 Jul 44	GO 55, 19 Jul 44, 2nd ID
Vallario, Oreste V.	PFC	17 Dec 44	GO 12, 1 Feb 45, 2nd ID
Vance, Russell W.	1/Sgt	16 Jun 44 – 25 Oct 44, 22 Dec 44 – 8 May 45	GO 77, 30 Jul 45, 8th AD
Vizacaro, Thomas J.	Sgt	17 Jun 44 – 8 May 45	GO 77, 30 Jul 45, 8th AD
Voich, Michael A.	T/4	6 Jun 44 – 8 May 45	GO 79, 23 Jun 45, 2nd ID
Wahler, William J.	Capt	15–20 Jan 45	GO 31, 22 Mar 45, 2nd ID
Walker, George H.	T/5	17 Jun 44 – 8 May 45	GO 77, 30 Jul 45, 8th AD
Walker, Sam E.	T/4	11 Jul 44 – 5 Aug 44	GO 95, 15 Aug 44, 2nd ID
Walker, Sam E.	T/4	17–19 Dec 44	GO 12, 1 Feb 45, 2nd ID
Walsdorf, Raymond J.	T/4	6–17 Jun 44	GO 55, 19 Jul 44, 2nd ID
Walsh, Joseph T.	Cpl	26 Jul 44 – 5 Aug 44	GO 95, 15 Aug 44, 2nd ID
Warren, Willis D.	Sgt	11 Jul 44	GO 71, 24 Jul 44, 2nd ID
Warren, Willis D.	Sgt	26 Jul 44	GO 95, 15 Aug 44, 2nd ID
Washut, Henry J.	S/Sgt	6–17 Jun 44	GO 55, 19 Jul 44, 2nd ID
Welch, John W.	S/Sgt	7 Jun 44 – 8 May 45	GO 79, 23 Jun 45, 2nd ID
White, Leonard B.	Sgt	17 Dec 44	GO 78, 31 Jul 45, 8th AD
White, Uleyse G.	Cpl	17–19 Dec 44	GO 12, 1 Feb 45, 2nd ID
Wilborn, Billie M.	PFC	17 Dec 44	GO 74, 11 Jun 45, 2nd ID
Wilson, George R.	T/4	16–20 Jul 44	GO 84, 31 Jul 44, 2nd ID

Wilson, George W.	2/Lt	11 Jul 44	GO 71, 24 Jul 44, 2nd ID
Winzenreid, Ernest C.	T/4	17 Jun 44 – 8 May 45	GO 77, 30 Jul 45, 8th AD
Wisecup, George A.	T/Sgt	15–20 Jan 45	GO 79, 23 Jun 45, 2nd ID
Wolfe, Robert E.	T/4	20 Jun 44 – 1 Aug 44	GO 95, 15 Aug 44, 2nd ID
Womack, Walter H.	PFC	6 Jun 44 – 22 Jul 44	GO 84, 31 Jul 44, 2nd ID
Woodward, Ralph A.	T/4	6 Jun 44 – 8 May 45	GO 77, 30 Jul 45, 8th AD
Young, Charles R.	Capt	15 Sep 44	GO 74, 17 Nov 44, V Corps
Young, Charles R.	Capt	15 Jun 44 – 15 Jul 44	GO 59, 20 Jul 44, 2nd ID
Young, Charles R.	Capt	17–19 Dec 44	GO 22, 1 Mar 45, 2nd ID
Youngblood, Mayne B.	Sgt	20 Sep 44	GO 70, 5 Nov 44, V Corps
Youngblood, Mayne B.	Sgt	7 Mar 45	GO 78, 31 Jul 45, 8th AD

All ranks reflect the soldier's grade at the time of his award. The BSM roster includes those who received the medal for "heroic achievement" and those who received it for "meritorious service." The latter awards usually encompass a broad date range, e.g., 6 Jun 44 – 8 May 45. The names in the BSM roster are not included in this book's index.

Appendix D

Prisoners of War

The place of internment listed for each prisoner came from records of the Prisoner of War Information Bureau, Office of the Provost Marshal, US Army. These records reside at the National Archives (RG 389).

Ayers, Edward L.	Sgt	A	19 Dec 44*	Stalag IID, Stargard
Belcher, John H.	1/Lt	D	22 Sep 44	Stalag IIIB, Fürstenberg
Calderaro, Daniel L.	Pvt	D	22 Sep 44	Stalag VIIA, Moosburg
Caughey, James P.	Pvt	D	22 Sep 44	Stalag XIID, Waldbreitbach
Chastain, Thomas D.	Pvt	A	17 Jun 44	Stalag XVIIB, Gneixendorf
Coffin, James P.	Pvt	B	18 Dec 44	Stalag IVB, Mühlberg
Crain, Von	Pvt	D	30 Sep 44	Stalag IIA, Neubrandenburg
Durrell, Robert E.	Pvt	D	30 Sep 44	Stalag XIIA, Limburg
Gross, Henry Jr.	Cpl	B	14 Apr 45	**
Gullatt, Harrell E.	Cpl	B	14 Apr 45	**
Hockenberry, Neal F.	Pvt	B	18 Dec 44	Stalag XIB, Fallingbostel
Howard, Clayton A.	T/5	B	18 Dec 44	Stalag XIB, Fallingbostel
Kammeyer, Albert H.	Sgt	B	14 Apr 45	**
Kassel, Gilbert R.	Pvt	D	30 Sep 44	Stalag IIB, Hammerstein
Klotz, Frank A.	1/Lt	A	29 Apr 45	Stalag IVC, Wistritz bei Teplitz
McMullen, Robert B.	Sgt	SV	17 Jun 44	Frontstalag 221, Rennes
Roach, Ernest B.	T/5	D	30 Sep 44	Stalag IIIC, Alt-Drewitz
Stover, Robert F.	Sgt	D	22 Sep 44	Stalag IIIC, Alt-Drewitz
Teffeteller, Edward	T/5	A	29 Apr 45	Stalag IVC, Wistritz bei Teplitz
Walsh, Joseph T.	Cpl	B	18 Dec 44	Stalag IID, Stargard
Williams, Elmer F.	Sgt	D	22 Sep 44	Stalag IIIC, Alt-Drewitz

* actual capture date may have been one or two days earlier
** held captive for ten hours at Hill 102 near Merseburg, Germany

Appendix E

Commanding Officers and Staff Officers
1 January 1944 – 27 October 1945

1 Jan 44 is the start date for this roster, not the date any of the following officers assumed their command or staff position. All subsequent changes were announced in "special orders" published by the battalion.

Battalion Commander	
Lt Col Jacob R. Moon - Detached service 3rd Armd Gp, 24 Dec 43 – 5 Jan 44 - Sick, to 2nd Gen Hosp, 11 Jan 44 - Dropped from assignment, 19 Feb 44	1 Jan 44 – 11 Jan 44
Maj (later Lt Col) Robert N. Skaggs	11 Jan 44 – 28 May 45
Maj Jack H. Browder	28 May 45 – Jul 45
Maj Cecil D. Thomas	1 Jul 45 – 3 Jul 45
Lt Col Preston R. Bishop	3 Jul 45 – 2 Sep 45
Maj Jack H. Browder	2 Sep 45 – 27 Oct 45
Executive Officer	
Maj Robert N. Skaggs - Acting battalion commander, 24 Dec 43 – 5 Jan 44	1 Jan 44 – 11 Jan 44
Capt (later Maj) Jack H. Browder	27 Feb 44 – 28 May 45
Maj Cecil D. Thomas	29 May 45 – 1 Jul 45
Maj Jack H. Browder - Temporary duty Paris, 1–11 Jul 45	11 Jul 45 – 2 Sep 45
Maj Cecil D. Thomas	2 Sep 45 – 27 Oct 45
Adjutant & Personnel Officer (S-1)	
1/Lt Ralph E. Lundquist	1 Jan 44 – 3 Apr 44
1/Lt William E. Park	3 Apr 44 – 29 May 45
1/Lt George M. Kovachy Jr.	29 May 45 – 1 Jul 45
1/Lt William E. Park - Simultaneously held S-2 position	1 Jul 45 – 19 Jul 45
1/Lt George M. Kovachy Jr.	19 Jul 45 – 27 Oct 45

Intelligence Officer (S-2)	
Capt Jack W. Duke	1 Jan 44 – 16 Feb 45
Capt James B. Eure	16 Feb 45 – 29 May 45
1/Lt William E. Park - Nobody replaced Park as S-2	29 May 45 – 4 Aug 45
Operations Officer (S-3)	
Capt Jack H. Browder	1 Jan 44 – 27 Feb 44
Capt (later Maj) William M. King	27 Feb 44 – 25 Sep 44
Capt (later Maj) Cecil D. Thomas	16 Oct 44 – 29 May 45
Capt Hurdle E. McDaniel Jr.	29 May 45 – 1 Jul 45
Maj Cecil D. Thomas	3 Jul 45 – 2 Sep 45
Maj Shepard J. Freed - S-3 position vacant until Freed joined Bn HQ	27 Sep 45 – 27 Oct 45
Supply Officer (S-4)	
1/Lt (later Capt) Albert O. G. Joven	1 Jan 44 – 27 Oct 45
Medical Detachment	
Capt Clarence J. Kocovsky	1 Jan 44 – 22 Jan 44
Capt Michael R. Godett	22 Jan 44 – 26 Mar 45
Capt Lawrence F. Barker	26 Mar 45 – 16 Jul 45
1/Lt William W. Campbell - Campbell returned from 191st Gen Hosp, 19 Jul 45	19 Jul 45 – 27 Oct 45
Headquarters Company	
Capt Hurdle E. McDaniel Jr.	1 Jan 44 – 24 Jan 44
1/Lt Earl H. Asker	24 Jan 44 – 25 Feb 44
Capt Hurdle E. McDaniel Jr.	25 Feb 44 – 29 May 45
Capt James B. Eure	29 May 45 – 27 Oct 45
Service Company	
Capt Douglas G. Mahon Jr.	1 Jan 44 – 27 Apr 45

1/Lt Wilbur T. Latshaw	27 Apr 45 – 20 Aug 45
1/Lt Walter R. Silbaugh	20 Aug 45 – 27 Oct 45
Company A	
Capt Cecil D. Thomas	1 Jan 44 – 16 Oct 44
Capt Edward S. Sledge II	16 Oct 44 – 7 Mar 45
1/Lt (later Capt) Roger J. McDonough	7 Mar 45 – 29 Apr 45
1/Lt Thomas A. Brooks	30 Apr 45 – 9 May 45
Capt Edward S. Sledge II	9 May 45 – 28 Jul 45
1/Lt Norbert F. Krob	28 Jul 45 – 20 Aug 45
2/Lt Donald W. Hecox	20 Aug 45 – 3 Oct 45
Capt Daniel C. Patterson	4 Oct 45 – 27 Oct 45
Company B	
Capt James G. Thornton Jr.	1 Jan 44 – 14 Sep 44
1/Lt (later Capt) Vincent A. Cinquina	14 Sep 44 – 20 Aug 45
2/Lt Turner G. Sheppard	20 Aug 45 – 1 Sep 45
Capt William J. Wahler	1 Sep 45 – 27 Oct 45
Company C	
Capt William M. King	1 Jan 44 – 23 Jan 44
1/Lt Thomas N. Snyder	23 Jan 44 – 26 Jan 44
Capt William M. King	26 Jan 44 – 27 Feb 44
1st Lt (later Capt) Charles R. Young	27 Feb 44 – 1 Feb 45
1/Lt (later Capt) John H. Covington Jr.	1 Feb 45 – 27 Oct 45
Company D	
1/Lt (later Capt) John F. Sicks	1 Jan 44 – 7 Aug 44
1/Lt (later Capt) Edward S. Sledge II	7 Aug 44 – 16 Oct 44
1/Lt (later Capt) Thomas N. Snyder	16 Oct 44 – 27 Oct 45

Appendix F

Postwar Rebirth

In February 1949, the Department of the Army redesignated the 741st Tank Battalion as the 396th Heavy Tank Battalion and allotted it to the Organized Reserve Corps. Activated the following month as a reserve unit, the battalion made its headquarters in Spokane, Washington, and joined the 96th Infantry Division, another reserve unit. Lieutenant Colonel Fulton G. Gale Jr. (1918–1996), a Spokane resident, was the battalion's first commander. His unit consisted of four companies: Headquarters & Service Company, Company A, Company B, and Company C. The men in Gale's battalion wore the polygon-shaped insignia of the 96th Division but not the distinctive insignia (*Strenue et Audacter*) worn by the 741st during the Second World War. The 396th moved its headquarters to Seattle in 1950 and remained there until redesignated as the 741st Tank Battalion and simultaneously inactivated in March 1952. It then became a paper unit assigned to the Army Reserve until 1953, when the phantom 741st joined the 11th Armored Division, a Regular Army formation that also existed only on paper. No activation ever occurred.

During its three-year existence, the 396th spent fourteen days on active duty. Those occurred during a summer training encampment held by the 96th Division in July 1951 at Camp W.G. Williams, Utah. At that time, the battalion commander was Lieutenant Colonel Edwin J. Beamer (1911–1983), a Seattle-area resident and veteran of the 1st Armored Division.

Appendix G

Coat of Arms

SPX 314.73 741st Tank Bn.
(9-10-42) LO

Coat of Arms for the
741st Tank Battalion (M)

September 10, 1942.

The Quartermaster General.

1. The following coat of arms for the 741st Tank Battalion (M), authorized under the provisions of paragraph 5, AR 260-10, November 20, 1931, is approved:

<u>SHIELD</u>: Per fess vert and argent, two swords in saltire, points in chief between as many griffin's heads erased palewise, all counter-changed.

<u>CREST</u>: None

<u>MOTTO</u>: Strenue et audacter (Strenuously and bravely).

The 741st Tank Battalion (M) was constituted 1 March, 1942, and made active at Fort George G. Meade, Maryland, 15 March 1942, pursuant to letter from the War Department, A.G. 320.2 (2-20-42), dated 1 March, 1942.

In the green and silver of the Armored Forces the griffin with its scaly armor sets forth the property of a valorous soldier whose magnamity is such that he will dare all dangers, even death itself rather than become captive. It is also a symbol of vigilancy. The two swords are representative of the military aspect of their functions.

The suggested motto: "Strenue et audacter" (Strenously and bravely) is expressive of the strenous and brave manner in which the personnel undertake its allotted tasks, and is being tentatively reserved for the exclusive use of the organization.

2. The above blazon and description are sent to you with instructions to have a drawing made indicating the tinctures, and to furnish the organization commander with a painting of the coat of arms.

By Command of Lieutenant General SOMERVELL:

/stamp/ J.S. Richards
Adjutant General

<u>"A TRUE COPY"</u>

Jack W. Duke

JACK W. DUKE
Captain, 741st Tank Battalion (M)
Adjutant

Distinctive Insignia

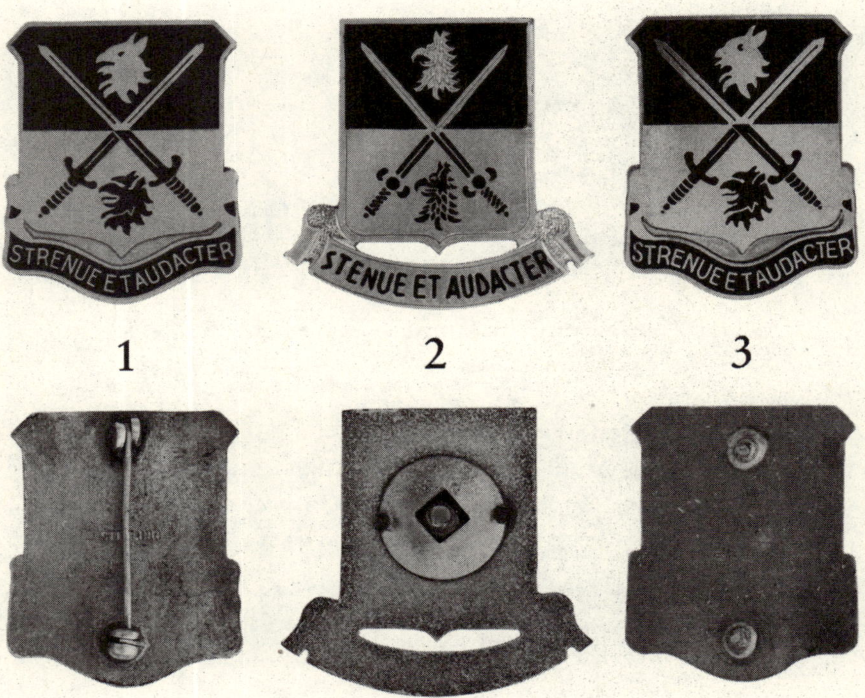

1 — Wartime US made; probably sold by A.H. Dondero, Inc., Wash. DC; no hallmark, incised "STERLING" on reverse; ASMIC number 741A1.

2 — Postwar "fantasy" insignia made in Austria; sold by Fessel B. Koepnick (1916-2004) of Houston, Texas; no hallmark, motto enameled in blue; not shown above: flat knurled 14 mm nut made of brass; ASMIC number Z741A1.

3 — Postwar US made copy; sold by Ernest E. Bora (1915-1984) of Brooklyn, New York; no hallmark; chrome finish; not shown above: two Ballou clutch fasteners; no ASMIC number.

ASMIC = American Society of Military Insignia Collectors.

Appendix H

Tanks Available, December 21, 1944

Document received by G-3, 2nd Infantry Division, at 00:20 hours, December 21, 1944 (Journal entry 9872). The number of Charley Company "tanks lost thru enemy action" was actually two, not three. Declassified by authority NND 735017 (NARA).

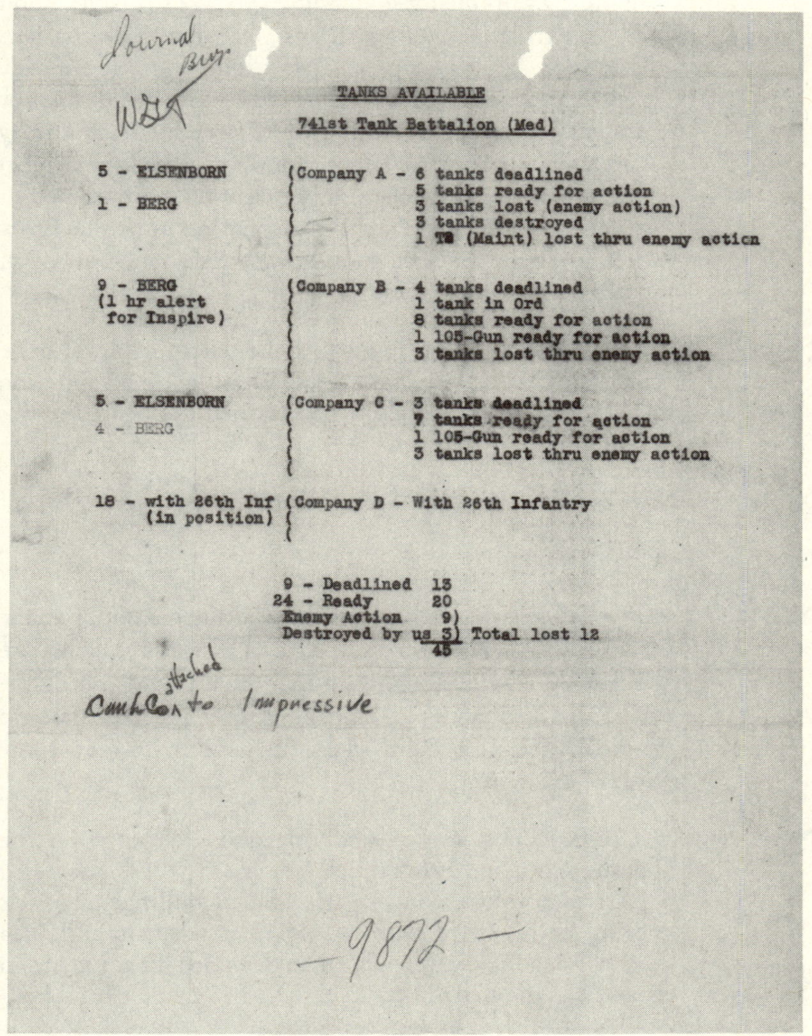

Journal Bur.

WGA

TANKS AVAILABLE

741st Tank Battalion (Med)

5 - ELSENBORN (Company A - 6 tanks deadlined
1 - BERG
5 tanks ready for action
3 tanks lost (enemy action)
3 tanks destroyed
1 T2 (Maint) lost thru enemy action

9 - BERG (Company B - 4 tanks deadlined
(1 hr alert for Inspire)
1 tank in Ord
8 tanks ready for action
1 105-Gun ready for action
3 tanks lost thru enemy action

5 - ELSENBORN (Company C - 3 tanks deadlined
4 - BERG
7 tanks ready for action
1 105-Gun ready for action
3 tanks lost thru enemy action

18 - with 26th Inf (in position) (Company D - With 26th Infantry

9 - Deadlined 13
24 - Ready 20
Enemy Action 9)
Destroyed by us 3) Total lost 12
45

Comb Co attached to Impressive

—9872—

Appendix I

Krinkelt-Rocherath Property Owners (1940–1944)

The following house numbers appear on maps in this book:

Krinkelt

Nr. 1 — Adolf Stoffels, a tanner by trade, sometimes self-employed but mostly working for Josef Kalpers (Nr. 92 Krinkelt). Adolf and his two sons, Peter and Klaus, were notorious poachers along with Adolf's brother Peter. Adolf committed suicide in 1955, and his son Peter died in 1956 due to an automobile accident. The rest of the family moved away, and the house lay abandoned for decades. The author often visited the derelict structure during the 1980s. Its brick walls bore numerous battle scars. The house remained standing until the 1990s, when it was demolished to make way for a new home that stands on the property today.

Nr. 37 — Sankt Johannes der Täufer Kirche, neo-Gothic church, cornerstone laid 1905, church dedicated 1907. Its demolition began in 1948 and ended in 1951, when the belltower was razed.

Nr. 38 — Fire Station, known as "Spitzenhaus."

Nr. 39 — Nikolaus Elsen.

Nr. 41a — Aloys Stoffels owned and operated a butcher shop on the premises, which opened in 1932. Destroyed during the "Bulge," Aloys rebuilt the shop after the war, and, during the 1970s, his nephew Felix Palm took over the business and named it Metzgerei Palm. The shop is still in business at the time of this writing.

Nr. 42 — Friedrich Rauw owned and operated a grocery on the premises. His son Franz (born January 15, 1917) was killed by German soldiers on May 10, 1940 at Büllingen. He was fleeing in his truck with a Belgian policeman, and, at a German roadblock, the policeman defended his country, and both Belgians died.

Nr. 87 Bernard Vilz.

Nr. 88 Peter Stoffels, brother of Adolf Stoffels. House served as CP for Recon Company, 644th TD Battalion, code-named "Hazard Roger" (CO, 1st Lt Harold L. Hoffer).

Nr. 89 Josef Palm.

Nr. 90 Robert Reuter's widow.

Nr. 91 Martin Küches (1905–1979) owned and operated a small grocery and butcher shop on the premises until 1934, when fire gutted the building. Martin rebuilt the place as a Gasthaus, and he also ran a small farm with several cows. On October 6, 1944, Martin relocated to Malmédy along with his wife and six children. The pyramidal roof of his house was (and still is) unique in Krinkelt-Rocherath.

Nr. 92 Nicolaus Balthasar Kalpers built the house in 1925–1927. As a successful businessman, he owned a tannery adjacent to the house. Nicolaus was the first mayor of Rocherath after its incorporation into Belgium following the First World War. He died in 1935, and his wife, Bertha, died in 1941. Their eldest son, Joseph, and his wife, Caroline, lived in the house during the war. With their two children, they evacuated to Malmédy on October 7, 1944.

Rocherath

Nr. 4 Ewald Henzen.

Nr. 5 Elisabeth Klinkhammer was the widow of Joseph Johann who died in 1928 after an automobile accident, leaving her with five children. Her son Franz died in Russia on May 16, 1943.

Nr. 53 Johann Rauw's house and barn contiguous with Nr. 54. Building totally demolished during the battle and never rebuilt. Two new homes stand today on the land where 53 and 54 once stood. Johann is not to be confused with the Johann Rauw who owned Nr. 61.

Nr. 54 Johann Vassen's house and barn were contiguous with Nr. 53. Building demolished during the battle and never rebuilt. On December 19, 1944, Second Platoon, Company A, 644th TD Battalion lost an M10 tank destroyer on the property occupied by 53 and 54. The gun commander, Sgt William E. Hughes, died along with Pvt Ignatz F. Bieniek. Three other crewmen were wounded: Cpl John Karpiak, T/5 Charlie Lee, Pvt Joseph W. Garpetti.

Nr. 60 Richard Palm was the husband of Thekla Palm, the woman who waved a white cloth at SS-Unterscharführer Heinz Freiberg on December 18, 1944.

Nr. 61 Johann Rauw (Fickers).

Nr. 62 Joseph Jansen's house sustained minor damage, and he repaired it after the war. One of his grandchildren demolished the house and built a new home on the property in 2015. Building served as CP for Antitank Company, 38th Infantry Regiment, December 17-18, 1944 (CO, Capt James W. Love).

Nr. 63 Maria Radermacher's house was destroyed during the battle and rebuilt after the war. During the battle, the building served as CP for Company C, 38th Infantry Regiment (CO, Capt Edward C. Rollings). Soldiers from AT/38 also garrisoned the house. After the battle, the house became Forward CP for 953rd Field Artillery Battalion.

Nr. 71 Albert and Franziska Palm, owners of the farm at Lausdell.

Nr. 74 Michael Schröder and Margaretha Jansen. Their son Aloys died in Russia on July 22, 1943. His twin, Arnold, a combat engineer in the German Army, surrendered near Rocherath in January 1945 to US soldiers. Paul Drösch claimed that Arnold was not a POW, and that he volunteered to clear minefields around Rocherath-Krinkelt. On a road shoulder between Krinkelt and Wirtzfeld, he died on April 14, 1945, while deactivating a mine.

Nr. 75 Alex Kreutz.

Nr. 76 Johann Steffens.

Nr. 77	Wilhelm Drösch, house and barn severely damaged on November 27, 1944 by V-1 flying bomb.
Nr. 80	Police station, a government-owned building originally built as a private residence. From October 25 to December 17, 1944, the house served as the CP for Battery B, 17th Field Artillery Observation Battalion, code named "Violet Baker." Immediately afterward, December 18–19, the house became the CP for I.Abteilung, SS-Panzer-Regiment 12 (Arnold Jürgensen). After the war, the house remained a police station until 1958, when the Belgian forestry service acquired the building. Albert Bettendorff was the first forester to reside there, and his successors have lived there ever since.
Nr. 90	Barbara Andres-Behrens was the widow of Mathias Behrens. Home occupied in 1944 by Maria Jost, a relative of Barbara.
Nr. 91	Bernard Pfeiffer.
Nr. 92	Mathias Schumacher (son-in-law of Nicolaus Kalpers) died of disease in February 1945 at Blumenthal, Germany.
Nr. 93	Mathias Brüls.
Nr. 94	Joseph Josten, 1944 tenant was Arnold Strong and his family, evacuees from Cologne, Germany.
Nr. 95	Mathias Faymonville was the mayor of Rocherath during the German occupation. He and his family fled to Germany with their possessions when U.S. troops arrived in September 1944. The family returned in 1946.
Nr. 96	Hubert Melchior, house named "Marxen."
Nr. 97	August Hönen, cousin of Mathias Hönen. House and barn contiguous with Nr. 98. The two properties were collectively known as "Lexandesch."
Nr. 98	Leonard Josten. House and barn contiguous with Nr. 97.

Nr. 99 Julius Rauw (1896–1944) remained in Rocherath after the civilian evacuation of the twin villages in October 1944. Julius and about twenty other people stayed behind to tend cattle. On the evening of December 17, while en route to visit his neighbor August Hönen, an American soldier mistakenly shot and killed Julius, probably thinking he was a German soldier. The house was named "Schwitze," most likely the surname of the family who built it during the eighteenth century. The building survived the war with only minor damage only to be demolished in 1992.

Nr. 100 Mathias Hönen remained in Rocherath to tend cattle after US troops arrived in September 1944, His house was named "Vorne Kättringen" (Nr. 90 Rocherath prior to 1940). It survived the war intact. The barn portion of the building was torn down in May 2011, leaving the structure less than half of its original size. House served as Forward CP for the 741st Tank Battalion from about 18:00, December 12, 1944 until 17:30, December 18, 1944, when the CP moved to Krinkelt.

Nr. 101 Joseph Hönen was the son of Mathias Hönen. His house was named "Hintern Kättringen."

Nr. 106 Primary School and Break Hall, the latter served as a motor-vehicle shed for the 393rd Infantry Regiment beginning in November 1944.

After the German invasion in 1940, the eastern cantons of Belgium became part of Germany, specifically the Köln-Aachen administrative division. As part of the transition, the government assigned new numbers to buildings in Krinkelt-Rocherath. Those numbers are represented in this appendix.

There was no town hall in Krinkelt or Rocherath during the war years. The nearest Rathaus was in Büllingen. The only government buildings in the twin villages were the fire house at Nr. 38 Krinkelt and the police station at Nr. 80 Rocherath.

Appendix J

Tactical Numbers, I. Abteilung, SS-Panzer-Regiment 12, Ardennes

Stubaf Arnold Jürgensen (died 23 Dec 44)					
Stabs-Kompanie	Abteilungsstab (Panther)	155	154		
	Panzer-Aufklärungszug	156	157	158	159

1. Kompanie, Panther – Hptm Walter Hils (died 21 Dec 44)					
		105	104		
I.Zug	115	116	117	118	
II.Zug	125	126	127	128	
III.Zug	135	136	137	138	

3. Kompanie, Panther – Hstuf Kurt Brödel (died 18 Dec 44)					
		305	304		Oscha Walter Freier (304)
I.Zug	315	316	317	318	Ustuf Karl Schittenhelm
II.Zug	325	326	327	328	Oscha Johann Beutelhauser
III.Zug	335	336	337	338	Ustuf Willi Engel

5. Kompanie, Panzer IV – Ustuf Eberhard Jeran (died 11 Mar 45)					
		505	504		
I.Zug	515	516	517	518	
II.Zug	525	526	527	528	
III.Zug	535	536	537	538	

6. Kompanie, Panzer IV – Hstuf Götz Grossjohann					
		605	604		Ustuf Karl Pucher (604)
I.Zug	615	616	617	618	
II.Zug	625	626	627	628	Ustuf Kurt Mühlhaus
III.Zug	635	636	637	638	

Gray numbers on the chart represent tanks destroyed December 17–19, 1944 and confirmed by photographic evidence. Three other Panthers appear in photos, but their tactical numbers remain unclear. The same is true of one Panzer IV lost near the Rocherath water tower.

Pages 71 and 150 of *Die 3.Kompanie* state that Brödel's company had fourteen Panthers. The book goes on to explain the situation: "In October 1944, the equipment for the combat echelons of tank companies was reduced to four tanks per platoon due to the diminished capacity of the armaments industry, owing to the Allied bombing campaign. Previously, the platoons had five tanks." Hubert Meyer's book affirms that each Panther and Panzer IV company had fourteen tanks (*Band II*, page 389). Meyer also states that the Abteilungsstab had two tanks but makes no mention of the Aufklärungszug, which may have possessed no armor in December 1944 (although, four machines are represented on the previous page to show the numbering scheme).

Appendix K

Organizational Diagrams

Headquarters and Headquarters Company

T/O & E 17-26, 18 November 1944 (with Change No. 1, 6 January 1945)

13 Commissioned Officers, 127 Enlisted Men

Battalion Headquarters

Tank Section

M1 bulldozer blade not in T/O & E

Headquarters Section

Company Headquarters

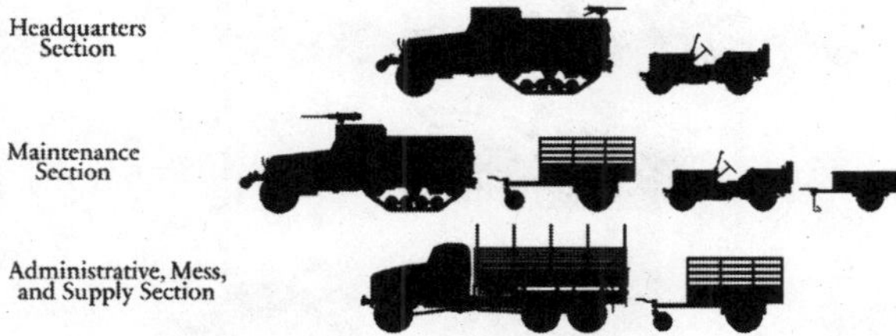

Mortar Platoon

Assault Gun Platoon

Battalion Reconnaissance Platoon

Medium Tank Company

T/O & E 17-27, 18 November 1944 (with Change No. 1, 6 January 1945)

5 Commissioned Officers, 112 Enlisted Men

Headquarters Section

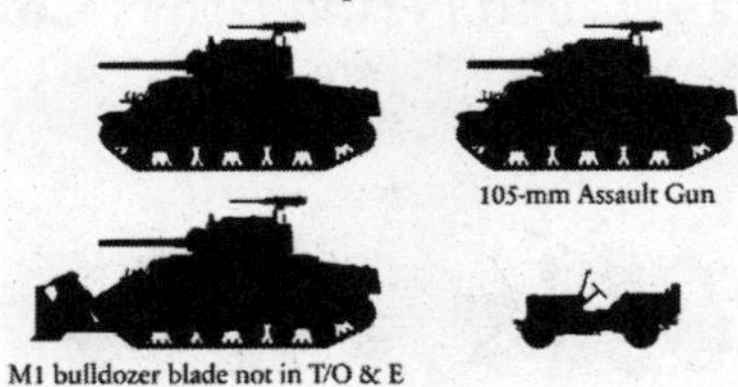

Administrative, Mess, and Supply Section

Maintenance Section

First Platoon

First Section

Second Section

Second Platoon

First Section

Second Section

Third Platoon

First Section

Second Section

Light Tank Company

T/O & E 17-17, 11 November 1944 (with Change No. 1, 13 January 1945)

5 Commissioned Officers, 89 Enlisted Men

Headquarters Section

Administrative, Mess, and Supply Section

Maintenance Section

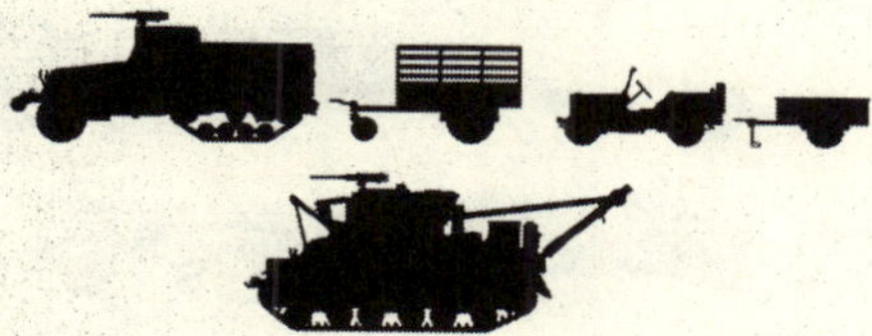

First Platoon

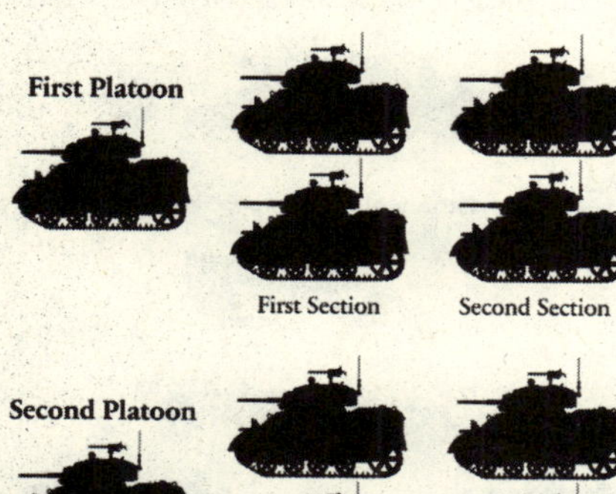

First Section Second Section

Second Platoon

First Section Second Section

Third Platoon

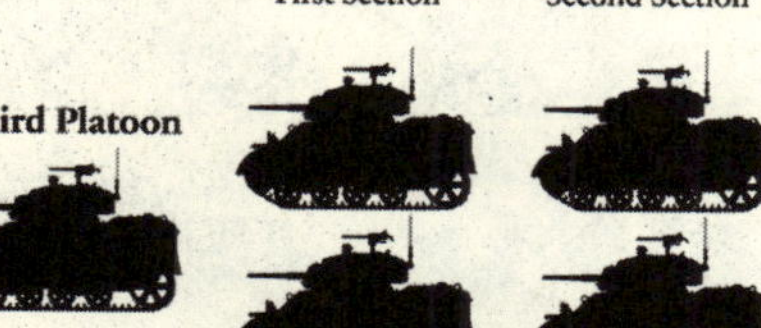

First Section Second Section

Medical Detachment

T/O & E 17-25, 18 November 1944

2 Commissioned Officers, 18 Enlisted Men

Service Company

T/O & E 17-29, 18 November 1944 (with Change No. 1, 6 January 1945)

4 Commissioned Officers, 3 Warrant Officers, 108 Enlisted Men

Company Headquarters

Battalion Administrative and Personnel Section

Battalion Maintenance Platoon

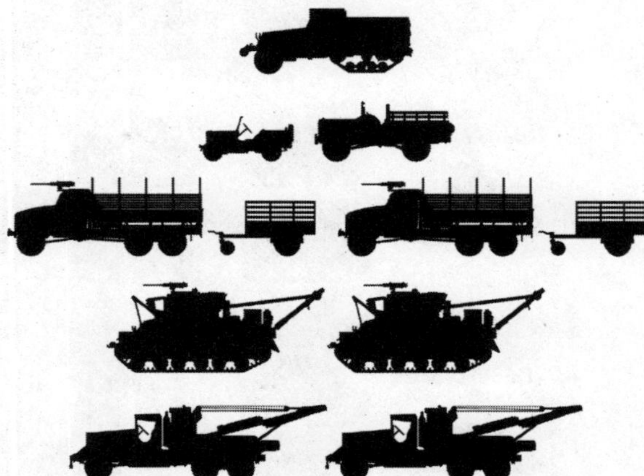

Service Company (continued)

Battalion Supply and Transportation Platoon

Tables of Organization and Equipment 17-26 and 17-27 allotted one caliber .50 machine gun for the turret of each medium tank and assault gun. This was for antiaircraft use, but with Allied air superiority after D-Day, there was little need for these guns. Most crews removed them altogether, though some tankers relocated them on the turret to serve in an anti-personnel role. No matter where the weapon was mounted, the operator exposed himself to enemy fire. The gun also tended to snag on tree branches and other obstructions. – Bill Warnock

Bibliography

Books, Manuscripts, and Journal Articles

Alford, Kenneth D., *American Crimes and the Liberation of Paris: Robbery, Rape and Murder by Renegade GIs, 1944–1947* (Jefferson, NC: McFarland & Company Inc., 2016).

Baily, Charles M., *Faint Praise: American Tanks and Tank Destroyers during World War II* (Hamden, CT: Archon Books, 1983).

Barkley, Cleve C., *In Death's Dark Shadow: A Soldier's Story* (Loraine, IL: Instantpublisher.com, 2006).

Beck, Alfred, et al., *The Corps of Engineers: The War Against Germany, U.S. ARMY IN WORLD WAR II* (Washington, DC: United States Government Printing Office, 1957).

Beeley, George E., *I Must Have Had A Guardian Angel* (N. Providence, RI: unpublished manuscript, 1996).

Blumenson, Martin, *Breakout and Pursuit, U.S. ARMY IN WORLD WAR II* (Washington, DC: United States Government Printing Office, 1961).

Boardman, Thayer M. and Edward Steere, *Final Disposition of World War II Dead 1945–51, QMC Historical Studies, Series II, No. 4* (Washington, DC: Historical Branch, Office of the Quartermaster General, 1957).

Balkoski, Joseph, *Omaha Beach: D-Day, June 6, 1944* (Mechanicsburg, PA: Stackpole Books, 2004).

Bradley, Omar N., *A Soldier's Story* (New York, NY: Henry Holt and Company, 1951).

Briggs, Elaine, *Joe Dew: A Glorious Life* (CreateSpace Independent Publishing, 2018).

Brown, James D., "Medium Tank Doctrine—The Sherman Era," *Armor*, November/December 1973, pp. 22-26.

Cannon, Paul R., Alvin W. Heintzleman and Lewis H. Lanier, *The Story of "Vitamin Baker": We'll Never Go Overseas* (Plzeň: Grafické Závody Pour a Spol., 1945).

Capa, Robert, *Slightly Out of Focus* (New York, NY: Henry Holt and Company, 1947).

Casely, Steve, "The Tank that Missed D-Day," *After The Battle*, Number 45 (1984), pp. 19-32.

Cavanagh, William C. C., *Krinkelt-Rocherath: The Battle for the Twin Villages* (Norwell, MA: The Christopher Publishing House, 1986).

Cluster, Herbert R., "D-Day: No Way to Storm a Beach," *Naval History*, May/June 1994, pp. 10-18.

Cole, Hugh M., *The Ardennes: Battle of the Bulge, U.S. ARMY IN WORLD WAR II* (Washington, DC: United States Government Printing Office, 1965).

Curley, Charles D., *How a Ninety-Day Wonder Survived the War: The Story of a Rifle Platoon Leader in the Second Indianhead Division during World War II* (Richmond, VA: Ashcraft Enterprises, 1998).

Dallas, John T., *Major Cecil D. Thomas Normandy Beach to the Rhine River with the 741st Tank Battalion World War II and Others from Rockingham County* (Greensboro, NC: Self-published, 2012).

Elliott, DeRonda E., "D-Day: What it Cost," *American Heritage*, May/June 1994, pp. 61-80.

Fane, Francis D., *Naked Warriors* (New York, NY: Prentice Hall, 1956).

Fardella, Peter J., *Combat History of the 741st Tank Battalion, July 26 1944 – May 7, 1945* (Brooklyn, NY: unpublished handwritten manuscript, July 28, 1948). Company C is the primary subject of this thirty-nine page document.

Folkestad, William B., *The View from the Turret: The 743rd Tank Battalion During World War II* (Shippensburg, PA: White Mane Publishing Company Inc, 1996).

Garth, David and Charles H. Taylor, *St-Lô, 7 July – 19 July 1944, AMERICAN FORCES IN ACTION SERIES* (Washington, DC: United States Government Printing Office, 1947).

Gawne, Jonathan, *Spearheading D-Day: American Special Units in Normandy* (Paris: Histoire & Collections, 1998).

Gillie, Mildred H., *Forging the Thunderbolt: A History of the Development of the Armored Force* (Harrisburg, PA: The Military Service Publishing Company, 1947).

Green, Michael, *M4 Sherman: Combat and Development History of the Sherman Tank and all Sherman Variants* (Osceola, WI: Motorbooks International, 1993).

Groß, Manfred, *Der Westwall zwischen Niederrhein und Schnee-Eifel* (Köln: Rheinland-Verlag GmbH, 1982).

Hancock, William F., *The 9th Infantry Regiment As I Knew It* (Fresno, CA: unpublished manuscript, November 2000).

Harrison, Gordon A., *Cross-Channel Attack, U.S. ARMY IN WORLD WAR II* (Washington, DC: United States Government Printing Office, 1951).

Heavey, William F., *Down Ramp!: The Story of the Army Amphibian Engineers* (Washington, DC: Infantry Journal Press, 1947).

Hemingway, Ernest, "Voyage to Victory," *Collier's*, July 22, 1944, pp. 11-13, 56-57.

Hewitt, Robert L., *Work Horse of the Western Front: The Story of the Thirtieth Infantry Division* (Washington, DC: Infantry Journal Press, 1946).

Hillner, George, Richard Meacham and George M. Smith, *Vitamin Charley: A History of Company C, 741st Tank Battalion in World War II* (New York, NY: Ferris Printing Company with binding by J.F. Tapley Company, 1946). George Cuthbert's father worked for Tapley, and he published this book in memory of his son and the other Charley Company soldiers who lost their lives in Europe.

Hunnicutt, Richard P., *Sherman: A History of the American Medium Tank* (Novato, CA: Presidio Press, 1978).

Irzyk, Albin F., "Tank versus Tank," *Military Review*, January 1946, pp. 11-16.

Jensen, Marvin G., *Strike Swiftly: The 70th Tank Battalion from North Africa to Normandy to Germany* (Novato, CA: Presidio Press, 1997).

Joyce, Carlton S., *Stand Where They Fought: 150 Normandy Battlefields of the 77-Day Campaign* (Bloomington, IN: AuthorHouse, 2006).

Karamales, Jay and Allyn R. Vannoy, *Against the Panzers: United States Infantry versus German Tanks, 1944–1945* (Jefferson, NC: McFarland & Company Inc, 1996).

Kemp, Harry M., *The Regiment: Let the Citizens Bear Arms!: A Narrative History of an American Infantry Regiment in World War II* (Austin, TX: Nortex Press, 1990).

Kleine, Egon and Volkmar Kühn, *Tiger: The History of a Legendary Weapon 1942–45*, Trans. David Johnston (Winnipeg, Manitoba: J. J. Fedorowicz Publishing Inc, 1989).

Lodieu, Didier, *Dying for Saint Lô: Hedgerow Hell, July 1944* (Paris: Histoire & Collections, 2007).

Lüttgens, Karl J., *Kriegsjahre, Kriegsende und Erste Neuanfänge im Kreis Schleiden 1939–1946* 2 vols. (Gemünd, Germany: Wallraf Druck + Design, 1997).

MacDonald, Charles B., *Company Commander* (Washington, DC: Infantry Journal Press, 1947).

———, *The Siegfried Line Campaign, U.S. ARMY IN WORLD WAR II* (Washington, DC: United States Government Printing Office, 1963).

———, *The Mighty Endeavor: American Armed Forces in the European Theater in World War II* (New York, NY: Oxford University Press, 1969).

———, *The Last Offensive, U.S. ARMY IN WORLD WAR II* (Washington, DC: United States Government Printing Office, 1973).

Major, Ben C. and Lois S. Montbertrand, *Unit Serial Numbers from the First U.S. Army Build-Up Priority Tables, List "A" D+1 through D+14* (United Kingdom: WW2 US Medical Research Centre, 2011). English edition

McGhee, Addison F. Jr., *He's In The Armored Force Now* (New York, NY: Robert M. McBride & Company, 1942).

Meyer, Hubert, *Kriegsgeschichte der 12.SS-Panzerdivision "Hitlerjugend," Band II* (Osnabrück: Munin Verlag GmbH, 1982).

Mittelman, Joseph B., *Eight Stars to Victory: A History of the Veteran Ninth U.S. Infantry Division* (Washington, DC: Ninth Infantry Division Assn, 1948).

Möller, Jürgen, *Flak im Endkampf – Leuna 1945: Die Besetzung des mitteldeutschen Chemiezentrums Schkopau-Merseburg-Leuna durch das V. US Corps im April 1945* (Bad Langensalza, Germany: Rockstuhl Verlag, 2013).

Möller, Jürgen, *Die letzte Schlacht – Leipzig 1945: Der Besetzung der Reichsmessestadt durch das V. US Corps der 1st US Army im April 1945. Kriegsende in Mitteldeutschland 1945* (Bad Langensalza, Germany: Rockstuhl Verlag, 2014).

Munoz, Antonio J., *The Last Levy: Waffen-SS Officer Roster, March 1st 1945* (Bayside, NY: Axis Europa Books, 2001).

Parker, Danny S., *Battle of the Bulge: Hitler's Ardennes Offensive, 1944–1945* (Conshohocken, PA: Combined Books, 1991).

Pederson, Maynard D., et al., *Armor in Operation Neptune* (Fort Knox, KY: The Armored School, May 1949). A research report prepared by Committee 10, Officers Advanced Course, The Armored School, 1948–49.

Robinson, Anthony W. III, *Move Out Verify: The Combat Story of the 743rd Tank Battalion* (Frankfurt am Main, Germany: 743rd Tank Battalion, July 1945).

Rolf, Rudi, *Der Atlantikwall: Die Bauten der deutschen Küstenbefestigungen 1940–1945, Teile 1-2* (Osnabrück: Biblio Verlag, 1998).

Roppelt, Fritz, *"Der Vergangenheit auf der Spur": 3.Fallschirmjäger-Division 1943–1945* (Hösbach: 1994).

Rupert, Barbara J., *The History of the Lima Army Tank Plant* (Lima, OH: August 1994).

Ryan, Cornelius, *The Longest Day* (New York, NY: Simon and Schuster, 1959).

Salomon, Gerhard, *Die 3.Fallschirmjägerdivision in der Normandie* (Neumünster: Selbstverlag, 1995).

Samsel, Harold J., *Operational History of the 102nd Cavalry Regiment* (Short Hills, NJ: ca. 1983).

Sawicki, James A., *Tank Battalions of the U.S. Army* (Dumfries, VA: Wyvern Publications, 1983).

Scherer, Wingolf, *Die 277.Volksgrenadierdivision in der Ardennenoffensive und im Untergang am Rhein 1944/45* (Düsseldorf, Germany: 1987).

Schneider, Wolfgang, *Tigers in Combat, Vol. I* (Winnipeg, Manitoba: J. J. Fedorowicz Publishing Inc, 1994).

Schreier, Konrad F. Jr., *Standard Guide to U.S. World War II Tanks and Artillery* (Iola, WI: Krause Publications, 1994).

Shirer, William L., *The Rise and Fall of the Third Reich* (New York, NY: Simon and Schuster, 1959).

Small, Kenneth, *The Forgotten Dead* (London: Bloomsbury Publishing Limited, 1988).

Steere, Edward, *The Graves Registration Service in World War II, QMC Historical Studies, No. 21* (Washington, DC: United States Government Printing Office, 1951).

Spielberger, Walter J., *Der Panzerkampfwagen Panther und Seine Abarten* (Stuttgart: Motorbuch Verlag, 1987).

———, *Der Panzerkampfwagen IV und Seine Abarten* (Stuttgart: Motorbuch Verlag, 1988).

Taylor, Charles H., *Omaha Beachhead, 6 June – 13 June 1944, AMERICAN FORCES IN ACTION SERIES* (Washington, DC: United States Government Printing Office, 1945).

Tessin, Georg, *Verbände und Truppen der deutschen Wehrmacht und Waffen-SS im Zweiten Weltkrieg 1939–1945*, Vols. 1-14 (Frankfurt/Main: Verlag E.S. Mittler & Sohn).

Thayer, Dale R., *The Famous 299th* (299th Engineer Combat Battalion, ca. 1950).

Thompson, Royce L., "Tank Fight of Rocherath-Krinkelt (Belgium), 17–19 December 1944" (OCMH Historical Manuscript File, call # 2-3.7 AE. P-12, February 1952).

Upham, John S., "DD Tanks," *Military Review*, February 1947, pp. 42-45.

Wertenbaker, Charles C., *Invasion!* (New York, NY: D. Appleton-Century Company, 1944).

Wiener, Alan D., "The First Wave," *American Heritage*, May/June 1987, pp. 136-138.

White, Isaac D., *United States vs. German Equipment* (Bennington, VT: Merriam Press, 1988)

White, William Lindsay, "Sergeant Culin Licks the Hedgerows," *The Reader's Digest*, February 1950, pp. 81-84.

Yeide, Harry, *The Infantry's Armor* (Mechanicsburg, PA: Stackpole Books, 2010)

Books and Journal Articles without Acknowledged Authors

A Manual for Courts-Martial U.S. Army – 1928 Corrected to April 20, 1943 (Washington, DC: United States Government Printing Office, 1943).

Combat History of the Second Infantry Division in World War II (Baton Rouge, LA: Army & Navy Publishing Company, 1946).

"DD Tanks: Straussler's Solution to Make 'Landships' Swim," *Wheels & Tracks*, Number 40 (1992), pp. 34-42.

Die 3.Kompanie (Preußisch Oldendorf: Eigenverlag Kompanie-Kameradschaft, 1978).

The author translated seven chapters from this book into English. These pages cover the period from August 1944 through January 1945.

First United States Army: Report of Operations, Vols. 1-4, *20 October 1943 – 1 August 1944* (First U.S. Army, 1944).

———, Vols. 5-8, *1 August 1944 – 22 February 1945* (First U.S. Army, 1945).

———, Vols. 9-11, *23 February 1945 – 8 May 1945* (First U.S. Army, 1945).

Handbook on German Military Forces (Washington, DC: United States Government Printing Office, 1945).

History, V Corps: Operations in the ETO 6 January 1942 – 9 May 1945 (U.S. Army, 1945).

Official Army Register, 1945 ed. (Washington, DC: United States Government Printing Office, January 1, 1945).

Operation "Neptune": Landings in Normandy, June 1944 (London: HMSO, Crown Copyright, 1994).

Register of Commissioned and Warrant Officers of the United States Naval Reserve (Washington, DC: United States Government Printing Office, July 31, 1944).

Service Instruction Book for Sherman III (Med.M4.A2) D.D. and Sherman V (Med.M4.A4) D.D. (Chilwell Catalogue No. 72/496, March 1944).

The Roerich Pact and The Banner of Peace (New York, NY: The Roerich Pact and Banner of Peace Committee, 1947).

"Salvaging the D-Day Beaches," *After The Battle*, Number 34 (1981), pp. 42-46.

A Short Guide to Great Britain (Washington, DC: United States Government Printing Office, 1942).

644th Tank Destroyer Battalion (Göttingen: Muster-Schmidt-Verlag, 1945).

741 D-Day to V-E Day (Paris: Paul Dupont, 1945).

"25 Years Ago, Firestone was at Normandy Beach," *Firestone Non-Skid*, June 3, 1969 (published by the Firestone Tire & Rubber Company, Akron, Ohio).

Field Manuals

Armored Force Field Manual 17-5: Armored Force Drill (Washington, DC: United States Government Printing Office, January 12, 1942).

Armored Force Field Manual 17-10: Tactics and Technique (Washington, DC: United States Government Printing Office, March 7, 1942).

Armored Force Field Manual 17-12: Tank Gunnery (Washington, DC: United States Government Printing Office, 10 July 1944).

Armored Force Field Manual 17-15: Combat Practice Firing, Armored Force Units (Washington, DC: United States Government Printing Office, May 15, 1942).

Armored Force Field Manual 17-25: Assault Gun Section and Platoon (Washington, DC: United States Government Printing Office, June 25, 1942).

Armored Force Field Manual 17-27: 81mm Mortar Squad and Platoon (Washington, DC: United States Government Printing Office, July 30, 1942).

Armored Force Field Manual 17-30: Tank Platoon (Washington, DC: United States Government Printing Office, October 22, 1942).

Armored Force Field Manual 17-32: The Tank Company, Light and Medium (Washington, DC: United States Government Printing Office, August 2, 1942).

Armored Force Field Manual 17-33: The Armored Battalion, Light and Medium (Washington, DC: United States Government Printing Office, September 18, 1942).

War Department Field Manual 17-36: Employment of Tanks with Infantry (Washington, DC: United States Government Printing Office, 13 March 1944).

Armored Force Field Manual 17-62: Fire Control and Coordination (Washington, DC: United States Government Printing Office, July 14, 1942).

Armored Force Field Manual 17-67: Crew Drill and Service of the Piece Medium Tank M4 (Washington, DC: United States Government Printing Office, April 9, 1943).

War Department Field Manual 17-68: Crew Drill, Light Tank M5 Series (Washington, DC: United States Government Printing Office, 24 May 1944).

Technical Manuals

War Department Technical Manual 9-731A: Medium Tanks M4 and M4A1 (Washington, DC: United States Government Printing Office, 23 December 1943).

War Department Technical Manual 9-731AA: Medium Tank M4 (105 mm Howitzer) and Medium Tank M4A1 (76 mm Gun) (Washington, DC: United States Government Printing Office, 23 June 1944).

War Department Technical Manual 9-732: Light Tanks M5 and M5A1 (Washington, DC: United States Government Printing Office, 27 November 1943).

War Department Technical Manual 9-2853: Preparation of Ordnance Material for Deep Water Fording (Washington, DC: United States Government Printing Office, 5 January 1944).

Audio-Visual

"D-Day Beneath the Waves," *Discovery Channel*, narrator John Badila, producer Will Aslett, BBC, first aired June 5, 2002 (aired in Great Britain on May 30 as "Journeys to the Bottom of the Sea—D-Day: The Untold Story")

> "D-Day Beneath the Waves" includes interviews with Phil Fitts (C/741) and Bill Merkert (B/741), both survivors of DD tanks that sank. During pre-production, Wanda Koscia and Bucy McDonald of the BBC attended the 741st Tank Battalion reunion held at the Ambassador Hotel in New Orleans, July 12–15, 2001. The two assistant producers conducted research at the reunion. The resulting film revolved around marine archeology conducted by Project Neptune 2K, an undertaking of the Institute of Nautical Archeology, headquartered at Texas A&M University. Brett A. Phaneuf served as project leader.

"The Lost Tanks of D-Day," Deep Sea Detectives, episode 37, *The History Channel*, narrator John Chatterton, produced by Lone Wolf Documentary Group and Liquid Pictures for A&E Television Networks, first aired June 1, 2004.

Microfiche

Rosters WW-II Dead (All Services) World War II (Ref. No. 601-07) 268 fiche sheets

Internet

Joseph DeMarco, Leife Hulbert, and Pierre-Olivier Buan, "Sherman Minutia," http://the.shadock.free.fr/sherman_minutia/

Cornelius Ryan Questionnaires

Questionnaires for *The Longest Day*, Ryan Collection, Alden Library, Ohio University:

William Friedman, questionnaire, 3 pages, undated (Hq, 16th Inf Rgt)

William M. King, letter and questionnaire, 5 pages, May 20, 1958 (Hq, 741st Tank Bn)

William D. McClintock, questionnaire, 4 pages, undated (Hq Co, 741st Tank Bn)

Robert N. Skaggs, questionnaire, 3 pages, undated (Hq, 741st Tank Bn)

Edward S. Sledge II, questionnaire, 3 pages, undated (Co A, 741st Tank Bn)

Primary Sources

National Archives and Records Administration

741st Tank Battalion Operational Records 1942–1945
Archives II, College Park, Maryland
NM-3 Entry 427, 407/270/58/35/4-5, Boxes 13502-13504
Declassified per NND 735017

BOX 13502 (ARBN-740-0.7 to ARBN-741-0.3)
NM-3 Entry 427, 407/270/58/35/4

Box 13502 also contains four folders of 740th Tank Battalion records.

FOLDER ARBN-741-0.1

» History of the 741st Tank Battalion (M), 15 May 1942 to 30 Sep 1943 (9 pages)
» Commendation from VII Corps, 9 Sep 1942 (2 pages)
» Letter of Appreciation from 75th Field Artillery Brigade, 16 Sep 1942 (1 page)
» Commendation from 7th Motorized Division, 26 Sep 1942 (1 page)
» Reorganization of GHQ Tank Battalions, 16 Mar 1942 (6 pages)
» Weapons Qualifications Statistics, 741st Tank Battalion, 1 Oct 1943 (2 pages)
» Names caption for panoramic photo, 741st Tank Battalion officers, Fort Dix, 18 Sep 1943 (1 page)
» Names caption for panoramic photo, Headquarters Company, Fort Dix, 18 Sep 1943 (1 page)
» Names caption for panoramic photo, Company A, Fort Dix, 18 Sep 1943 (1 page)
» Training Memorandum Number 43, 751st Tank Battalion, 19 Mar 1942 (8 pages)
» Special Orders Number 53, Headquarters 751st Tank Battalion, 18 Mar 42 (7 pages)
» Special Orders Number 54, Headquarters 751st Tank Battalion, 19 Mar 42 (6 pages)
» Special Orders Number 75, Headquarters Armored Force, 16 Mar 42 (1 page)
» Special Orders Number 80, Headquarters Armored Force, 21 Mar 42 (2 pages)
» General Orders Number 14, Headquarters Armored Force, 15 Mar 42 (1 page)

FOLDER ARBN-741-MD-0.2

» History of the Medical Detachment, 741st Tank Battalion (M), dated 1 Jan 1943 (3 pages)

FOLDER ARBN-741-0.2

» Battalion History for March 1945, dated 31 Mar 1945 (23 pages)
» Battle Casualty Reports, BCR Control Numbers 70-73, dated 7 to 13 Mar 45 (1 page)
» Battle Casualty Reports, BCR Control Numbers 74-76, dated 25 to 29 Mar 45 (1 page)

FOLDER ARBN-741-0.3

» After-Action Report, January 1945, dated 4 Feb 1945 (6 pages)
» Unit Journal, 1–31 January 1945 (17 pages)

» Commendation for C/741 from 2nd Inf Div, dated 14 Feb 1945, for actions 15-20 Jan 1945 (2 pages)
» After-Action Report, February 1945, dated 3 Mar 1945 (5 pages)
» After-Action Report, March 1945, dated 6 Apr 1945 (7 pages)
» Unit Journal, 1–31 March 1945 (17 pages)
» After-Action Report, April 1945, dated 4 May 1945 (10 pages)
» Unit Journal, 1–30 April 1945 (12 pages)
» After-Action Report, May 1945, dated 27 May 1945 (2 pages)
» Battle Casualty Reports, BCR Control Numbers 59-62, dated 2 to 5 Feb 1945 (1 page)
» Battle Casualty Reports, BCR Control Numbers 70-94, dated 7 Mar 45 to 25 May 45 (8 pages)
» folder contains carbon copies of several battle casualty reports

BOX 13503 (ARBN-741-0.3 to ARBN-741-3.7)
NM-3 Entry 427, 407/270/58/35/5

FOLDER ARBN-741-0.3
» After-Action Report, June 1944, dated 19 Jul 1944 (18 pages incl. 4 appendixes)
» Vehicle Losses for June 1944, dated 20 Jul 1944 (1 page)
» After-Action Report, July 1944, dated 7 Aug 1944 (26 pages incl. 3 appendixes and 9 map overlays)
» Unit Journal, 1–31 July 1944 (20 pages incl. 4-page addendum)
» After-Action Report, August 1944, misdated 2 Aug 1944, probably 2 Sep 1944 (7 pages)
» Unit Journal, 1–31 August 1944 (8 pages)
» After-Action Report, September 1944, dated 5 Oct 1944 (6 pages)
» Unit Journal, 1–30 September 1944 (36 pages incl. 15-page addendum)
» Unit Journal, Military Government Detachment, 1–30 September 1944 (8 pages)
» Statistical Armored Report for September 1944, dated 5 Oct 1944 (1 page)
» After-Action Report, October 1944, dated 1 Nov 1944 (4 pages)
» After-Action Report, November 1944, dated 2 Dec 1944 (2 pages)
» Unit Journal, 1–30 November 1944 (7 pages)
» Unit Reports 28-57, daily reports, November 1944 (30 pages)
» After-Action Report, December 1944, dated 4 Jan 1945 (6 pages)
» Statistical Armored Report for December 1944, dated 4 Jan 1945 (1 page)
» Unit Journal, 1–31 December 1944 (13 pages)
» Non Battle Casualty Report, Robert B. McMullen, dated 13 Jul 1944
» Non Battle Casualty Report, Thomas D. Chastain, dated 13 Jul 1944
» Battle Casualty Reports, BCR Control Numbers 5-16, dated 11 Jul to 4 Aug 1944 (6 pages)
» Battle Casualty Reports, BCR Control Numbers 22-30, dated 15 Sept to 5 Oct 1944 (4 pages)
» Battle Casualty Report, BCR Control Number 37, dated 21 Nov 1944 (1 page)
» Battle Casualty Reports, BCR Control Numbers 48-58, dated 2 to 29 Jan 1945 (4 pages)
» Battle Casualty Reports, BCR Control Numbers 63-69, dated 6 to 26 Feb 1945 (2 pages)

FOLDER ARBN-741-0.7

- » Unit Journal, 6–30 June 1944 (16 pages)
- » Unit Journal, 1–31 July 1944 (23 pages incl. 4-page addendum)
- » Unit Journal, 1–31 August 1944 (23 pages)
- » Unit Journal, 1–30 September 1944 (36 pages incl. 15-page addendum)
- » Unit Journal, Military Government Detachment, 1–30 September 1944 (8 pages)
- » Unit Journal, 1–31 October 1944 (9 pages incl. 2-page addendum)
- » Unit Journal, 1–30 November 1944 (7 pages)
- » Unit Journal, 1–31 December 1944 (12 pages)
- » Unit Journal, 1–31 January 1945 (22 pages incl. 6-page January After-Action Report Addendum)
- » Unit Journal, 1–28 February 1945 (12 pages)
- » Unit Journal, 1–31 March 1945 (17 pages)
- » Unit Journal, 1–30 April 1945 (14 pages)
- » Unit Journal, 1–31 May 1945, actually 1–8 May (4 pages)

FOLDER ARBN-741-0.9

- » Unit Reports 80-110, daily reports, January 1945 (31 pages)
- » Unit Reports 169-198, daily reports, April 1945 (30 pages)

FOLDER ARBN-741-0.16

- » Coat of Arms for the 741st Tank Battalion, 10 Sep 1942 (3 pages)

FOLDER ARBN-741-1.13

- » General Orders, Headquarters, 741st Tank Battalion

1942

- » GO 1, 15 Mar; GO 2, 19 Mar; GO 3, 20 Mar; GO 4, 24 Mar; GO 5, 27 Mar; GO 6, 8 Apr; GO 7, 22 Apr; GO 8, 23 Apr; GO 9, 5 Jun; GO 10, 5 Aug; GO 11, 8 Aug; GO 12, 15 Aug GO 13, 3 Sep; GO 14, 14 Sep; GO 15, 27 Oct; GO 16, 29 Oct; GO 17, 3 Nov; GO 18, 5 Nov; GO 19, 12 Nov; GO 20, 16 Nov; GO 21, 30 Nov; GO 22, 1 Dec; GO 23, 10 Dec; GO 24, 16 Dec; GO 25, 17 Dec; GO 26, 18 Dec;

1943

- » GO 1, 18 Jan; GO 2, 22 Jan; GO 3, 19 Mar; GO 4, 21 Mar; GO 5, 4 Apr; GO 6, 10 Apr; GO 7, 10 May; GO 8, 19 Jun; GO 9, 23 Jun; GO 10, 14 Jul; GO 11, 25 Jul; GO 12, 7 Aug; GO 13, 16 Aug; GO 14, missing; GO 15, 9 Sep; GO 16, 14 Sep

1944

- » GO 1, 6 Jan; GO 2, 7 Jan; GO 3, 11 Jan; GO 4, 18 Apr; GO 5, 8 Jul; GO 6, 18 Sep; GO 7, 18 Oct; GO 8, 19 Oct; GO 9, 9 Nov; GO 10, 22 Nov; GO 11, 26 Nov; GO 12, 8 Dec

1945

- » GO 1, 5 Jan; GO 2, 10 Jan; GO 3, 11 Jan; GO 4, 18 Jan; GO 5, 24 Jan; GO 6, 18 Mar; GO 7, 21 Mar; GO 8, 5 Apr; GO 9, 13 May; GO 10, 17 May; GO 11, 23 May (Bronze Service Arrowheads awarded for D-Day); GO 12, 27 May; GO 13, 28 May; GO 14, 30

May (Bronze Service Arrowhead for Charles F. Urbanas); GO 15, 4 Jun; GO 16, 12 Jun; GO 17, 18 Jun; GO 18, 1 Jul; GO 19, 3 Jul; GO 20, 4 Jul; GO 21, 10 Jul; GO 22, 31 Jul; GO 23, 2 Aug; GO 24, 15 Aug; GO 25, 1 Sep; GO 26, 9 Sep

FOLDER ARBN-741-3.1

» S-3 Periodic Report, 5 Oct 1944 (1 page and 3 overlays)
» Unit Reports 2-28, 6-31 Oct 1944 (28 pages plus 3 overlays for Report 4, 072400 – 082400 Oct 1944)
» The title S-3 Periodic Report was superseded after the first report by the title Unit Report.

FOLDER ARBN-741-3.18

» Movement Order Number 1, 1 Sep 1944 (1 page)
» Movement Order Number 3, 10 Sep 1944 (1 page)
» Movement Order Number 4, 4 Oct 1944 (1 page)
» Movement Order Number 5, 10 Dec 1944 (3 pages)
» Movement Order Number 6, 1 May 1945 (1 page)

FOLDER ARBN-741-3.7

» Terrain Study Hill 192, Sheet 1, 1:10,000, no handwritten markings or notations
» Assault Map, Elsenborn and Vicinity, Sheet 10 of 14, 1:10,000 (pencil-written notations showing 741 positions in and around twin villages, 16-19 Dec 1944)
» Assault Map, Elsenborn and Vicinity, Sheet 9 of 14, 1:10,000 (pencil-written notations showing 741 positions around Wirtzfeld, 17-19 Dec 1944)
» Assault Map, Sheet 1, 1:10,000 (pencil-written notations showing 741 positions at Wahlerscheid, 16 Dec 1944, artillery concentrations printed in blue)
» Krinkelt-Rocherath map on vellum, 1:10:000, no handwritten markings
» Krinkelt-Rocherath map on vellum, 1:10:000, pen and pencil markings showing 741 positions, 17-19 Dec 1944
» Krinkelt-Rocherath map on vellum, 1:10:000, pen markings showing knocked out enemy trucks and armored vehicles, 17-19 Dec 1944

BOX 13504 (ARBN-741-3.9 to ARBN-743-0.1)

NM-3 Entry 427, 407/270/58/35/5

FOLDER ARBN-741-3.9, F.O. #1

» Field Order #1, 741st Tank Bn, "Neptune," 21 May 1944 (3 pages)
» Operations Overlay, Annex #1 to FO #1, undated, 1:25,000
» S-2 Estimate of the Enemy Situation, Annex #2 to FO #1, 21 May 1944 (3 pages)
» Appendix (map overlay of enemy HQ locations in Normandy) to Annex #2, 12 May 1944, 1:500,000
» Tactical Study of Terrain, Annex #3 to FO 1, 21 May 1944 (1 page)
» Signal Annex to FO #1, 21 May 1944 (1 page)
» Administrative Annex to FO #1, 21 May 1944 (3 pages)
» Operations Overlay, Engineer Plan, Provisional Engineer Group, 22 May 1944, 1:25,000

- Obstacles Offshore, map locating Hedgerow, Element C, and Ramp-Type obstacles in vic. E-1 and E-3, no scale indicated
- Vehicle Losses for June 1944, dated 20 Jul 1944 (1 page)
- Non Battle Casualty Report, Robert B. McMullen, dated 13 Jul 1944
- Non Battle Casualty Report, Thomas D. Chastain, dated 13 Jul 1944
- Battle Casualty Reports, BCR Control Numbers 1-8, dated 4 to 17 Jul 1944 (5 pages)

FOLDER ARBN-741-3.9, Field Orders #2 and #3

- Field Order #2, 741st Tank Bn, 3 Jul 1944 (2 pages)
- Operations Overlay, Annex #1 to FO #2, undated, 1:25,000
- Intelligence Annex, Annex #2 to FO #2, 9 Jul 1944 (1 page)
- Situation Overlay for Annex #2 to FO #2, undated, 1:25,000
- Annex #3 to FO #2, Traffic Circulation Plan, undated, 1:25,000
- Field Order #3, 741st Tank Bn, 21 Jul 1944 (5 pages)
- Intelligence Annex, Annex #1 to FO #3, 21 Jul 1944 (1 page)
- Situation Overlay for Annex #1 to FO #2, 21 July, 1:25,000
- Companies A and B Operations Overlay for FO #3, 21 Jul 1944, 1:10,000
- Overlay of Tank Sorties, Companies A and B, Annex #3 to FO #3, 21 July 1944, 1:10,000
- Tank Plan Company C, Annex #4 to FO #3, 21 July 1944, 1:10,000

FOLDER ARBN-741-3.9, F.O. #1 (duplicate records)

- Field Order #1, 741st Tank Bn, "Neptune," 21 May 1944 (3 pages)
- S-2 Estimate of the Enemy Situation, Annex #2 to FO #1, 21 May 1944 (2 pages)
- Tactical Study of Terrain, Annex #3 to FO 1, 21 May 1944 (1 page)
- Signal Annex to FO #1, 21 May 1944 (1 page)
- Administrative Annex to FO #1, 21 May 1944 (3 pages)
- Operations Overlay, Annex #1 to FO #1, undated, 1:25,000
- Appendix (map overlay of enemy HQ locations in Normandy) to Annex #2, 12 May 1944, 1:500,000
- Operations Overlay, Engineer Plan, Provisional Engineer Group, 22 May 1944, 1:25,000
- Obstacles Off Shore, map locating Hedgerow, Element C, and Ramp-Type obstacles in vic. E-1 and E-3, no scale indicated

Box 13504 has three folders containing 742nd Tank Battalion records
Box 13504 has three folders containing 743rd Tank Battalion records

Signal Corps Motion-Picture Photography

Archives II, College Park, Maryland
35 mm film reels

NARA Ref #	Location	Date
111-ADC-1288	Hill 192	July 12, 1944
111-ADC-1334	Hill 192	July 11, 1944
111-ADC-2232	Paris	August 29, 1944
111-ADC-2233	Paris	August 29, 1944
111-ADC-2235	Paris	August 29, 1944
111-ADC-2237	Paris	August 29, 1944
111-ADC-2239	Paris	August 29, 1944
111-ADC-2262	Paris	August 29, 1944
111-ADC-3238	Wahlerscheid	February 1, 1945
111-ADC-3578	Königsfeld and Dümpelfeld	March 9, 1945
111-ADC-4046	Leipzig	April 18, 1945

111 = Record Group 111 (US Army Signal Corps)
ADC = Army Depository Copy

Morning Reports, 741st Tank Battalion

National Personnel Records Center, St. Louis, Missouri
16 mm microfilm

Month, Year	Item Numbers	Box Numbers	Reel Numbers
Mar 1942	not applicable	2647	8.198
Apr 1942	not applicable	2647	8.198
May 1942	not applicable	2647	8.198
Jun 1942	not applicable	2647	8.198
Jul 1942	not applicable	2647	8.198
Aug 1942	not applicable	2647	8.198
Sep 1942	not applicable	2647	8.198
Oct 1942	not applicable	2647	8.198
Nov 1942	not applicable	2647	8.198
Dec 1942	not applicable	2647	8.198
Jan 1943	not applicable	2647	8.198
Feb 1943	not applicable	2647	8.198
Mar 1943	not applicable	2647	8.198
Apr 1943*	not applicable	2647	8.198
May 1943*	not applicable	2647	8.198
Jun 1943*	not applicable	2647	8.198
Jul 1943*	not applicable	2647	8.198
Aug 1943	07285	190	19.16
Sep 1943	13915, 03730	415, 136	9.99, 10.3
Oct 1943	03855, 22102, 31586	122, 492, 696	7.64, 3.404, 11.950
Nov 1943	40046	673	7.45
Dec 1943	31643	542	5.64
Jan 1944	32483	664	2.84
Feb 1944	25631	592	9.135
Mar 1944	17559	519	17.53
Apr 1944	27200	635	5.134
May 1944	13827	488	14.50
Jun 1944	26105	639	3.140

Jul 1944	16237	540	10.147
Aug 1944	25288	732	20.103
Sep 1944	11173	444	8.132
Oct 1944	15795, 15797**	560, 560	17.196
Nov 1944	09869	400	10.226
Dec 1944	23037	670	5.343
Jan 1945	11091	426	16.356
Feb 1945	20070	544	15.182
Mar 1945	07374	317	7.428
Apr 1945	11012	413	4.527
May 1945	14884	548	18.495
Jun 1945	13194	496	13.389
Jul 1945	13424	491	8.487
Aug 1945	23519	743	1.833
Sep 1945	22367	717	20.415
Oct 1945	00819	31	11.610

* A/741 and C/741 morning reports missing
** B/741 morning reports included with 771 FA Bn reports (Item # 15797)

Beginning in 2024, NARA began posting morning reports online. These records reference the box numbers shown above, but they are called roll numbers. To locate a specific report, one needs to know the month and year as well as the roll number.

Monthly Rosters for Officers

National Personnel Records Center, St. Louis, Missouri
16 mm microfilm

Month(s), Year	Item Numbers	Box Number	Reel Number
Mar–Dec 1942	58603	4581	15.74
Jan–Nov 1943	58603	4581	15.74
Dec 1943	58602	4581	15.74

Payroll Records & Monthly Rosters for Enlisted Men

National Personnel Records Center, St. Louis, Missouri
16 mm microfilm

Company, Year	Item Number	Box Number	Reel Numbers
Hq Co (1943) [1]	27929	3473	2.120
Serv Co (1943)	27929	3473	2.120
Serv Co (1942)	27929	3473	2.120
Co A (1943)	27929	3473	2.120
Co A (1942)	27929	3473	2.120
Co B (1943)	27929	3473	2.120
Co B (1942)	27929	3473	2.120
Co C (1942)	27929	3474	2.121
Co C (1943)	27929	3473	2.120
Co D (1943) [2]	27929	3474	2.121
Med Det (1943)	27929	3474	2.121
Med Det (1942)	27929	3474	2.121

1 Hq Co records missing for 1942
2 D/741 activated per WD letter file AG 322, 4 October 1943

The microfilmed rosters (WD AGO Form No. 305A) and the payrolls (WD Form No. 366/366a) for 1942 and 1943 have box numbers and item numbers that start with the latest date and progress toward the earliest date, the order being counterintuitive.

The rosters and payrolls for 1944 through 1946 no longer exist. The original paper documents were reportedly destroyed prior to the catastrophic fire at the National Personnel Records Center which broke out on July 12, 1973. Apparently, the US Army never microfilmed these rosters and payrolls before the disposal of the paper originals, but that remains an open question. It is known that the General Services Administration accessioned no such film among the sizeable collection it received in 1975, which included all the morning reports from the Second World War plus the rosters and payrolls for 1940 through 1943. The army created this collection after the war by using 16 mm nitrate film that was surplus from the defunct V-Mail service. The collection existed well before the NPRC fire, and the blaze affected none of the film. The full story behind the missing rosters and payrolls remains elusive, though rumors have abounded. While investigating this matter, the author located a memorandum sent to FBI Director Clarence M. Kelley. It stated that on July 17, 1973, an officer of the Florida Highway Patrol temporarily detained two men on a traffic violation, one of whom possessed shredded army payroll records. This man also admitted to being at the NPRC shortly before the fire, and he joked about the calamity. The officer felt that both detainees were being "evasive and holding back information." The FBI never found sufficient evidence for the US Government to prosecute anyone for arson or the unlawful destruction of government property. Aside from the Florida incident, there is another clue in this mystery. During the mid-1960s, author and historian George E. Koskimaki obtained photostatic copies of May 1944 payrolls for the 101st Airborne Division, and these copies survive today at the US Army Heritage and Education Center. Koskimaki acquired them from the NPRC, and they appear to have been made from the paper originals. As of this writing, the enigma of the missing rosters and payrolls elicits more questions than answers.

Note: The General Services Administration oversaw the National Archives from 1949 until the formation of the National Archives and Records Administration in 1984.

Acknowledgments

This book would have been a futile undertaking without the contributions of the people named below. My sincere thanks to all of them.

741st veterans: Lloyd C. Ball, John H. Barner, George E. Beeley, Roger V. Bigham, Jack B. Boardman, John R. Brewer, Jack H. Browder, Charles R. Buchanan, Robert C. Buckholtz, Craig K. Carter (a.k.a Kelly B. Layman), Thomas D. Chastain, Alphonse D'Ambrose, Louis J. DiClementi Jr., Andrew W. Divers, Charles W. Eubanks, Valentine Fister Jr., Philip L. Fitts, Frank F. Frimel, Ralph R. Garber, George R. Geddes Jr., Verdie M. Guest Jr., Harrell E. Gullatt, Nelson W. Hall, Leonard E. Hatfield, Walter R. Jutkins, Glenn W. Kieffer, Anthony E. Kroeger, Virgil H. Lohman, Lytle C. Morton, Wilbert C. Mortzfield, Niles N. Nutter, John Onuschak, Paul W. Ragan, Herman W. Sendelbach, John G. Suydam, Edward Teffeteller, Leonard H. Trimpe, Joseph T. Walsh, Edward W. Warnock, Willis D. Warren, Vernon C. Wilson, Ralph A. Woodward, Mayne B. Youngblood, Philip J. Zodda

741st widows and family members: Douglas R. Anderson, Joan Barcellona, Elaine Briggs, Jeffrey B. Brown, Dorothy T. Carpenter, Mary Louise Clark, Eva Covington, Rosemary Crisler, John T. Dallas, Joshua A. Dew, DeRonda E. Elliott, Antonio R. Jean, Ruth M. Kieffer, Pamela Krunsberg-Pitcher, Anna Kadlecova-Lucas, Robert C. Legnini, Paul J. McDaniel, Karen A. Quill, Dorothy Rutyna, Mary O'Shaughnessy, Edward S. Sledge III, Edward S. Sledge IV, Peggie J. Smith, Connie M. Starcher, C. Arthur Steineker, George H. West, David G. Wiltrout

Other veterans: Donald G. Adkins, Odis Bone, Joseph Busi, Richard H. Byers, Harlin E. Coffinger, Charles D. Curley, Kenneth J. Kurtenbach, James W. Love, William C. Mitchell, Tom C. Morris, Joseph Laurence Noël, Rudolf von Ribbentrop, Gerhard Salomon, Anthony C. Sienkiewich, Edgar M. Stuart, Charles Warax, Ralph W. Widener Jr.

National Archives employees: Richard L. Boylan, Sharon A. Culley, Theresa A. Fitzgerald, Martin A. Gedra, Lynn A. Goodsell, Thomas C. Hayes, James M. Hébert, James K. Kelling, James S. Konicek, Christina M. Kovac, Andrew Knight, Daria M. Labinsky, Sarah E. LeRoy, Mark P. Meader, Susan K. Nash, Timothy K. Nenninger, Donna L. Noelken, Holly L. Reed, Eric S. VanSlander, Victoria S. Washington

Mark A. Bando, Cleve C. Barkley, Peter W. Brown, Helmut Brüls, William C. C. Cavanagh, Bernard Collard, Thomas S. Colones, Daniel Fong, Geoff D. Gentilini, Raphaël d'Amico-Gérard, Joseph L. DeMarco, Dale Edwards, Jonathan G. Gawne, Christian Habisohn, Erich Hönen, Edwin F. Huffine, Dale Johnson, Norbert Koch, Guy Lejoly, Charles R. Lemons, Marcel Leveel, Marc Marique, Douglas McCabe, Jean-Luc Menestrey, Holger Menrad, Laurence I. Munnikhuysen III, Douglas E. Nash, Darren R. Neely, Christopher Nesbit, Joseph P. Oney, Jaclyn A. Ostrowski, Katherine H. Rasdorf, Michael F. Reynolds, David A. Schwind, Jean-Louis Seel, Jo Anna Shipley, Brian N. Siddall, Anthony L. Sienkiewich, Jean-Philippe Speder, Remy M. Spezzano, Hanno Spoelstra, Deborah P. Tamulis, John W. Tengea, Jean Torchio, Jos Vogel

Special thanks to my creative team: Laura A. Johnson, Craig W. H. Luther, Ian Pamplona, Michael J. Rivilis, Felipe Rodna, and Kelly A. Tooman

Notes and Citations

1 Jacob Robert Moon retired as a lieutenant colonel in 1948 and became president of Tomoka Brick & Tile Company in Florida. While in Florida, he established an equestrian academy at Ormand Beach in the former stables of John D. Rockefeller. Moon's love for horses extended back to his boyhood in Goodwater, Alabama, where he learned to ride his father's mules and horses. While attending West Point, he assisted with cadet riding instruction and played on the polo team. He continued riding while posted to The Infantry School, The Cavalry, and later the Philippine Scouts. The war interrupted his constant contact with horses and turned his attention toward the Armored Force. Once hostilities ended, he returned to his beloved animals. He not only taught riding and jumping but also bred and trained purebloods. After disposing of Tomoka Brick & Tile, he and his wife retired to Hillsborough, North Carolina. In 1969, he became president of Moon-Web Products Incorporated. The company provided an outlet for his inventive spirit, which led to patents for a self-adhesive bandage for horses, a horse blanket intended to stay in place, and a smokeless burner. He also held a series of patents for a water-conservation device to reduce the water volume used by bathroom toilets. The patents for the latter invention were infringed upon, and litigation was underway at the time of his death. He was seventy-four years old when he died on December 5, 1973, at the VA hospital in Durham, North Carolina. He left behind a wife, son, daughter, six grandchildren, and five great-grandchildren. They interred him at Hillsborough's old Saint Matthew's Episcopal Church cemetery.

2 Cannon, Heintzleman, and Lanier, *The Story of "Vitamin Baker,"* p. 2.

3 The Mail Steamship (MS) *Capetown Castle* was built in 1938 at the Belfast ship works of Harland & Wolff Limited, the same firm that built the RMS *Titanic*. The Union-Castle Mail Steamship Company Limited (more commonly called the Union-Castle Line) owned and operated the ship. After its conversion to a troopship during the war, the vessel returned to civilian passenger service in January 1947.

4 The collision sent thirty-five USS *Murphy* sailors to their deaths: two ensigns and thirty-three enlisted men. The USS *Jeffers* rescued three survivors from the water. Some accounts erroneously state that thirty-six men perished at sea, mistakenly including Seaman First Class Ernest S. Kelly. He died the previous day aboard the SS *Gulfbelle* when it collided with the SS *Gulfland* along the Florida coast. The ABMC erroneously lists his date of death as October 21, 1943.

5 During the autumn of 1944, the US Army officially adopted the name "General Sherman" for the M4 series and "General Stuart" for the M5 series. The Ordnance Department announced the name adoptions in Research and Development Service Order No. 13-44 ("Nicknames for Ordnance") dated November 24, 1944. Bearing the signature of Major General Gladeon M. Barnes, the order indicated that the names Sherman and Stuart were "in current use by men in the field." Newspapers in the United States had also published the names, but they didn't gain widespread usage by Americans until the postwar era. It's worth noting that

German troops used both names during the war. Thanks to Nicholas Moran for locating an original copy of Barnes's order at the US National Archives.

6 Letter from Frank M. Elliott to his wife, November 12, 1943.

7 Ladock became the temporary home for Companies A and B as well as Headquarters Company, Service Company, and the Medical Detachment. Only Company C settled at Brixham.

8 Luftwaffe aircraft struck Bryanston Square exactly two weeks after Moon joined First US Army Group (FUSAG). Incendiary bombs destroyed three buildings used for billets and five office buildings, including the one used by Headquarters Special Troops. Moon was uninjured. There were no fatalities during the nighttime raid, part of the so-called "Baby Blitz" from January through May 1944.

9 Cannon, Heintzleman, and Lanier, *The Story of "Vitamin Baker,"* p. 15.

10 Four 741st companies (A, D, Service, and Headquarters) trained at Camp Merrion from January 29 to February 23, 1944. The Medical Detachment was also there throughout the period. Company D and Service Company returned in March, and so did the Medical Detachment.

11 The US military conducted two major rehearsals designed to duplicate the events and conditions expected during the actual invasion: Exercise Tiger (April 22-30) and Exercise Fabius I (May 2-6). Tiger involved the units scheduled to storm the Normandy beach code-named Utah, and Fabius I involved the units destined for the beach code-named Omaha.

12 For the journey to Normandy, the support echelon had space aboard nine LCTs, three LSTs, two attack transport ships, and one infantry landing ship.

13 The 610th Engineer Light Equipment Company largely consisted of draftees from Florida and Tennessee. The unit was activated in California per General Orders #5, Headquarters, II Armored Corps, 13 January 1943. The 610th reached the United Kingdom in December 1943. Captain Robert B. Shanks commanded the company until a landmine wounded him during August 1944. First Lieutenant Lloyd E. York succeeded Shanks.

14 Harrison, *Cross-Channel Attack*, p. 274.

15 Valentine Fister's oral history, recorded March 1993.

16 Author's interview with Valentine Fister, January 23, 1998.

17 Fister told the author that the tank had a pin-up girl shellacked to its sponson alongside the name "Always Smiling." He said the girl was a blonde, who looked something like Marilyn Monroe.

18 Raymond E. Eberhardt's Action Report dated July 21, 1944, and submitted to the Commander, US Naval Forces, Europe.

19 Valentine Fister's oral history, recorded March 1993.

20 Ibid.

21 Ibid.

22 Author's interview with Paul J. McDaniel, July 12, 2001.

23 Robert A. Rowe's interview with Ralph A. Woodward, July 17, 1987, p. 54 (Rowe Collection, USAHEC).

24 Ibid., p. 55.

[25] Brett A. Phaneuf, marine archeologist and leader of Project Neptune 2K, was the first person to identify poor seamanship as a factor in the disaster. His team surveyed nineteen sunken DD tanks during the summer months of 2000 and 2001. The orientation of the wrecks on the seabed was valuable evidence. All pointed toward the same location on land, the church steeple at Colleville-sur-Mer. The church was the most visible landmark from the sea, a landmark known to all the tank crews. As Phaneuf explained, "What makes sense is that they were trying to fight the current to stay in this area and were heading toward the steeple." While struggling to stay on course, they turned sideways to the swells, which made them vulnerable to swamping. The DD tank research conducted by Project Neptune 2K served as the basis for a 2002 documentary film called "D-Day Beneath the Waves." The project was an undertaking of the Institute of Nautical Archaeology, headquartered at Texas A&M University.

[26] Author's interview with Philip L. Fitts, December 17, 1995.

[27] "Launching of tanks for the assault on D-Day," by Ens. Henry Patrick Sullivan, June 15, 1944, p. 1 (RG 38, NARA).

[28] Cannon, Heintzleman, and Lanier, *The Story of "Vitamin Baker,"* p. 26.

[29] Robert A. Rowe's interview with Jack B. Boardman and Millard I. Case, July 18, 1987, p. 4 (Rowe Collection, USAHEC).

[30] Three members of Millard Case's crew died, namely Merrill E. Chapman (driver), Ralph E. Kunkle (assistant driver), Joseph B. Kaminesky (loader).

[31] Cannon, Heintzleman, and Lanier, *The Story of "Vitamin Baker,"* p. 28.

[32] "Introduction to the 741st Tank Battalion," by Joseph "Larry" Noël, October 1997, p. 1. Noël served as executive officer aboard LCT(HE) 2425 on D-Day.

[33] S/Sgt. Fair's & Sgt. Larsen's Report, Unit Journal, 741st Tank Battalion, June 1944 (RG 407, NARA).

[34] Ibid.

[35] Author's interview with Donald G. Adkins, February 24, 1996.

[36] Author's interview with William C. Mitchell, February 28, 1996.

[37] Shortly before the invasion, LCT 210 replaced LCT 2228 due to a mechanical failure. Both craft were Mark 5 LCTs, though 210 was unarmored, unlike 2228.

[38] Edward L. Ayers served with Company A, 91st Armored Reconnaissance Battalion (11th Armored Division) before joining Headquarters Company, 741st in May 1943. He received an Honorable Discharge in September 1945 but returned to active duty in January 1948 by joining the US Air Force as a clerk typist. Ayers soon faced a fraudulent-enlistment charge when an investigation revealed his civilian arrest record. He had divulged this information when entering the military in 1942, and there seemed no need to mention the matter again when reenlisting. His unblemished combat record helped mitigate the situation, and his commanding officer waived the charge. Afterward, alcoholism and unexcused absences ended his military career in March 1950, when the Air Force cast him out with a General Discharge (after his squadron commander first sought an "Undesirable Discharge").

[39] Report by Pvt's Arthur Barker, John Froehlich, and Charles D. Anderson, Unit

Journal, 741st Tank Battalion, June 1944 (RG 407, NARA).

40 Robert F. Hunter from Tulsa, Oklahoma, commanded the 336th Engineer Combat Battalion.

41 Besides John Froehlich, Privates Arthur W. Barker and Charles D. Anderson received Bronze Star Medals for heroic achievement.

42 Habib's crew consisted of Corporal Anthony E. Annunziata, Technician Fourth Grade Lynn E. Hurte, Private Arthur W. Mains, and Private Charles Lombardo.

43 Spooner's crew consisted of Corporal Lawrence E. Meade, Technician Fifth Grade James C. Boyett, Technician Fifth Grade John W. Farr Jr., and Private Franklin E. Stallings.

44 Author's interview with Herman W. Sendelbach, February 21, 1999.

45 Ibid.

46 S/Sgt. Fair's & Sgt. Larsen's Report, Unit Journal, 741st Tank Battalion, June 1944 (RG 407, NARA).

47 Sledge's M4A1 carried artwork alongside the name "Alabama II." The art consisted of the letter "A" with an elephant's head peeking out from under it. The composition was a tribute to the University of Alabama. The "A" included a scroll with writing that may have read "Bama" or "Crimson Tide." It's not legible in the only known photograph, a still image from motion-picture film published online by Critical Past. Although a Citadel graduate, Sledge identified with the Crimson Tide as did many Alabamians. He grew up in Mobile. The registration number on his tank was USA-3036825. The registration number on Roger McDonough's tank ("Ace of Spades") was USA-3036894.

48 Cannon, Heintzleman, and Lanier, *The Story of "Vitamin Baker,"* p. 28.

49 Ibid.

50 Author's interview with George R. Geddes, November 30, 1995.

51 Cannon, Heintzleman, and Lanier, *The Story of "Vitamin Baker,"* pp. 24-25.

52 Robert A. Rowe's interview with Jack B. Boardman and Millard I. Case, July 18, 1987, p. 51 (Rowe Collection, USAHEC).

53 Cornelius Ryan's questionnaire completed by William D. McClintock, undated but probably 1958 (Ryan Collection, Ohio University).

54 Letter from Leonard H. Trimpe to the author, undated.

55 Letter from Lytle C. Morton to the author, undated but 1997.

56 "H-Hour Plus 60 Minutes, Omaha Beach," Leonard "Dick" Trimpe, 2001, p. 38.

57 Cornelius Ryan's questionnaire completed by William Friedman, undated. Friedman served as S-1 for the 16th Infantry (Ryan Collection, Ohio University).

58 Valentine Fister's oral history, recorded March 1993.

59 Ibid.

60 Ibid.

61 Ibid.

62 Letter from Pauline L. Elliott to Frank M. Elliott, June 6, 1944.

63 Author's interview with Leonard E. Hatfield, August 15, 1997.

64 Ibid.

65 Letter from Leonard H. Trimpe to the author, undated.

66 On June 15, the 2nd Infantry Division moved its command post from le Molay-Littry to a wooded grove near the Château de la Boulaye. The former CP closed at 14:08, by which time the new location was set up.

67 Some postwar historians have rated the 3.Fallschirmjäger-Division as Germany's best division in Normandy, albeit a subjective appraisal.

68 Author's interview with John H. Barner, April 13, 1996.

69 Unit Journal, 741st Tank Battalion, June 16, 1944 (RG 407, NARA).

70 Hurley Fuller received a second chance at command, taking over the 110th Infantry Regiment (28th Division) on November 24, 1944. He and many others in his regiment became prisoners of war on December 18, 1944, at Clervaux, Luxembourg.

71 Letter from Tom C. Morris to the author, March 16, 1996.

72 The Château de la Commanderie near Balleroy served as the 1st Infantry Division command post.

73 Frontstalag 221 consisted of the main camp established in 1940 at Saint-Médard-en-Jalles in Bordeaux and several subcamps, including a far-flung location opened in 1943 at Rennes in Brittany. Though nearly 500 kilometers apart, Rennes and Saint-Médard were both known as Frontstalag 221 and utilized the same Feldpost number (17427A).

74 Author's interview with Jack B. Boardman, February 22, 1996.

75 Ibid.

76 Letter from Charles D. Curley to the author, April 19, 1996.

77 Historian Kenneth W. Hechler's narrative, Hill 192, 11 July 1944 (Combat Interviews, 2nd Infantry Division, RG 407, NARA).

78 Letter from Tom C. Morris to the author, March 16, 1996.

79 Ibid.

80 Unit Journal, 741st Tank Battalion, 1 July 1944 (RG 407, NARA).

81 *History, V Corps*, p. 104.

82 Annex #2, Field Order #2, 741st Tank Battalion, July 9, 1944 (RG 407, NARA).

83 Lieutenant Colonel Richard J. Hunt joined the team at Vitamin Forward. The visitor commanded the 744th Light Tank Battalion, which had recently arrived in France. Hunt and a lieutenant from his unit planned to study the July 11 attack as it unfolded.

84 Garth and Taylor, *St-Lô (7 July – 19 July 1944)*, p. 65.

85 On July 11, 1944, the Company C platoon leaders were: Edwin H. Jacobs, First Platoon; Jerry B. Brown, Second Platoon; Quentin J. Swan, Third Platoon.

86 Unlike its sister companies, Company C committed all three of its tank platoons during the initial assault on July 11, 1944. The company had no reserve platoon due to the challenges posed by Purple Heart Draw and another draw farther south. Those terrain obstacles necessitated additional armor and an elaborate plan for tanks to circumvent the two gullies. The "Tank Scheme of Maneuvers," drafted on July 7 (Annex No. 6 to Field Order No. 5, Headquarters 23rd Infantry), diagrammed the movements. The additional armor gave Company A, 23rd four more tanks. Company C, 23rd had two additional tanks. Captain Ken Hechler

lacked the Tank Scheme of Maneuvers document when he conducted "combat interviews" five months after the battle (a deficiency he noted). The tankers he interviewed had difficulty recalling the plan and the moves they actually executed. Minor errors and omissions thus appeared in Hechler's manuscript. The War Department Historical Division repeated these errors in its 1946 publication, *St-Lô (7 July – 19 July 1944)*.

87 Author's interview with Philip L. Fitts, December 19, 1995.

88 Fallschirm-Sturmgeschütz-Brigade 12, commanded by Hauptmann Günther Gersteuer.

89 Author's interview with Philip L. Fitts, December 19, 1995.

90 Jerry Brown (born Benjamin Brown in Cleveland, Ohio) changed his name to Jerry Ben Brown before volunteering for military service. He joined the 741st on December 28, 1943, and transferred to Company C on January 6, 1944. Brown returned to civilian life in November 1945 and worked in the wholesale lighting business, eventually starting his own company, J.B. Brown Electric. The former platoon leader switched careers in mid-life and entered the employment-services industry. In 1973, he and his son, Jeffrey, established J.B. Brown and Associates, an employment agency in Cleveland. In later years, Jerry and his wife, Viola, retired to Tamarac, Florida, where he pursued photography as a hobby, specializing in object photography, e.g., art objects such as sculptures. He died at age seventy in 1989 and rests at Beth David Memorial Gardens in Hollywood, Florida.

91 Ibid.

92 Author's interview with Charles R. Buchanan, April 13, 1996.

93 Ibid.

94 Company B's Third Platoon supported Company E, 38th Infantry, while the Second Platoon supported Company F. The First Platoon was in reserve. Company E attacked at H-hour on July 11, and Company F attacked twenty minutes later. Company E had more ground to cover, hence the earlier start.

95 Letter from Charles D. Curley to the author, May 12, 1996.

96 Curley, *How a Ninety-Day Wonder Survived the War*, p.83.

97 Ibid.

98 Letter from Charles D. Curley to the author, April 19, 1996.

99 Curley, *How a Ninety-Day Wonder Survived the War*, p.86.

100 Ragan's crew destroyed two assault guns on July 11. Contemporary accounts are imprecise, but the guns were apparently of different types. Given the composition of Fallschirm-Sturmgeschütz-Brigade 12, one gun may have been a StuG III and the other a StuH 42.

101 Author's interview with Wilbert C. Mortzfield, November 30, 1995.

102 Ibid.

103 Curley, *How a Ninety-Day Wonder Survived the War*, p.94.

104 White, "Sgt. Culin Licks the Hedgerows," p. 82.

105 Culin and Litton and two other collaborators each received the Legion of Merit.

106 V-Mail letter from Edward S. Sledge II to his father, July 12, 1944.

107 V-Mail letter from Edward S. Sledge II to his father, July 17, 1944.

108 *The Mobile Press,* Mobile, Alabama, July 27, 1944, p.1.

109 Airmail letter from George R. Coleman to his parents, July 14, 1944.

110 Letter from Roy D. Coleman to George R. Coleman, May 2, 1944. Roy served in combat as a private with Company K, 130th Infantry Regiment (33rd Division) until wounded in the upper right arm by a rifle bullet on October 14, 1918, during the Meuse-Argonne offensive.

111 Cannon, Heintzleman, and Lanier, *The Story of "Vitamin Baker,"* p. 38.

112 Ibid.

113 Ibid.

114 Ibid.

115 Ibid., p. 39.

116 Ibid.

117 Ibid., pp. 39-40.

118 Author's interview with Willis D. Warren, July 21, 1996.

119 Author's interview with Philip L. Fitts, December 17, 1995.

120 First Army records confirm the destruction by fire of Malcolm Reynolds's tank (USA W-3036860) and Willis Warren's tank (USA 3022788). It is also possible that Calvin Meyers's tank was a total loss, but the author found no corroborating evidence. (*Intelligence Report of Tanks Rendered Inoperative Due to Enemy Action, First United States Army, Armored Section, July 1944,* RG 331, NARA)

121 After-Action Report, 741st Tank Battalion, July 1944, p. 7 (RG 407, NARA).

122 Fardella, *Combat History of the 741st Tank Battalion*, p. 3.

123 The railway gun was a 24cm "Theodor Bruno" (Deutsche Reichsbahn number 919084) from Eisenbahn-Batterie 722. The unit also abandoned three guns at Vire.

124 The 741st Tank Battalion After-Action Report for August 1944 states that Company A attacked astride the Guilberville-Cabotière road, and that the core of enemy resistance centered on a crossroads near la Renevière (mistakenly called Larveniere on wartime maps). The report goes on to indicate that the tanks commanded by Hecox and Smulik were lost at this crossroads. That is incorrect. La Renevière lies along N174 and the attack route of Company C, which supported Second Battalion, 23rd Infantry. Company A supported Third Battalion, 23rd and attacked down N175 from la Couaille toward le Poteau. Both tanks met harm at or near the crossroads called la Croix Étêtée, referred to as Pt. Aunay in wartime documents. Le Petit Aunay is actually a farmstead close to the crossroads. (The same documents refer to le Poteau as la Cabotière. The name la Cabotière actually applies to a nearby farm.)

125 The projectile that struck Smulik's tankdozer may have been a Stielgranate 41. However, First Army personnel examined the tank (USA 3039419) and recorded that an 88 mm round hit the rear motor compartment at a thirty-degree angle, causing total destruction of the vehicle by fire. (*Intelligence Report of Tanks Rendered Inoperative Due to Enemy Action, First United States Army, Armored Section, August 1944,* RG 331, NARA)

126 Lt. Covington, Individual Reports, Company Commanders and Platoon Leaders, Addenda to Unit Journal, 741st Tank Battalion, August 1944, p. 3 (RG 407, NARA).

127 Unit Journal, 741st Tank Battalion, August 3, 1944 (RG 407, NARA).

128 Mr. Swan, Individual Reports, Company Commanders and Platoon Leaders, Addenda to Unit Journal, 741st Tank Battalion, August 1944, p. 3 (RG 407, NARA).

129 Lt. Covington, Individual Reports, Company Commanders and Platoon Leaders, Addenda to Unit Journal, 741st Tank Battalion, August 1944, p. 3 (RG 407, NARA).

130 V-Mail letter from Edward S. Sledge II to his mother, August 8, 1944. Captain Sicks returned to duty and became commander of Company A, 19th Tank Battalion (9th Armored Division).

131 Medics transported Ragan to the 109th Evacuation Hospital before removing him to a general hospital in the United Kingdom. He spent sixty-nine days hospitalized before rejoining Company B on November 18. On December 14, he became first sergeant, replacing Robert E. McLendon, who moved up to battalion headquarters and became a master sergeant.

132 Letter from Charles D. Curley to the author, June 2, 1996.

133 Ibid.

134 Unit Journal, 741st Tank Battalion, August 18, 1944 (RG 407, NARA).

135 Cannon, Heintzleman, and Lanier, *The Story of "Vitamin Baker,"* p. 48.

136 V-Mail letter from George R. Coleman to his family, August 15, 1944.

137 V-Mail letter from George R. Coleman to his parents, August 31, 1944.

138 Hillner, Meacham, and Smith, *Vitamin Charley*, p. 33.

139 Fardella, *Combat History of the 741st Tank Battalion*, pp. 9-10.

140 Hillner, Meacham, and Smith, *Vitamin Charley*, p. 34.

141 Cannon, Heintzleman, and Lanier, *The Story of "Vitamin Baker,"* p. 50.

142 *History, V Corps*, p. 235.

143 Fardella, *Combat History of the 741st Tank Battalion*, p. 11.

144 In October 2022, the author visited the site of Dew's breakthrough and located a gap in the dragon's teeth at 50° 9'12.52" N, 6°13'8.20" E. The gap measured 114 inches (290 cm) in width at its narrowest point, just large enough for an M4 or M4A1 medium tank to pass through. Teeth on either side of the gap exhibited scrape marks likely caused by armored vehicles. The gap itself was created by eliminating four concrete teeth; its location roughly corresponds to that shown on Lieutenant O'Shaughnessy's battle map, which the author discovered in 2010 (see note 174). Among the 28th Infantry Division's combat interviews at the US National Archives, the author found a map overlay labeled "Attack—2nd Bn, 110th, 14 Sept 44." The point of penetration circled on the overlay roughly corresponds to the 114 inch gap. Some one hundred meters southwest of this area, the author located a section of missing teeth measuring forty feet (twelve meters) wide. This was too large for Dew and his crew to have created with AP shells, but could have been made by combat engineers using high explosives. It's also possible that local residents removed the teeth after the war to allow for the passage of farm machin-

ery. While combing through textual records at the National Archives, the author found a relevant entry in the unit journal of the 110th Infantry Regiment for 14 September 1944: "Tanks knocked out two dragon's teeth—are making path for themselves." But alas, the journal entry provides no grid coordinates for the path. The aforementioned map and overlay are the only known primary source documents that locate Dew's gap.

145 Lt. Cinquina, After-Action Report, Company B, Addenda to Unit Journal, 741st Tank Battalion, September 1944 (RG 407, NARA).

146 Lt. O'Shaughnessy, After-Action Report, Company B, Addenda to Unit Journal, 741st Tank Battalion, September 1944 (RG 407, NARA).

147 Ibid.

148 Cannon, Heintzleman, and Lanier, *The Story of "Vitamin Baker,"* pp. 51-52.

149 Leonard's statement, Missing Report for Capt. James G. Thornton Jr., dated September 18, 1944 (Individual Deceased Personnel File for Thornton).

150 Letter from Charles R. Thomas to James G. Thornton Sr., June 8, 1945 (Individual Deceased Personnel File for Thornton).

151 Ibid.

152 Historian John S. Howe's narrative, "Into Germany, 110th Infantry Regiment, 17-30 September 1944," insert for p. 19 (Combat Interviews, 28th Infantry Division, RG 407, NARA).

153 PW Recapitulation, G-2 Periodic Report No. 50, 28th Infantry Division, 15-16 September 1944 (RG 407, NARA).

154 The captured commander was Captain Robert H. Shultz (1919–1998), CO of Company F, 110th Infantry. In addition to the thirty-six prisoners, seven men from his company died.

155 Author's interview with Philip J. Zodda, July 20, 1996.

156 Ibid.

157 Ibid.

158 Ibid.

159 Ibid.

160 Fardella, *Combat History of the 741st Tank Battalion*, p. 14.

161 Capt. Young, After-Action Report, Company C, Addenda to Unit Journal, 741st Tank Battalion, September 1944 (RG 407, NARA).

162 Steineker joined the 741st from the 41st Replacement Battalion. He previously served with the 9th Tank Battalion, 20th Armored Division. In January 1944, the War Department pulled him from the 9th (along with other officers such as Lt. George Coleman) and dropped him into the replacement system.

163 Beck's IDPF contains the following statement: "An examination by 463rd Ordnance Evacuation Company of the interior of subject tank (USA W-3039916) at the time it was evacuated failed to disclose any evidence of human remains inside."

164 Author's interview with Frank F. Frimel, February 20, 1997.

165 Ibid.

166 On October 4, 1944, Roger McDonough recommended Steineker for the Silver Star Medal, citing his gallantry on September 20. Cecil Thomas and Robert Skaggs

endorsed the recommendation. The Awards and Decorations Board at Headquarters, 28th Infantry Division concurred, but two staff officers disagreed, the G-1 and Chief of Staff. They considered the Bronze Star Medal more appropriate, and the Commanding General, Norman Cota, felt the same or at least deferred to them. McDonough's recommendation ultimately reached V Corps headquarters, and Steineker received the BSM on October 27.

167 Author's interview with Frank F. Frimel, February 20, 1997.

168 Cannon, Heintzleman, and Lanier, *The Story of "Vitamin Baker,"* p. 53.

169 Ibid., p. 54.

170 Letter from Nelson W. Hall to the author, August 30, 1996.

171 Ibid.

172 Unit Journal, 741st Tank Battalion, September 22, 1944 (RG 407, NARA).

173 G-2 Periodic Report No. 56, 28th Infantry Division, 21-22 September 1944 (RG 407, NARA).

174 After-Action Report, 741st Tank Battalion, September 1944, p. 6 (RG 407, NARA).

175 Historian John S. Howe's narrative, "Into Germany, 110th Infantry Regiment, 17-30 September 1944," p. 37 (Combat Interviews, 28th Infantry Division, RG 407, NARA).

176 Patrick O'Shaughnessy (1919–2008) retained a map that he used in September 1944. It's a soiled and battle-worn copy of AMS 841, Sheet 5803, Leidenborn, 1:25,000. The author visited his widow, Mary, on October 9, 2010, and scanned portions of the map. This rare document has a red overprint showing German defense works. It also exhibits handwritten markings made by O'Shaughnessy, including the location of Joe Dew's path through the dragon's teeth.

177 Letter from Robert N. Skaggs to William C. C. Cavanagh, undated but probably 1983.

178 Father Poepperling earned two Bronze Star Medals during the Second World War—one for aiding wounded soldiers on D-Day and another for similar actions during the "Bulge." He retired in 1979 and died in May 1983 at age seventy-eight. Over twenty years later, after the child sexual abuse scandal surfaced in the Roman Church, two female plaintiffs identified Poepperling as a pedophile, who repeatedly assaulted them as young children during the 1950s.

179 Company C occupied the bivouac from 19:30 hours on September 16 until 11:00 hours on December 11. The men built log cabins and used tanks to furnish electricity for lights inside the dwellings. Many soldiers called the place "Snow Mountain," a name partly derived from the German word Schnee-Eifel.

180 The Cafe Henkes no longer exists in Schönberg. The building was torn down after the war and replaced by the Zum Burghof café, a hotel and restaurant located at K.F.-Schinkel Straße 11.

181 Letter from Nelson W. Hall to the author, December 21, 1996.

182 Airmail letter from George R. Coleman to his parents, October 9, 1944.

183 Airmail letter from George R. Coleman to his parents, October 20, 1944.

184 Major King left the 741st on September 25, 1944. He traveled to the 67th Evacuation Hospital and then to a general hospital, spending eighty-three days recuper-

ating before returning to duty. His arthritic condition prevented him from returning to the 741st or any combat unit.

185 Letter from Nelson W. Hall to the author, December 21, 1996.

186 Letter from Robert N. Skaggs to William C. C. Cavanagh, undated but probably 1983.

187 Letter from Nelson W. Hall to the author, date unknown.

188 Fardella, *Combat History of the 741st Tank Battalion*, p. 15.

189 *Combat History of the Second Infantry Division in World War II*, p. 81.

190 Airmail letter from George R. Coleman to his parents, November 16, 1944. Bob Connell and Chick Bennorth both survived the war. Bob piloted a B-24J with the 372nd Bomb Squadron, 307th Bomb Group. Gunfire from a Japanese Nakajima Ki-43 (Oscar) brought down his ship on October 3, 1944, while on a mission to bomb oil refineries at Balikpapan, Borneo. Ten crewmen survived the calamity, and a US Navy submarine later rescued them. Chick served with Company C, 745th Tank Battalion and spent fifty-seven days hospitalized with burns suffered on October 13, 1944, during the ferocious house-to-house battle for Aachen, Germany.

191 Donald Kepplin (1920–2010) joined Company A on June 19, 1944, from the 16th Replacement Depot.

192 Airmail letter from George R. Coleman to his parents and sister, November 28, 1944.

193 The army deemed the criteria for home furloughs to be unfair because soldiers longest in combat and most in need of rest were often denied priority. A policy change followed, and the sole criterion became the length of time overseas. Although the number of men granted furloughs in each unit was small, the resulting morale boost spawned a plan to limit frontline service. A soldier would thus have a goal toward which to fight. About 200 combat days was the desired period before a reprieve, preferably stateside duty, but the program remained nothing more than a grand notion. Army planners could not solve the monumental problem of replacing departed combatants. US Army soldiers never benefited from such a program during the Second World War.

194 Their ranks included the younger brother of the hospitalized Lieutenant O'Shaughnessy. Private First Class Vincent J. O'Shaughnessy served with the 423rd Infantry Regiment and became a prisoner of war on December 21, 1944. The Germans sent him to Stalag VIIIA Görlitz.

195 Location of Vitamin Forward CP (Nr. 100 Rocherath) shown on Assault Map, Elsenborn and Vicinity, Sheet 10, 1:10,000 (NARA, Box 13503, 407/270/58/35/5, NM-3 Entry 427). The map also locates Hall and Neidrich's tanks. The battalion's after-action report for December corroborates their position, stating the tanks were in "a lane just east of the CP."

196 Letter from Nelson W. Hall to the author, August 8, 1996.

197 Airmail letter from George R. Coleman to his parents, December 13, 1944.

198 Letter from Nelson W. Hall to the author, August 8, 1996.

199 Beeley, *I Must Have Had A Guardian Angel*, p. 63.

200 G-3 Journal, 2nd Infantry Division, 12:57, December 16, 1944. The journal entry reads "Inspire Hqs," which translates to 23rd headquarters.
201 Cannon, Heintzleman, and Lanier, *The Story of "Vitamin Baker,"* p. 62.
202 Unit Journal, 741st Tank Battalion, December 1944, p. 3.
203 Cannon, Heintzleman, and Lanier, *The Story of "Vitamin Baker,"* p. 62.
204 McDonough also received credit for destroying an armored car, but no photographic evidence has ever surfaced to corroborate this claim. Furthermore, Peiper's vanguard is not known to have employed armored cars. The vanguard did possess halftracks, and the one lost to McDonough and the 644th TD Battalion appears on page 45 of the latter's unit history published in 1945 by Muster-Schmidt-Verlag of Göttingen, Germany. The destroyed halftrack rests in a meadow overlooking Holzwarche Creek.
205 Company D helped defend Domäne Bütgenbach on December 17, along with three 76 mm guns from the Third Platoon of Company C, 801st Tank Destroyer Battalion and soldiers from the 254th Engineer Combat Battalion. The next day, the TD platoon and Company D were attached to Company C, 612th Tank Destroyer Battalion also now defending Domäne Bütgenbach alongside the Second Battalion, 26th Infantry Regiment.
206 History of the 23rd US Infantry, December 1944, p. 7 (RG 407, NARA).
207 Gasthaus zur Waldeslust stood in open pastureland on the road between Lausdell and Ruppenvenn (50°26'13.50"N, 6°19'34.75"E). Mathias Rauw and his wife, Johanna Catharina, owned and operated the inn, which catered to local hunters. Prior to the battle, the 324th Engineer Combat Battalion (99th Division) maintained its CP there. The building was destroyed during the battle, never rebuilt, and no trace of it remains today. It had a Krinkelt house number: Nr. 96. Part of Krinkelt extended far to the northeast beyond Rocherath, a little-known fact.
208 Author's interview with Charles B. MacDonald, October 12, 1984.
209 MacDonald, *Company Commander*, p. 90 (1947 edition).
210 Born on October 18, 1914, Siegfried Müller grew up without his father who was an architect and also served in the military. Paul Müller died in 1916 at age twenty-eight during the battle for Verdun. As a teenager, Siegfried followed in his father's footsteps, entering into an apprenticeship with an architect named Wilhelm Eggeling in Essen-Bredeney. Also like his father, Siegfried's life turned toward military service when he joined the German Army in 1935 as a member of an engineer battalion located at Königsberg. He rose from the enlisted ranks to become a reserve officer in 1938. That same year, he left the army and transferred as an officer to the SS-Totenkopfverbände, serving with SS-Totenkopf-Standarte 2 "Brandenburg," which oversaw the Sachsenhausen Concentration Camp. During the Polish campaign, the men of Brandenburg conducted "police and security" operations behind the front, arresting and executing hundreds of Polish Jews, intellectuals, and other perceived enemies. Afterward, the unit furnished troops for the newly formed SS-Totenkopf-Division, and so the Brandenburgers transitioned from camp guards to combat soldiers. Müller gravitated to the division's engineer battalion, where he gained his first command in January 1940, taking charge of the battal-

ion's 1.Kompanie. He led that unit in combat across Belgium and France, earning the Iron Cross Second Class and the Iron Cross First Class. Promoted to SS-Hauptsturmführer in early 1941, Müller briefly moved to Norway, where he helped SS-Kampfgruppe "Nord" build an engineer company, an assignment lasting only four months. His frontline service resumed with the Totenkopf-Division, where his engineer company saw heavy combat in the Soviet Union south of Lake Ilmen. He subsequently received the German Cross in Gold and led an engineer training battalion for the Totenkopf-Division, which withdrew from Russia to repair its losses and reorganize as a Panzer-Grenadier-Division. During this period, Müller departed Totenkopf and attended a battalion commander's school in Paris before being posted to the Hitlerjugend Division in 1943 as the first commander of its engineer battalion. Rising to SS-Sturmbannführer in January 1944, his performance during the Normandy campaign marked him for greater responsibility, though members of his battalion were later accused of murdering Canadian prisoners of war at Mouen, France. After surviving the shambolic retreat from France, Müller became the commander of SS-Panzer-Grenadier-Regiment 25 on November 15, 1944. This unit would form the nucleus of Kampfgruppe Müller. For his "outstanding leadership" during the battle for Krinkelt-Rocherath, Müller received the Knight's Cross on December 19, though the official typewritten recommendation was dated February 28, 1945. He remained in charge of the regiment until the war ended. After a term in captivity, Müller returned to his wife and young daughter. He died in Hamburg at age fifty-nine on April 7, 1974.

211 During the early 1990s, the author communicated with Rudolf von Ribbentrop, who lived in Ratingen, Germany. He stated that the 1.Kompanie was the only element of SS-Panzer-Regiment 12 to reach the battle area on December 17, and its tanks joined the fight that day. Hubert Meyer missed this fact in his two-volume history of the Hitlerjugend Division. While researching his tome, Meyer never located a single veteran of the 1.Kompanie who served in the Ardennes, and this helped create a gap in his understanding. Meyer erroneously claimed that the 1.Kompanie first entered combat on the morning of December 18, an assertion that has led subsequent writers astray. Ribbentrop served as regimental adjutant during the Krinkelt-Rocherath battle, a position that afforded him knowledge of the company's whereabouts. He entered combat himself at Domäne Bütgenbach when he took over the I.Abteilung after a white-phosphorus shell mortally wounded Arnold Jürgensen.

212 MacDonald, *Company Commander*, p. 93 (1947 edition).

213 Unit Journal, 741st Tank Battalion, December 1944, p. 4.

214 Author's interview with Louis J. DiClementi Jr., July 20, 1996.

215 Richard Smith joined Company C on August 16, 1944. Michael Fox joined on September 18. Both men came from the 3rd Replacement Depot. Fox died on June 30, 1984, at age seventy-three and rests today at Calverton National Cemetery, Long Island, New York. Richard Smith died of pneumonia on August 1, 1991, at age sixty-seven. He rests at Jefferson Barracks National Cemetery, Missouri. Alvin Lanier (1925–1944) was the loader in Miller's tank, and Guy A. Dedmon (1917–

1944) was the gunner. Both men perished with Miller.

216 Ibid.

217 Ibid.

218 Author's interview with Andrew W. Divers, August 15, 1997.

219 Letter from Helmut Zeiner to Jean-Philippe Speder, July 8, 1997. Several authors have asserted that Zeiner's Jagdpanzers destroyed Victor Miller's tanks at Ruppenvenn. Zeiner himself denied it in the aforementioned letter. There were actually two US tanks there, but we can forgive Zeiner for only remembering one after more than fifty years. These tanks fell victim to the 1.Kompanie of SS-Panzer-Regiment 12. Unfortunately, no author ever located a survivor from this Panther company to obtain a firsthand account. Even the HJ Division historian, Hubert Meyer, failed to find one. The full story of the 1.Kompanie remains largely untold.

220 Meyer, *Kriegsgeschichte der 12.SS-Panzerdivision "Hitlerjugend," Band II*, p. 421. English translation by H. Harri Henschler for J.J. Fedorowicz Publishing with subsequent translation corrections by the author.

221 Cannon, Heintzleman, and Lanier, *The Story of "Vitamin Baker,"* p. 63.

222 Meyer, *Kriegsgeschichte der 12.SS-Panzerdivision "Hitlerjugend," Band II*, p. 422.

223 Sometime after Theodore Parker lost his tank, he single-handedly captured an enemy tank and killed one member of its crew in the process. He relinquished the German machine to infantrymen who relieved him. The story first appeared in *The Stars and Stripes* Paris edition on January 25, 1945. Titled "Lone GI Captures Nazi Tank Intact," the article lacks sufficient detail to determine where the incident fits into the event chronology for the Krinkelt-Rocherath battle. No corroborating source is known. *Stars and Stripes* provided no author credit for the article.

224 Cannon, Heintzleman, and Lanier, *The Story of "Vitamin Baker,"* p. 67.

225 Author's interview with Kelly B. Layman (a.k.a. Craig K. Carter), April 5, 1997.

226 After the battle, Kelly Layman returned to the Krinkelt church and found Charles Ireson's body. Val Fister did the same and checked Ireson's dog tags. "I knew him," Fister recalled. "He had on his tanker's helmet and was lying under some bricks. We went to battalion headquarters and told them about this body because he was one of our men." (Valentine Fister's oral history, recorded March 1993)

227 Sergeant Carlson's tankdozer was an M4A1(75) built in May 1943 by the Pressed Steel Car Company in Chicago. The vehicle bore US Army registration number USA-W-3036870. The tank was a total loss except for its bulldozer blade, which was salvaged.

228 Meyer, *Kriegsgeschichte der 12.SS-Panzerdivision "Hitlerjugend," Band II*, p. 422.

229 Beeley, *I Must Have Had A Guardian Angel*, p. 64.

230 Letter from Ray M. Wilson to George E. Beeley, June 6, 1994.

231 Beeley, *I Must Have Had A Guardian Angel*, p. 66.

232 Ibid., p. 65.

233 Aloys Stoffels (1904–1984) owned the butcher shop (Nr. 41a Krinkelt) that Frank Mildren used as his CP. The building was destroyed during the battle. Stoffels rebuilt the shop after the war, and, during the 1970s, his nephew Felix Palm took over the business and named it Metzgerei Palm. The shop is still in operation at

the time of this writing. The M8 armored car belonged to the Reconnaissance Company of the 644th TD Battalion. First Lieutenant Harold L. Hoffer commanded the company and established his CP at Nr. 88 Krinkelt, a house owned by Peter Stoffels. The M8 crew named their vehicle "SKYROCKET." The 644th is known to have lost an M20 armored command car elsewhere in the twin villages.

234 Meyer, *Kriegsgeschichte der 12.SS-Panzerdivision "Hitlerjugend," Band II*, p. 429. "*Wir igelten uns ein.*" translates to "We set up an all-around defense."

235 The lead machine, Panther 135, struck an antitank mine laid by PFC Harlin E. Coffinger (1921–2001) of Company D, 9th Infantry Regiment. The three noncoms who conspired to throw an open can of gasoline on the tank were Sergeant Charley L. Roberts (1914–1975) of Company D, 9th Infantry; Staff Sergeant Odis Bone (1921–2011) of Company B, 9th Infantry; and Sergeant Joseph Busi (1920–2015), also of Company D. The trio approached the tank from the rear with Bone and Busi lifting the heavy can onto the machine's aft end. When Roberts began climbing onto the tank to reposition the can, one of the Germans inside tossed out a grenade that wounded Roberts in the hand. Though injured, he threw a phosphorus grenade toward the can and ignited the gasoline. The fire blazed brightly but failed to spread and engulf the tank. Had Roberts succeeded in moving the can, the results may have been different. Amid the melee, Panther 127 bypassed 135 and began turning left toward Rocherath. Private William A. Soderman (1912–1980) of Company K, 9th Infantry rose from behind a hedgerow and launched a bazooka rocket. The projectile shattered a track link, and 127 rolled to a halt near Soderman's position. The crew kept firing their main gun. Intent on ending the menace, Busi launched rifle grenades at 127. One of them detonated at the base of the turret, sending out a long tongue of fire. "The turret stopped rotating and didn't move again," Busi remembered. Crewmen quickly emerged from 127, and GIs everywhere opened up on them. In 1987, the author began corresponding with Joe Busi and Odis Bone. Both men pinpointed on a Lausdell map the location where a mine stopped the lead German tank. Bone and Busi independently identified the same spot, the very place where the wreck of Panther 135 appears in wartime and postwar photographs as well as Harrison S. Standley's pen-and-ink sketch. Careful study of the photos reveals that the Panther's left track was shattered, a detail that Harlin Coffinger shared with the author. Coffinger also corroborated the location of the tank. In January 1988, the author visited Odis Bone at his home in Marble Falls, Texas. In his driveway, Odis reenacted the assault on 135, using his automobile as a stand-in for the enemy tank. The author also corresponded with John Stanley Milesnick (1921–2013), who commanded Company B at Lausdell. He confirmed the details provided by Bone, Busi, and Coffinger. All the aforementioned veterans were emphatic about the time and date—a couple of hours after dark on December 17. In October 1989, the author traveled to Cincinnati for a joint reunion of Companies A and C, 9th Infantry. The Lausdell veterans in attendance provided accounts that jibed with the recollections of Bone, Busi, and Coffinger. Leonel G. Vidaurri (1915–2001), the former first sergeant for Company A, gave an especially vivid description of the battle. Situated at the Company

A command post, he had a ringside seat for the fireworks that night. The next morning, he became a prisoner of war along with eighty-one other members of his company. As for wartime documentation, Harold N. Denny of *The New York Times* wrote the most accurate contemporary account. Denny interviewed at least ten men, including Bone and Busi. Conversely, major discrepancies exist in the two relevant combat interviews conducted by an army historian in March 1945 (interviewees Major William F. Hancock, Staff Sergeant Norman R. Bernstein, and First Lieutenant Wesley E. Knutsen). The same is true of First Lieutenant Roy E. Allen's dictated statement made within a month of the battle. In 1947, Charles B. MacDonald (CO, Company I, 23rd Infantry Regiment) published a memoir that included his perspective on the epic clash at Lausdell. His story meshes with the information presented above.

236 The Company A morning report for June 15, 1945, reported Edward Ayers as a prisoner of war. The same report also stated he "returned to military control" on May 2, 1945.

237 The author found no evidence that corroborates Lohman's claim that an American TD (presumably self-propelled) knocked out his tank. One TD unit was at Lausdell, namely Third Platoon, Company B, 801st Tank Destroyer Battalion. It is unclear whether this unit deployed any of its towed antitank guns. Its men did provide antitank mines used to defend Lausdell. Elements of the 612th and 644th Tank Destroyer Battalions were in the Krinkelt-Rocherath area, but none of their self-propelled TDs are known to have participated in the Lausdell action. Lohman's claim cannot be refuted or verified.

238 Beeley, *I Must Have Had A Guardian Angel*, p. 66.

239 Ibid., p. 68.

240 Meyer, *Kriegsgeschichte der 12.SS-Panzerdivision "Hitlerjugend," Band II*, p. 429.

241 Parker, *Battle of the Bulge*, p. 128.

242 Meyer, *Kriegsgeschichte der 12.SS-Panzerdivision "Hitlerjugend," Band II*, p. 430.

243 Hancock, *The 9th Infantry Regiment As I Knew It*, p. 29.

244 Silver Star Medal recommendation for Gaetano R. Barcellona, dated 8 March 1945 (RG 338, NARA).

245 Ibid.

246 Born on May 24, 1919, at Oberndorf am Neckar, Walter Erich Hils first experienced combat in May 1940 as a platoon leader with the 6.Kompanie of Panzer-Regiment 7 (10.Panzer-Division). His unit sliced through Luxembourg, cracked the French lines at Sedan, and penetrated to the English Channel. He earned the Iron Cross Second Class and the Wound Badge in Black, the latter for a back injury caused by an antitank shell. The following year, he became adjutant for the I.Abteilung of Panzer-Regiment 7 and participated in Operation Barbarossa, the attack on the Soviet Union. In early December, amid the frigid grip of winter, his regiment reached Lenino, a settlement twenty-four miles from the Kremlin, but days later the Red Army launched a counteroffensive that ended the German campaign against Moscow. Hils received the Iron Cross First Class and Wound Badge in Silver later that month. In April 1942, his regiment relocated to

France for recuperation, but after just two months, the I.Abteilung returned to the Eastern Front. The battalion underwent two reorganizations, finally becoming the III.Abteilung of Panzer-Regiment 36 (14.Panzer-Division). Its men and armor joined the German summer offensive and fought across the vast grasslands of the Eurasian Steppe all the way to Stalingrad. Fate intervened to spare Hils from the Stalingrad catastrophe. Recalled to Germany in October 1942, he became a training officer, ultimately serving as a company commander with Panzer-Lehrgänge "Panther," established at Erlangen in March 1943. He instructed Panther crews until joining the I.Abteilung of SS-Panzer-Regiment 12 as a replacement officer. Before the Ardennes Offensive, Hils succeeded SS-Hauptsturmführer Walter Bormuth as commander of the 1.Kompanie. The only known account of subsequent events appeared in a 1978 book titled *Die 3.Kompanie*. The book described changes within the I.Abteilung after the Krinkelt-Rocherath battle: "The commanding officer, Sturmbannführer Jürgensen, was forced to regroup in view of the severe Panther losses. The remnants of the combat elements of the 1st and 3rd Companies were consolidated, but their supply trains maintained their functions. On the evening of December 20, the newly reorganized battalion was ready to attack on the road between Losheimergraben and Büllingen. Panzers 315, 325, and 335 of the 3rd Company belonged to this unit, all of them platoon leader's vehicles; Panzer 335 having returned from the workshop. Hauptmann Hils, commander of 1st Company, had been appointed to lead the unit. Since he had lost his tank [Panzer 105] in Krinkelt, he chose 325 as his command vehicle." The history goes on to record that Hils died on December 21 when American fire destroyed his Panther at Domäne Bütgenbach. Two men from the 3.Kompanie perished with him: SS-Sturmmann Nanning Lorenzen, the gunner, and SS-Rottenführer Hermann Krieg, the loader. On January 13, 1945, SS-Obersturmführer Helmut Gaede succeeded Hils as commander of the 1.Kompanie (per an Ergänzung or supplement to Gaede's Soldbuch). After the war, Hils's parents contacted the Deutsches Rotes Kreuz (DRK) in an effort to clarify the fate of their son. His name and photograph appeared in one of 225 volumes published by the DRK during the 1950s, the so-called Vermisstenbildlisten or Missing Persons Picture Lists: Walter Hils, Field-Post Number 59043B, Volume WC, p. 324 (FPN 59043B = 1.Kompanie, SS-Panzer-Regiment 12). The remains of the twenty-five-year-old Panzer officer have never been recovered or positively identified.

247 The intelligence officer was First Lieutenant Sidney P. Dane and the sergeant was Charles White. Captain Francis W. Phelps Jr. recorded their actions in a combat interview he conducted on February 25, 1945. The interviewees were Captain Fred L. Rumsey, Major Martin B. Coopersmith, and Sergeant Grover C. Farrell of the First Battalion, 38th Infantry Regiment. In addition, Hal Boyle of the Associated Press mentioned Dane and the bovine roadblock in a dispatch dated on January 15, 1945.

248 Kepinski and his crew belonged to the Third Platoon of Company C, 644th TD Battalion. As they ravaged Panther 126, three GIs from Detachment E, 165th Signal Photographic Company were on hand. The trio included a still photographer,

a motion-picture cameraman, and a jeep driver. Staff Sergeant Bernard J. Cook, the cinematographer and ranking member of the photo team, headed toward the Panther, and the last shot from the TD nearly wounded him. "A captain [most likely John J. Murphy, CO, L/38) and a private called down to me from a foxhole on my right," Cook remembered. "They wanted to know if I was all right." Just then, "two Germans got out of the tank on my left, and both were shot by the captain and private. A third man got out on the [opposite side], using the tank for cover." Cook tracked down the SS tanker after he dashed uphill and sought cover at a spring-fed well for cattle. "He was wounded, so, I took him back to the road and down past the tank where James Clancy, my still-photo man, took a photo." Clancy photographed Cook marching the prisoner at pistol point as blood streamed down the German's face. Moments later, the sergeant turned back to film the burning tank with his 35 mm motion-picture camera. He immortalized the last moments of 126 on celluloid. Cook's jeep driver, Campbell G. Snowden, escorted the PW to a nearby MP post, while Cook and Clancy headed for the center of town. Both men photographed two Panthers destroyed in front of Nr. 92 Krinkelt, a prominent house built during the 1920s by local businessman Nicolaus Balthasar Kalpers (1875–1935). Cook also filmed another nearby Panther while it burned. All of Clancy's photos from December 18 were later miscaptioned as being made the previous day. Looking back on the affair, Cook provided his recollections in a letter to William C.C. Cavanagh, dated July 29, 1990. At the US National Archives, the Still Pictures Branch maintains Clancy's original 4x5 negative of Cook and his prisoner (111-SC-198468). Careful study of the negative reveals that the prisoner wore a pair of Drillich-Panzerhose and a Drillich-Panzerbluse without shoulder boards or other insignia. The jacket and pants were German Army issue rather than Waffen-SS, though some HJ-Division tank crewmen received these items. Both garments appear water soaked, especially the pants. The negative also reveals wooden fencing around the well and pond in the background. Also at the National Archives, the combat interviews of the 2nd Infantry Division contain the story of Panther 126 and its rampage through Krinkelt. Captain Francis W. Phelps Jr. gleaned part of the story during an interview he conducted on February 25, 1945, with Captain Fred L. Rumsey, Major Martin B. Coopersmith, and Sergeant Grover C. Farrell of First Battalion, 38th Infantry Regiment. The typewritten narrative prepared by Phelps includes two map overlays and three sketches. Phelps learned about the demise of 126 when he interviewed Lieutenant Colonel Olinto M. Barsanti and Captain John L. Murphy of Third Battalion, 38th Infantry. The interview occurred on March 15, 1945, and the resulting narrative includes two map overlays.

249 Cannon, Heintzleman, and Lanier, *The Story of "Vitamin Baker,"* p. 67.

250 Layman and Williams may have sought shelter at Nr. 99 Rocherath, the home of Julius Rauw, a traveling butter-and-egg salesman. He died on the evening of December 17, when a GI mistakenly shot him, believing he was a German soldier. Rauw's house survived the battle with little damage but was ultimately torn down in 1992 after sitting derelict for over a decade. The infantry CP across the street

was at Nr. 91 Krinkelt, a Gasthaus and farm operated by Martin Küches. As an aside, Nr. 99 did have a coal cellar and a basement storage space. This seems inconsistent with the story told by Layman who said that he and Williams "tried to find a cellar but couldn't." Maybe they just failed to locate it or were in a different house altogether.

251 Cannon, Heintzleman, and Lanier, *The Story of "Vitamin Baker,"* p. 68.

252 The Normandy campaign decimated SS-Panzer-Regiment 12. For example, the 3.Kompanie lost all of its Panthers. The surviving crewmen, now little more than foot soldiers, withdrew to Belgium after the collapse of the Falaise pocket. The company crossed into Germany on September 10, 1944, and settled in Wilden, a village eight kilometers southeast of Siegen. The company remained there until October 5, when it moved to Hassbergen, eight kilometers north of Nienburg. Willi Gundlach of the 3.Kompanie described the refitting process in a unit history published in 1978. He stated that an Abhol-Kommando (procurement detail) led by SS-Untersturmführer Willi Engel departed by train on October 20 to receive new vehicles at the Heeres-Panzer-Zeugamt Breslau-Masselwitz. Five days later, the men headed west with eleven Panthers (Ausführung G) loaded on flatcars. Presumably a detail from the 1.Kompanie also journeyed to the same tank depot. Photographic evidence shows that the Maschinenfabrik Augsburg-Nürnberg (MAN) manufactured the Panthers procured by SS-Panzer-Regiment 12. Fahrgestellnummer (chassis number) 121137 appears in a postwar picture of Panther 125. MAN completed this particular vehicle in October 1944. In another photo, the number 121146 appears on Panther 327, indicating another vehicle completed in October 1944 by MAN.

253 Kurt Brödel commanded 3.Kompanie, Heeres-Panzerjäger-Abteilung 743 before transferring to the Waffen-SS. In September 1944, his former battalion commander recommended him for the German Cross in Gold, citing Brödel's actions against Soviet forces on December 6, 1941, December 4-5, 1942, February 19, 1943, February 21, 1943, and February 25-26, 1943. Unfortunately, most of those dates were incorrect; the 1942 action occurred in 1943, and the 1943 actions took place in 1944. Less than a week before Brödel died, the decorations officer on Heinrich Himmler's personal staff rejected the recommendation on a technicality. More than a year had elapsed between the actions and the date of the recommendation, something forbidden by an Army High Command (OKH) decree. The decorations officer added that if the recommending officer updated the proposal with more recent accomplishments, nothing was likely to stand in the way of an award. In actuality, the proposal only needed the years corrected and the 1941 material removed, but that never happened. It is unknown whether Brödel knew about the status of his German Cross in Gold at the time he fell in battle. His remains were never positively identified and presumably rest in an "unknown" grave at a German military cemetery, perhaps Lommel. As an aside, Brödel's highest award was the Führer's Certificate of Recognition (Anerkennungsurkunde des Führers), a rare accolade bestowed on January 7, 1944. This entitled Brödel to wear the Army Honor Roll Clasp (Ehrenblatt Spange des Heeres). The aforementioned

decorations officer mentioned that the grounds for the Führer's certificate could not form part of an updated German Cross in Gold proposal. The basis for the certificate *may* have been Brödel's action on December 4-5, 1943.

254 *Die 3.Kompanie*, p. 87. English translation by the author.

255 Ibid., p. 91.

256 Hal Boyle of the Associated Press interviewed Staff Sergeant Parker and included his sniper story in a dispatch dated January 13, 1945. Boyle wrote that Parker "knocked off" a dozen Germans, an amazing claim that the author could not corroborate. Parker received a Silver Star Medal for his actions in Krinkelt-Rocherath. His combat service ended on January 1, 1945, when an enemy artillery shell wounded him (and killed William Myers). Steel fragments penetrated the sergeant's abdomen and hospitalized him for 226 days. He received a disability discharge in August 1945. Parker hailed from Counce, Tennessee, and died in Memphis at age sixty-four on March 21, 1984.

257 Letter from James W. Love to the author, April 18, 1997.

258 Letter from Helmut Zeiner to Jean-Philippe Speder, April 26, 1996.

259 Author's interview with Jack B. Boardman, March 1997.

260 Captain Cinquina's command tank "B-1" was an M4(75) built in April 1943 by the American Locomotive Company in Schenectady, New York. The vehicle bore US Army registration number USA-W-3066160.

261 Author's interview with Jack B. Boardman, February 1997.

262 Ibid.

263 Western Union telegram to Irma S. Coleman, mother of 1st Lt. George R. Coleman, time stamped 12:24 P.M., February 19, 1945.

264 Jürgensen established his command post in Nr. 80 Rocherath, the local police station, originally built as a private residence. On page 90 of Die 3.Kompanie, Willi Engel described the place as a kind of "Verwaltungsgebäude" [administrative building]. The police station was the only structure matching that description on the main street in Rocherath. Page 89 of the same book includes a sketch by Engel showing the approximate location of Jürgensen's CP. This is also consistent with Nr. 80, except the sketch incorrectly shows the building on the west side of the main street. (The sketch also shows the church on the wrong side of the street.) Prior to Jürgensen's arrival at the building, American artillerymen had used the police station as their CP. These were men from Battery B, 17th Field Artillery Observation Battalion (code name "Violet Baker"), who vacated the area at 10:00 P.M. on December 17. After the war, the police station remained in operation until 1958, when ownership passed to the Belgian Forestry Service. Since that time, the head forest ranger for the area has resided there.

265 Author's interview with Ralph A. Woodward, April 20, 1997.

266 The four 741st members evacuated to the 622nd Clearing Company in December were: Newell F. Goff, T/4, cook, SV/741, 20 Dec 1944; Antonio Grossi, S/Sgt, communications chief, B/741, 21 Dec 1944; Roy C. Nobles, T/4, tank driver, A/741, 25

Dec 1944; William M. Polyak, Pfc, tank driver, C/741, 21 Dec 1944.

267 Author's interview with Charles W. Eubanks, April 20, 1997.

268 Ibid.

269 *The Chicago Tribune,* Chicago, Illinois, December 30, 1944, p.2.

270 Unit Journal, 741st Tank Battalion, December 1944, p. 7.

271 Cannon, Heintzleman, and Lanier, *The Story of "Vitamin Baker,"* p. 64.

272 The Krinkelt-Rocherath church bore the name of Sankt Johannes der Täufer (Saint John the Baptist). The architect was Heinrich Forthmann (1880–1943) from Cologne. Workers laid the church's cornerstone on June 25, 1905, and construction ended in 1907. The bell tower housed three bronze bells until the Germans confiscated the two largest ones in March 1943 and melted them down for the war effort. The sole remaining bell miraculously survived the Battle of the Bulge. After the war, between 1945 and 1954, the parish held church services in a wooden facility constructed by a local joiner. Demolition of the severely damaged church began in 1950. Architects Gaston Marchot and Robert Busch designed the new church in the neo-Roman style, and the consecration ceremony took place on September 29, 1954. The sole surviving original bell hangs in the bell tower of the new church.

273 Letter from Nelson W. Hall to the author, August 8, 1996.

274 The destruction was only a partial loss for posterity. Most 741st records remained safe at the rear CP in Robertville, the rear CP having moved there from Ovifat on December 14. The battalion staff located their rear CP at Nr. 17 Robertville, a boarding house known as the "Pension Bardenheuer." Marie Fagnoul-Bardenheuer had purchased the house in 1922, nine years before her marriage to Franz Josef Bardenheuer. She operated the establishment until her death in 1967 at age eighty-two. Today it's a vacation property known as la Maison de la Warche (Rue du Thier 14, Robertville).

275 Letter from Nelson W. Hall to the author, August 8, 1996.

276 Valentine Fister's oral history, recorded March 1993.

277 In its After-Action Report for December 1944, the 741st Tank Battalion recorded the loss of eleven Shermans and one tank-recovery vehicle. This author has corroborated those numbers and determined the US Army registration numbers for eight of the Shermans. In the same after-action report, the 741st claimed to have knocked out twenty-seven enemy tanks. Hubert Meyer found this number credible and stated so in his two-volume history of the Hitlerjugend Division (*Band II*, p. 435). Over a forty-year period, the author studied every known photograph of German armor destroyed in the Krinkelt-Rocherath area. These images reveal sixteen machines: eleven Panthers, three Panzer IVs, and two Jagdpanzer IV/70s. There is no way to determine the number of vehicles knocked out in the twin villages and repaired by the Germans. The absence of daily strength and loss reports for SS-Panzer-Regiment 12 and Panzerjäger-Abteilung 12 during the Ardennes Offensive makes a precise accounting impossible. Meyer attempted to fill this gap in the historical record when he discovered strength statistics written on OKH situation maps from the Führerhauptquartier. These 1:300,000 maps reside at the

Bundesarchiv in Freiburg. The map titled *3.Lage Frankreich - Nord* and dated December 17, 1944 (RH 2-KART/11324) gives the following numbers for the Hitlerjugend Division:

39 Panzer IVs—combined total, operational and out of service for repair
38 Panthers—combined total, operational and out of service for repair
53 Jagdpanzers—combined total, operational and out of service for repair

Another map with the same date (*2.Lage Frankreich - Nord*, RH 2-KART/11323) provides conflicting numbers:

20 Panzer IVs, including 15 out of service for repair (short term)
26 Panthers, including 7 out of service for repair (short term)
58 Jagdpanzers, including 4 out of service for repair (short term)

The same conflicting numbers appear on a map dated December 18 (RH 2-KART/11325), but another map for that date (RH 2-KART/11326) reflects the following machines:

30 Panzer IVs, including 9 out of service for repair (short term)
34 Panthers, including 4 out of service for repair (short term)
57 Jagdpanzers, including 4 out of service for repair (short term)

The OKH maps for December 19-21 are consistent:

26 Panzer IVs, including 13 out of service for repair (short term)
21 Panthers, including 14 out of service for repair (short term)
33 Jagdpanzers, including 17 out of service for repair (short term)

The Bundesarchiv also possesses a "Materielle Lage" report for the HJ Division dated December 31 and signed by Hugo Kraas (RH 10/321). The report shows the following:

7 Panthers, battle ready
10 Panthers, repairable within three weeks
13 Panzer IVs, battle ready
14 Panzer IVs, repairable within three weeks
15 Jagdpanzers, battle ready
3 Jagdpanzers, repairable within three weeks

The OKH map for December 31 (RH 2-KART/11362) offers different numbers:

6 Panzer IVs, including 7 out of service for repair (short term)
12 Panthers, including 7 out of service for repair (short term)
8 Jagdpanzers, including 28 out of service for repair (short term)

On all the maps (except 11362), the number of Jagdpanzers includes IV/70s along with Jagdpanthers from schwere Panzerjäger-Abteilung 560, which entered combat after the battle for Krinkelt-Rocherath. During that battle, the 277.Volksgrenadier-Division possessed Jagdpanzers of the 38(t) variety. These belonged to Panzerjäger-Kompanie 1277, but what, if any, role that unit played in the battle remains unknown. Both of the aforementioned maps for December 17 report the unit had eleven 38(t) Jagdpanzers. On December 21, the number was down to seven.

278 Beeley, *I Must Have Had A Guardian Angel*, p. 69.

279 Valentine Fister's oral history, recorded March 1993.

280 The allotment of US medium tanks received from British sources detailed in a

memorandum dated December 30, 1944, from Colonel Peter C. Haines III, Chief, Armored Section, First US Army to the Ordnance Officer, First US Army (Colonel John B. Medaris).

281 The Tiger mounted a Turmzielfernrohr 9d, featuring variable magnification: 2.5-power (25° field of view) and 5-power (14° field of view). The Panther had a Turmzielfernrohr 12a with the same variable magnification and fields of view. US telescopic sights lacked variable magnification, but American tank gunners had a supplemental sighting device, a non-rotating periscope with a wide field of view. This scope also had a built-in telescope with a narrower, 9° field of view and 1.44-power magnification. The aforementioned German optics were notable for their clarity and anti-reflective characteristics. During the glass-making process, the manufacturer, Carl Zeiss, added lanthanum to improve clarity. Zeiss also perfected the application of thin-film coatings to reduce glare.

282 Extended-end connectors also proved an utter folly on light tanks, throwing tracks and damaging bogie wheels.

283 *Armored Force Field Manual 17-33, The Armored Battalion, Light and Medium* (dated September 18, 1942) addressed "Tank versus Tank Action" on p. 107. The methods described there stemmed from *Armored Force Field Manual 17-10, Tactics and Technique* (dated March 7, 1942), pp. 144-147.

284 Lt. Covington, After-Action Report, Platoon Leader Company C, Unit Journal, 741st Tank Battalion, January 1945 (RG 407, NARA). Wartime documents also refer to Point 68 as Crossroads 68.

285 Author's interview with Philip L. Fitts, December 19, 1995.

286 Lt. Covington, After-Action Report, Platoon Leader Company C, Unit Journal, 741st Tank Battalion, January 1945 (RG 407, NARA).

287 MacDonald, *Company Commander*, p. 133 (1947 edition). Captain Roy G. McCracken commanded the Antitank Company of the 23rd Infantry Regiment.

288 Frederick Neidrich was promoted to staff sergeant on January 11, 1945, and later earned a battlefield commission to second lieutenant.

289 Cannon, Heintzleman, and Lanier, *The Story of "Vitamin Baker,"* p. 73.

290 Ibid.

291 The exact location of Young's CP remains unclear. As an aside, according to Leonard Trimpe, a veteran of Service Company, Colonel Skaggs located his Forward CP at Nr. 81 Rocherath upon returning to the twin villages after the "Bulge." Johann Jansen owned this property during the war. Trimpe supplied the author with a 1987 photograph of the house made during a battlefield tour. Partly corroborating Trimpe's assertion, the author obtained a February 1945 photograph made in the twin villages by James W. Love of the 38th Infantry Regiment. The photo shows a wooden road sign pointing northeast along the main street toward the Forward CP for the 741st. Although the sign doesn't confirm the CP location, it jibes with Trimpe's memory. However, the Unit Journal for the 741st offers contradictory information. It states that the Forward CP moved from Berg to Krinkelt (not Rocherath) on February 1, 1945, and it remained in Krinkelt until February 20. During that period, the Rear CP also moved to Krinkelt and four days later co-located with

the Forward CP. The latter shifted to the building occupied by the former.

292 Author's interview with Philip J. Zodda, November 9, 1998.

293 Author's interview with Jack B. Boardman, March 1997.

294 Major Robert L. Utley (Operations Officer, 38th Infantry Regiment) created a map covering Lausdell and Krinkelt-Rocherath on which he located enemy tanks destroyed and probably destroyed, December 17–19, 1944. His total was 78 tanks (and 77 on a second version of the map). Utley derived these numbers from after-action claims, but he never undertook a battlefield survey to corroborate them. Only sixteen of the tanks have been confirmed through photographic evidence, despite an abundance of pictures made by soldiers and civilians. One must take Utley's work with a grain of salt. The maps reside today among the operational records of the 2nd Infantry Division in RG 407 at Archives II. One map accompanies the December 1944 After-Action Report for the 38th Infantry; the other map (badly deteriorated) is filed with the 2nd Division's combat interviews. Both documents are available for free download on the author's website (etohistory.com).

295 Letter from Philip L. Fitts to LTC J. C. Herbert, Chief, U.S Army Personnel Services Division, US Army Reserve Personnel Center, St. Louis, MO, September 10, 1991, p. 2.

296 Author's interview with Philip L. Fitts, December 17, 1995.

297 As Fitts traveled to his embarkation port, his right elbow became infected and swollen from the injury he suffered on February 1. The infection forced him to visit an aid station at LeHavre, France, where a doctor squeezed pus from the wound and gave him sulfanilamide powder and Epsom salts. The medical officer could do no more without making Fitts a patient, an act that would end his furlough. Fitts soaked his elbow in a helmet filled with hot water and Epsom salts while steaming across the Atlantic aboard the RMS *Queen Elizabeth*. After reaching home, the sergeant's wife eyed the still-infected wound with alarm and insisted he visit Walter Reed Hospital, where another doctor examined the injury but refused to believe it was combat related. The cynical physician thought his patient had tangled with other GIs in a barroom brawl and hurt himself, an abuse of "Government property." The doctor said that a greenstick fracture caused the wound but abstained from documenting it. He doled out penicillin tablets as well as medicine for scabies that Fitts had around his waist. The penicillin and more Epsom salts cured the infection, and the wound healed. Fitts returned to Company C on May 6, 1945. It was not until 1990 that he realized no official report existed of his elbow injury, and thus no authority for an oak leaf cluster to his Purple Heart. Fitts started writing letters, hoping to correct the record. His effort met with skepticism from the US Army's Personnel Services Division in Saint Louis, but the veteran prevailed after his former company commander, Herb Covington, submitted a notarized statement corroborating the claim. On August 3, 1993, the Army Board for Correction (Military Records Branch) authorized an oak leaf cluster for his Purple Heart.

298 Air mail letter from Edward S. Sledge II to his mother, February 24, 1945.

299 Shirer, *The Rise and Fall of the Third Reich*, p. 919. "When even Zeitzler got up

enough nerve to suggest to the Führer that in view of the danger to the long northern flank along the Don the Sixth Army should be withdrawn from Stalingrad to the elbow of the Don, Hitler flew into a fury. 'Where the German soldier sets foot, there he remains!' he stormed." (Adolf Hitler, September 27, 1942).

300 *Litzmannstädter Zeitung*, Litzmannstadt, Germany, October 19, 1944, p. 2 (excerpt from Himmler's radio address at Bartenstein, East Prussia, during the first induction ceremony for *Volkssturm* troops on October 18, 1944: *"Unsere verfluchten Feinde werden es feststellen und sehen müssen, daß ein Einbruch in Deutschland, selbst wenn er irgendwo gelänge, für Angreifer Opfer kostet, die für ihn dem nationalen Selbstmord gleichkommen."*)

301 Author's interview with Mayne B. Youngblood, May 10, 1998.

302 Ibid.

303 As of this writing, the treadway bridge over Adenauer Creek remains in place, but it's only open to foot traffic. The bridge is adjacent to the Gasthaus Strohe (known as the Gasthof Strohe during the war). The author visited the bridge on two occasions—July 1996 and June 2011. In Louis Nemeth's photo, Mayne Youngblood and his driver, John Onuschak, wear ETO contract tanker helmets. After the Normandy campaign, the U.S. Army in Europe sought to replace the American-made M38 tank helmet due to supply shortages and the helmet's lack of ballistic properties. First Army contracted with the Paris firm Guéneau et Cie to supply French Army radio operator helmets (*Casque d'Opérateur Radio*) painted olive drab and retrofitted with American electronics, but widespread adoption never occurred. The manufacturer's lack of materials and electrical power curtailed output.

304 Fardella, *Combat History of the 741st Tank Battalion*, p. 28.

305 Ibid., pp. 28-29.

306 Cannon, Heintzleman, and Lanier, *The Story of "Vitamin Baker,"* p. 84.

307 Fardella, *Combat History of the 741st Tank Battalion*, p. 29.

308 Arthur Clifford Shirk was from Dubuque, Iowa, and joined the 741st in July 1944 as a replacement officer. He previously served as a platoon leader with Company D, 5th Tank Battalion (16th Armored Division) at Camp Chaffee, Arkansas.

309 The dead pilot was Captain Sherman N. Crocker, Commanding Officer of the 507th Fighter Squadron. He flew P-47D, 44-19772, named "Harriett" in honor of his fiancée Harriett Jey Jones. Hit by enemy Flak on February 13, 1945, his aircraft crashed and exploded on Hill 480 west of Ramersbach, Germany (Kreis Ahrweiler). Local residents know the hill as the Hilgesberg. Luftwaffe troops recovered one of the pilot's identification tags a month before Lieutenant Dew's men found the wrecked aircraft. On March 15, Dew reported the crash site at grid coordinate wF527101. During his combat career, Crocker shot down three FW-190s, two of them in one day. He rests today at Marstons Mills Cemetery, Marstons Mills, Massachusetts.

310 Air mail letter from Edward S. Sledge II to his parents, March 16, 1945.

311 "From the Rhine to Leipzig, 21 March - 20 April 1945, 38th Infantry Regiment," T/3 Jose M. Topete's interview with Maj. James W. Love, Capt. Edward L. Farrell, and Captain Edward O. Ethell, p. 3 (Combat Interviews, 2nd Infantry Division, RG 407, NARA).

312 Author's interview with Harrell E. Gullatt, August 16, 1997.

313 Author's interview with Willis D. Warren, July 21, 1996.

314 MacDonald, *Company Commander*, p. 146 (1947 edition).

315 Ibid., p. 147.

316 Ibid., p. 151.

317 Ibid., p. 152.

318 Ibid., p. 156.

319 Author's interview with Jack H. Browder, August 16, 1997.

320 After hostilities ended, Godett remained at the central blood bank and helped facilitate publication of *741 D-Day to V-E Day*, a pocket-sized history of the battalion printed by Paul Dupont of Paris. The battalion reconnaissance officer, Rom Legnini, owned his own printing company in civilian life and helped with the publication. In a letter dated July 8, 1945, the battalion adjutant, William E. Park, stated, "It is the intention of this headquarters to mail a copy to every individual that has ever been a member of this organization and to the next-of-kin of all deceased members. There will be no cost for this booklet." Park wrote the letter to the mother of George Coleman, the Company B platoon leader who died in Rocherath during the Bulge.

321 Fardella, *Combat History of the 741st Tank Battalion*, p. 31.

322 The two riflemen who died aboard Henkelman's tank were Privates First Class William H. Creasey (35906437) and Herman D. Taylor (36586909). Both belonged to Company L, 23rd Infantry. The Germans repaired Sauertalstrasse and closed it to vehicular traffic, the roadbed too weak for anything heavy.

323 Fardella, *Combat History of the 741st Tank Battalion*, pp. 31-32.

324 Annuity payments were $55.10 each month for twenty years for the beneficiary of a $10,000 National Service Life Insurance policy.

325 Veterans Administration Claim File XC 4099822 contains letters from Gordon's widow as well as an investigation report detailing the circumstances of his death plus numerous other documents (Federal Records Center, Eastpoint, Georgia).

326 Air mail letter from Edward S. Sledge II to his mother, April 5, 1945.

327 SS-Panzerbrigade „Westfalen" was formed in March 1945 under the command of SS-Obersturmbannführer Hans Stern. Its nucleus consisted of two SS training and replacement regiments from Sennelager near Paderborn. At peak strength, the brigade consisted of fifteen obsolete Panzer IIIs, twenty King Tigers, and three thousand troops. The Tigers were attached to the brigade on March 29 and belonged to schwere Panzerabteilung 507 (Heer), commanded by Major Fritz Schöck. The Panzer IIIs belonged to a tank company commanded by SS-Obersturmführer Max Tietz (born February 14, 1920 in Wanne-Eickel). His vehicles were training machines from Sennelager, and they became part of I.Abteilung, SS-Regiment „Holzer" (formerly I./SS-Panzer-Ausbildung-und Ersatz-Regiment).

328 Author's interview with Jack B. Boardman, October 16, 1997.

329 These tanks belonged to schwere Panzerabteilung 502 and schwere Panzerabteilung 510. Each battalion had dispatched its 3.Kompanie from the Eastern Front to procure new tanks from Henschel und Sohn AG, a manufacturer in Kassel that

produced all Tiger tanks. The Panzer men from 502 received eight Tigers, and the men from 510 procured six. They entered combat outside Kassel and retreated north across the area facing the 741st.

330 Hal Boyle of the Associated Press interviewed David Neill and included his story in a dispatch dated April 27, 1945. American newspapers across the country published it.

331 Cannon, Heintzleman, and Lanier, *The Story of "Vitamin Baker,"* p. 96. Author's translation of English into German.

332 MacDonald, *Company Commander*, p. 192 (1947 edition).

333 Fardella, *Combat History of the 741st Tank Battalion*, p. 35.

334 Author's interview with Harrell E. Gullatt, August 16, 1997.

335 MacDonald, *The Last Offensive*, p. 410.

336 The antiaircraft position at Schotterey was Flakstellung Nr. 103, six 88 mm guns and six 105 mm guns manned by 6./s.Flak-Abteilung 540 and Batterie zum besonderen Verwendung 1712. The position was just north of the village on Sandstrasse.

337 Cannon, Heintzleman, and Lanier, *The Story of "Vitamin Baker,"* p. 98.

338 Ibid., pp. 99-100.

339 Author's interview with Harrell E. Gullatt, August 16, 1997.

340 Author's interview with Jack B. Boardman, April 5, 1997.

341 Cannon, Heintzleman, and Lanier, *The Story of "Vitamin Baker,"* p. 102.

342 Clarence Gallow and Merlin LeBon remained in the ditch until nighttime when they crawled away.

343 The deceased soldier was Private First Class Emuel Whitfield, an ammunition handler from Company D, 23rd Infantry. His body ended up in a nearby "foreigner's cemetery" at Korbetha, Germany, where French prisoners of war interred him. The American Graves Registration Command identified his body in 1948. Private First Class Henry Lipfield (1916–2007) was the other mortarman from Company D held captive on Hill 102. Earlier, when the Flak guns fired on the five jeeps, one man died, Corporal Talmadge M. Sims, and three others were slightly injured, including First Lieutenant Charles J. Barclay (1917–1988), a section leader with the Mortar Platoon of Company D. These three soldiers avoided German captivity.

344 Author's interview with Harrell E. Gullatt, August 16, 1997.

345 Cannon, Heintzleman, and Lanier, *The Story of "Vitamin Baker,"* pp. 103-104.

346 HASAG was the acronym for munitions manufacturer Hugo Schneider Aktiengesellschaft. The firm's general manager, Paul Budin, became the subject of an American intelligence report: "Herr Bundis [Budin], director of a factory, which was one of the main suppliers of Panzerfausts, was also an ardent Nazi. Living in Leipzig, he felt the conquest of the city was only a matter of days. Being afraid of the fate a Nazi would suffer when captured, he decided upon a theatrical farewell gesture. He invited his closest associates to a formal banquet in a room rigged with mines. After dining and wining his guests, he set off the charges and blew himself as well as his friends into oblivion." (Annex No. 1, Consolidated Interrogation Report, G-2 Periodic Report No. 38, 9th Armored Division, 17 April 1945, RG 407, NARA).

347 The credit union was an office of the Allgemeine Deutsche Credit Anstalt (ADCA). Wilhelm Postler owned the photography studio. The address for both was Karl-Heine-Strasse 30.

348 Two Reports of Burial (GR Form 1) in Cuthbert's IDPF indicate he died from a gunshot wound to the head. He apparently lost his life to small-arms fire after the destruction of his tank. He was temporarily interred at Breuna #1 on April 21, 1945. Cuthbert was reburied in Plot OO at US Military Cemetery #1, Margraten, Holland, on July 27, 1945. He was always interred separately from the rest of his crew, their bodies recovered from inside the tank, his outside.

349 Richard W. Crampton and Jack H. Hutton served with the 165th Signal Photo Company. Edward R. Murrow mentioned Cuthbert's tank during a radio program broadcast nationwide on Sunday, April 22, 1945.

350 Author's interview with Verdie M. Guest Jr., August 15, 1997.

351 Verdie Guest provided the story of Norman Schneider's tank in Leipzig. The author interviewed Guest at the 741st Tank Battalion Reunion held at Paris Landing State Park, Tennessee, August 14-16, 1997. In addition to describing what happened, Guest drew a sketch showing the key locations in his story and the damage that his tank sustained from the Panzerfaust hit.

352 Parker spent thirty-eight days hospitalized, Neill twenty-eight days, and Suster thirty days. All three returned to duty. As for their tank, it burned and was a total loss. The vehicle carried US Army registration number 3051720 painted on its hull. It was an M5A1 (serial number 8609) built in August 1943 at the South Gate Assembly Plant, Southern California Division of General Motors. Presumably, this same light tank was the one that destroyed a King Tiger on April 8.

353 Unit Journal, 741st Tank Battalion, April 1945, p. 11 (RG 407, NARA).

354 Author's interview with Edward Teffeteller, January 7, 1996.

355 Sledge remained with Company A until relieved from that assignment on July 28, 1945, and reassigned to the 701st Tank Battalion. He had an ASR score of 108, more than sufficient for discharge. He departed Europe on September 28 aboard the RMS *Aquitania* and arrived at New York City on October 4. He traveled on to the Separation Center at Fort McPherson, Georgia, where he received an honorable discharge from active duty.

356 *Holdings and Opinions, Board of Review, Branch Office of the Judge Advocate General, European Theater of Operations, Vol 27 B.R. ETO*: Board of Review No. 3, CM ETO 13638, United States v. Donald A. Kepplin (36684359), July 5, 1945, pp. 2-3 (US Army JAG School Library).

357 Author's interview with Thomas D. Chastain, May 10, 1998.

358 Letter from Mr. & Mrs. Martin M. McDonough to Sen. James M. Mead, August 28, 1946 (Individual Deceased Personnel File for Roger J. McDonough).

359 Letter from Col. William J. McDonald to Sen. James M. Mead, September 7, 1946 (Individual Deceased Personnel File for McDonough).

360 Narrative of Investigation at Stollberg, Germany, by 1st Lt. Owen L. Nethery, American Graves Registration Command, p.1 (Individual Deceased Personnel File for McDonough).

361 Ibid., pp. 1-2.

362 Report of Official Travel by Lt. Col. E. M. Brown, June 9, 1950 (Individual Deceased Personnel File for Russell M. Bradsher).

363 Elliott, *D-Day: What it Cost*, p. 80.

364 Letter from Dorothy T. Carpenter to the author, February 25, 2001.

365 *Bergen Evening Record*, November 26, 1947, p. 3

366 *The Tribune*, Seymour, Indiana, March 13, 1995, p. 3B.

367 On June 17, 1945, the 741st transferred from V Corps to XXII Corps. The 741st joined the 8th Armored Division effective June 20, 1945, per verbal order of CG XXII Corps.

368 Preston Bishop (1904–1985) commanded the 778th Tank Battalion before joining the 741st. Earlier in the war, he served as a captain with the 741st but departed in February 1943 from Camp Polk to join the 738th Tank Battalion. He retired in 1964 as a full colonel.

369 At the Riviera Recreational Area, enlisted men quartered in Nice and officers in Cannes.

370 Author's interview with Eva Covington, July 13, 2001.

371 Author's interview with Jack B. Boardman, March 9, 1998.

372 Author's interview with Thomas D. Chastain, December 8, 1996.

373 Author's interview with Leonard H. Trimpe, August 25, 2008.

374 Dick Trimpe married a war widow. Leona E. Eckelman lost her husband in Normandy, a rifleman with Company K, 175th Infantry Regiment. She and Dick married in 1950 and had one son. On multiple occasions, the couple visited the Normandy American Cemetery, the final resting place of Leona's first husband.

375 First Lieutenant Albert I. Shapiro deliberately injured his knee during the battle for Hill 192. Jack Boardman recalled the incident and added that "Shapiro was not liked by the men or the other officers. ... It was a relief when he left." Author's interview with Jack B. Boardman, February 22, 1996.

376 Author's interview with Ralph A. Woodward, April 20, 1997.

377 Wiener, "The First Wave," *American Heritage*, May/June 1987, p. 138.

378 The 70th Tank Battalion lost two M4A1 DD tanks during Exercise Tiger. Kenneth Small (1930–2004) led an operation that recovered one of them in May 1984 from the bottom of Start Bay. Today the tank stands at a roadside memorial at Torcross.

379 "Salvaging the D-Day Beaches," *After the Battle*, Number 34, p. 46. Jacques Lemonchois died suddenly from a cerebral aneurysm on July 14, 2007. He left a wife and daughter.

380 Letter from Lt. Col. Lawrence J. Gomez to Philip L. Fitts, September 10, 1988, p. 2.

Index

People

Places